CELLULAR MODIFICATION AND GENETIC TRANSFORMATION BY EXOGENOUS NUCLEIC ACIDS

SIXTH INTERNATIONAL SYMPOSIUM ON MOLECULAR BIOLOGY

Sponsored by Miles Laboratories, Inc.

The Thomas B. Turner Auditorium
The Johns Hopkins Medical Institutions
Baltimore, Maryland
June 8-9, 1972

Edited by

Roland F. Beers, Jr. and R. Carmichael Tilghman

Supplement Number 2
The Johns Hopkins Medical Journal
by
The Johns Hopkins University Press
Baltimore

Manufactured in the United States of America

The Johns Hopkins University Press, Baltimore, Maryland 21218

The Johns Hopkins University Press Ltd., London

Library of Congress Catalog Card Number 73-5909

ISBN 0-8018-1511-8

Library of Congress Cataloging in Publication data will be found on the last printed page of this book.

The following figures and tables are used with the permission of the publications in which they first appeared:

Chapter	Figure or Table	Publication
1	Fig. 2	*Nature New Biol.* 233 (1971): 35, Fig. 2.
	Figs. 4 & 6	*J. Mol. Biol.* 72 (1972): 583-22, Figs. 4 & 6.
2	Figs. 2, 4, 6, 11, 12, 14, 16	*Proc. Nat. Acad. Sci.* 58 (1967): 650, Fig. 4; 68 (1971): 210, Figs. 1 & 3; 69 (1972): 2518, Figs. 2, 3, 4, & 5.
	Fig. 5	*Biochemistry* 9 (1970): 4428, Fig. 18.
3	Fig. 1	*J. Virol.* 10 (1972): 344, Fig. 3.
7	Figs. 1, 2, 3, 4 & Tables I & V	*C. R. Acad. Sc.* Paris, Serie D, 274 (1972): Fig. 1–p. 3560; Figs. 2 & 3, & Tables I & V–p. 2801.
	Figs. 4, 8, & Table III	*Proc. Nat. Acad. Sci.* USA 69 (1972): 191.
12	Figs. 7, 8, 9, 10	*Molecular Studies in Viral Neoplasia*, 25th Annual Sym. on Fundamental Cancer Research, Univ. Of Texas M. D. Anderson Hospital and Tumor Institute, Houston, 1972, Figs. 4, 5, 6, & 7. Also, *Molecular Studies in Viral Neoplasia*, Allen et al., Baltimore: Williams & Wilkins Co., in press, Figs. 4, 5, 6, & 7.
14	Figs. 1 & 2	*J. Virol.* 10 (1972): 622-27, Figs. 1 & 2.
	Fig. 3	*J. Biol. Chem.* 247 (1972): 7282-87, Fig. 3.
16	Fig. 1	*C. R. Acad. Sc.* Paris 274 (1972): 1977-80, Fig. 1.
18	Fig. 2 & Table 4	*Nature New Biol.* 240 (1972): 157, Fig. 1, Table 1.
	Fig. 3	*Proc. Natl. Acad. Sci.* USA 69 (1972): 1009, Fig. 1.
	Fig. 4, Tables 2,3,5	*Perspectives in Virology*, ed. M. Pollard. New York: Academic Press, 1972, pp. 81-99, Fig. 7, Tables 5, 6, 7.
19	Figs. 1, 2, & Table IV	*Federation Proc.* 31 (1972): 1655-57, Figs. 1 & 2, Table 5.
	Table III	*Proc. Soc. Expt. Biol. Med.* 140 (1972): 404-8, Table IV.
	Table VI	*Am. J. Epidemiol.* 91 (1970): 531-38, Table 4.
20	Figs. 8a,c	*Nature* 236 (1972): 103, Figs. 3a & 4b.
	Fig. 9	*Cancer Res.* 33 (1973): 626, chart 1.
21	Figs. 1-5, 8, 10, 14, 15, 16, 17, 18	*Proc. Nat. Acad. Sci.* USA 69: 3 (1972): 535, Figs. 1 through 5; 69:2 (1972): 435, Fig. 5; 69:7 (1972): 1727, Fig. 2; 69:8 (1972): 2020, Figs. 1 through 6.
	Figs. 6 & 7	*Nature* 235 (1972): 32, Figs. 1A & B, & 4.
	Figs. 9, 11, 12, 13	*Science* 175 (1972): 182-85, Fig. 3; 542-44. Fig. 1A; 174 (1971): 840-43, Figs. 1 & 2A. Copyrights 1971 and 1972 by the American Association for the Advancement of Science.
	Figs. 19, 20, 21	*Nature New Biol.* 240 (1972): 72, Figs. 1 through 3.

These Proceedings are dedicated to the memory of one of its participants, the late Professor Tsutomu Watanabe of Keio University School of Medicine, Tokyo, Japan

Tsutomu Watanabe

Professor of Microbiology
Keio University School of Medicine
Tokyo, Japan

Born: October 6, 1923
Died: November 4, 1972

Birthplace: Gifu, Japan

Nationality: Japanese

Academic Career:

1948 Graduated Keio University School of Medicine

1948–57 Department of Microbiology (Bacteriology), Keio University School of Medicine

1951–52 Exchange Student at Radiology Laboratory (under Prof. John Z. Bowers) and Department of Bacteriology (under Prof. Louis P. Gebhardt), University of Utah College of Medicine

1957–63 Assistant Professor of Bacteriology, Keio University School of Medicine

1957–58 Research Associate, Department of Zoology, Columbia University (under Prof. Francis J. Ryan)

1957 Participant in Phage Course, Bacterial Genetics Course, and Fungus Genetics Course, Cold Spring Harbor

1963–71 Associate Professor of Microbiology, Keio Univeristy School of Medicine

1971–72 Professor of Microbiology, Keio University School of Medicine

Honors:

1961 Dr. Kitasato Memorial Award (Japan) for "The Genetic Studies of Bacterial Drug Resistance, Especially of the Mechanism of Transfer of Multiple Drug Resistance"

1969 Dr. Asakawa Memorial Award of the Japan Bacteriological Society for "The Genetic Studies of Bacterial Drug Resistance with Special Reference to Transferable Drug Resistance"

1971 Medal of Honor, J. E. Purkyne Award (Czechoslovakia) for "The Studies of Antibiotic Resistance with Special Reference to Bacterial Plasmids"

THE JOHNS HOPKINS MEDICAL JOURNAL

Established in 1889 as The Johns Hopkins Hospital Bulletin

Published monthly under the sponsorship of
The Johns Hopkins University School of Medicine, The Johns Hopkins Hospital
and The Johns Hopkins Medical and Surgical Association

The Johns Hopkins Medical Journal, established in 1889 as The Johns Hopkins Hospital Bulletin, is dedicated to the publication of results of original research and reviews, especially of interdisciplinary interest to those working in the biomedical sciences, and to those in the broadest sense involved in the practice of medicine. It is designed particularly to reflect the contributions of present and former faculty, staff and students of The Johns Hopkins Medical Institutions. In addition, contributions from others are welcome. The circulation of The Journal includes libraries and individuals throughout the world.

THE JOHNS HOPKINS UNIVERSITY PRESS

BALTIMORE, MARYLAND 21218 U.S.A.

PREFACE

Molecular biology, despite its youth in comparison with older disciplines, today occupies a position in the center of the scientific stage. It continues to be an exciting venture in basic problems, and the degree of interest in the subject is attested to by the ever-increasing attendance at the international symposia dealing with selected phases of molecular biology. Miles Laboratories, Inc., under the guidance of its president, Dr. Walter Compton, inaugurated and has sponsored six programs in as many years. In 1971, The Fifth International Symposium was held at The Johns Hopkins Medical Institutions in Baltimore, Maryland, and the proceedings constituted a bound supplement of The Johns Hopkins Medical Journal. In June 1972 The Sixth Miles International Symposium was held again at Johns Hopkins and the proceedings are presented herewith by The Johns Hopkins Medical Journal as Supplement Number 2.

To the many people who have contributed to the success of the symposium and of this volume, the editors wish to express sincere thanks. Every effort has been made in editing the discussions, transcribed from tapes, to preserve the accuracy and intent of the participants. Permission was given the editors to do so without sending proof to the discussants. If error or misinterpretation of remarks have resulted, the editors take full responsibility.

On the morning of November 4, 1972, Professor Tsutomu Watanabe died after a losing struggle with carcinoma of the stomach. He had undergone surgery earlier in the year and despite the gravity of his illness made the arduous trip to the United States to participate in the Symposium. We are most grateful to Dr. Watanabe and to his former student, Dr. Arai, for the completion of the final draft of this presentation. It is our honor and pleasure to dedicate these proceedings to the memory of Dr. Watanabe, whose premature death comes as a shock to those who have come to know this remarkable scientist and man.

November 1972

Roland F. Beers, Jr.
R. Carmichael Tilghman

LIST OF AUTHORS

CONTENTS

Part Four

AUTHORS, CO-AUTHORS AND DISCUSSANTS

S. Aaronson, National Cancer Institute, Bethesda, Maryland

M. Abou-Sabe, Rutgers University, New Brunswick, New Jersey

P. T. Allen, M. D. Anderson Hospital and Tumor Institute, Houston, Texas

H. Vasken Aposhian, University of Maryland School of Medicine, Baltimore, Maryland

E. Applebaum, University of Wisconsin, Madison, Wisconsin

Domenica Attardi, St. Louis University School of Medicine, St. Louis, Missouri

G. S. Aulakh, Howard University Medical School, Washington, D.C.

Laure Aurelian, The Johns Hopkins University School of Medicine, Baltimore, Maryland

R. Axel, Institute for Cancer Research, Columbia University, New York, New York

John P. Bader, National Cancer Institute, Bethesda, Maryland

David Baltimore, Massachusetts Institute of Technology, Cambridge, Massachusetts

Alok K. Bandyopadhyay, Baylor College of Medicine, Houston, Texas

W. Baxt, Institute for Cancer Research, Columbia University, New York, New York

Roland F. Beers, Jr., The Johns Hopkins University, Baltimore, Maryland

M. Beljanski, Institut Pasteur, Paris, France

Sauyma Bhaduri, St. Louis University School of Medicine, St. Louis, Missouri

Richard Bockrath, Indiana University Medical School, Indianapolis, Indiana

J. M. Bowen, M. D. Anderson Hospital and Tumor Institute, Houston, Texas

Marvin H. Caruthers, Massachusetts Institute of Technology, Cambridge, Massachusetts

K. A. Chandrabose, Imperial Cancer Research Fund Laboratories, London, England

J. Christopher Cordaro, The Johns Hopkins University, Baltimore, Maryland

D. R. Critchley, Imperial Cancer Research Fund Laboratories, London, England

W. Howard Cyr, Armed Forces Institute of Pathology, Washington, D.C.

Edward C. DeFabo, Smithsonian Radiation Biology Laboratory, Rockville, Maryland

Arnold Dion, Institute for Medical Research, Camden, New Jersey

Leon Dmochowski, M. D. Anderson Hospital and Tumor Institute, Houston, Texas

James L. East, M. D. Anderson Hospital and Tumor Institute, Houston, Texas

Howard Elford, Duke University Medical School, Durham, North Carolina

James G. Gallagher, University of Vermont College of Medicine, Burlington, Vermont

Robert C. Gallo, National Cancer Institute, Bethesda, Maryland

J. Georgiades, M. D. Anderson Hospital and Tumor Institute, Houston, Texas

David H. Gillespie, Baltimore City Hospitals, Baltimore, Maryland

Robert Grafstrom, Hershey Medical Center, Hershey, Pennsylvania

Maurice Green, St. Louis University School of Medicine, St. Louis, Missouri

S. C. Gulati, Institute for Cancer Research, Columbia University, New York, New York

Richard Hackel, Brooklyn College, Brooklyn, New York

Eugene Hamori, University of Delaware, Newark, Delaware

H. Hashimoto, University of Wisconsin, Madison, Wisconsin

Stanley Hattman, University of Rochester, Rochester, New York

R. Hehlmann, Institute for Cancer Research, Columbia University, New York, New York

Donald R. Helinski, University of California, La Jolla, California

Thomas R. Henderson, Lovelace Foundation, Albuquerque, New Mexico

Paul Homsher, Old Dominion University, Norfolk, Virginia

Diane Humphrey, Imperial Cancer Research Fund Laboratories, London, England

Roland G. Kallen, University of Pennsylvania School of Medicine, Philadelphia, Pennsylvania

Chil-Yong Kang, University of Wisconsin Medical Center, Madison, Wisconsin

Mitchell Kanter, Duke University, Durham, North Carolina

D. T. Kingsbury, University of California, La Jolla, California

LeRoy Kuehl, University of Utah, Salt Lake City, Utah

D. Kufe, Institute for Cancer Research, Columbia University, New York, New York

Y. Kupersztoch, University of California, La Jolla, California

F. H. Lin, New York State Institute for Research in Mental Retardation, Staten Island, New York

Amnon Liphshitz, Syracuse University, Syracuse, New York

David Livingston, National Cancer Institute, Bethesda, Maryland

M. Lovett, University of California, La Jolla, California

I. A. MacPherson, Imperial Cancer Research Fund Laboratories, London, England

P. Manigault, Institut Pasteur, Paris, France

Koslir Maruyama, M. D. Anderson Hospital and Tumor Institute, Houston, Texas

Arnold Meisler, University of Rochester Medical Center, Rochester, New York

Martin L. Meltz, Southwest Foundation for Research and Education, San Antonio, Texas

Carl R. Merril, National Institute of Mental Health, Bethesda, Maryland

Nora L. Meuth, Massachusetts Institute of Technology, Cambridge, Massachusetts

S. Michel, University of Wisconsin, Madison, Wisconsin

Satoshi Mizatani, University of Wisconsin Medical Center, Madison, Wisconsin

Luc Montagnier, Institut du Radium, Orsay, France

Dan H. Moore, Institute for Medical Research, Camden, New Jersey

William A. Newton, Jr., M. D. Anderson Hospital and Tumor Institute, Houston, Texas

Sam K. Ng, Indiana University, Bloomington, Indiana

Alexander L. Nussbaum, Hoffman-LaRoche, Inc., Nutley, New Jersey

Kazuyo Oda, Keio University School of Medicine, Tokyo, Japan

Yasuko Ogata, Keio University School of Medicine, Tokyo, Japan

Samuel Oleinick, Duke University Medical Center, Durham, North Carolina

Wade Parks, National Cancer Institute, Bethesda, Maryland

D. Perlman, University of Wisconsin, Madison, Wisconsin

Quentin A. Pletsch, University of Maryland School of Medicine, Baltimore, Maryland

Morris Pollard, University of Notre Dame, Notre Dame, Indiana

Pradman K. Qasba, University of Maryland School of Medicine, Baltimore, Maryland

Marvin S. Reitz, Bionetics Research Laboratories, Inc., Bethesda, Maryland

Philip Roane, Howard University School of Medicine, Washington, D.C.

Martin S. Robin, St. Louis University School of Medicine, St. Louis, Missouri

Robert H. Rownd, University of Wisconsin, Madison, Wisconsin

H. J-P Ryser, University of Maryland School of Medicine, Baltimore, Maryland

Tatsuo Saito, Keio University School of Medicine, Tokyo, Japan

Samuel Salzberg, St. Louis University School of Medicine, St. Louis, Missouri

Leo T. Samuels, University of Utah, Salt Lake City, Utah

P. S. Sarin, National Cancer Institute, Bethesda, Maryland

Nural H. Sarkar, Institute for Medical Research, Camden, New Jersey

M. C. Sarngadharan, Bionetics Research Laboratories, Inc., Bethesda, Maryland

J. Schlom, Institute for Cancer Research, Columbia University, New York, New York

Edward Scolnick, National Cancer Institute, Bethesda, Maryland

V. S. Sethi, University of Maryland School of Medicine, Baltimore, Maryland

G. Shanmugam, St. Louis University School of Medicine, St. Louis, Missouri

D. J. Sherratt, University of California, La Jolla, California

Robert Silber, New York University Medical Center, New York, New York

R. Graham Smith, National Cancer Institute, Bethesda, Maryland

Donna F. Smoler, Massachusetts Institute of Technology, Cambridge, Massachusetts

Sol Spiegelman, Institute for Cancer Research, Columbia University, New York, New York

Katomi Sugawara, Keio University School of Medicine, Tokyo, Japan

D. Taylor, University of Wisconsin, Madison, Wisconsin

Howard M. Temin, University of Wisconsin Medical Center, Madison, Wisconsin

Frances M. Thompson, University of California, Berkeley, California

George J. Todaro, National Cancer Institute, Bethesda, Maryland

Nobuo Tsuchida, St. Louis University School of Medicine, St. Louis, Missouri

Giancarlo Vecchio, St. Louis University School of Medicine, St. Louis, Missouri

Inder M. Verma, Massachusetts Institute of Technology, Cambridge, Massachusetts

P. Vigier, Institut du Radium Orsay, France

Tsutomu Watanabe, Keio University School of Medicine, Tokyo, Japan

John W. Winkert, Howard University Medical School, Washington, D.C.

D. B. Yelton, University of Maryland Medical School, Baltimore, Maryland

Yoshitsuru Yokoyama, Keio University School of Medicine, Tokyo, Japan

INTRODUCTION

Roland F. Beers, Jr.

The Johns Hopkins University School of Hygiene and Public Health
Baltimore, Maryland

The concept of cell transformation by an exogenous agent originated at the turn of the century. The French investigators, Borrel (1) and Bosc (2), in 1903 proposed the theory that the etiological agent for cancer was a virus. Eight years later Rous and his associates (3) demonstrated the viral etiology of spontaneous chicken sarcomas. In the decades to follow, partly, perhaps, because of the propensity of most investigators to search for a common etiological mechanism for malignant cell transformation, the viral hypothesis has had a very controversial history. Fortunately, Peyton Rous was blessed with longevity and was awarded the Nobel Prize shortly before his death at the age of ninety-one.

The fact that virus infections involved a form of transformation of the infected cell was not appreciated for many decades after their discovery by several French investigators in 1892 (4–6), although it was clearly recognized by the late thirties that viruses could not carry out metabolic processes independent of the host cell. It is of interest to note that as recently as 1939 the major chemical constituent of the virus was believed to be its protein, with little or no significance attached to its nucleic acid component. Perhaps this view was influenced by the metabolic activity of protein as both enzymes and antigens. The information about bacteriophages was even more fragmentary.

The ability of nucleic acids to enter the cell and serve as exogenous genomes was first demonstrated by Avery (7) in 1944, at a time when molecular biology was just coming of age. Cell transformation became a concept distinct from or, perhaps, distinguished as a special kind of cell mutation. Within a very short time the phenomenon of pneumococcal transformation by DNA was seen to have many characteristics in common with cell responses to phage and viral infections. It was clear that the metabolic machinery of the host cell was "directed" by the exogenous agent to carry out functions that were not dictated by the genetic determinants of the cell.

At about the same time the concept of a plasmagene was introduced to account for cytoplasmic inheritance (cf., Brachet, J. Embryologic Clinique. Masson, Paris, 1954.) It was a term that quickly fell into disrepute because it was used to account for a variety of cytoplasmic functions and was associated with particulates of the cytoplasm, such as chromidia and microsomes. In an interesting theory put forth at this time, Spiegelman (8) proposed that the synthesis of adaptive enzymes, that is, enzymes which appeared in the cell only in the presence of a specific substrate, occurred by means of a replicating enzyme-substrate complex which he called a plasmagene.

Others attempted to identify plasmagenes according to their nucleic acid content. The issue of replicating bodies in the cytoplasm still remains a major question today, enhanced, perhaps, by the discovery of mitochondrial DNA.

It is of interest today to reflect on one or two related questions raised about the time molecular biology came into its own. For example, in 1943 Earle and Nettleship (9) demonstrated the transformation of normal mouse fibroblasts into malignant cells when grown in tissue cultures. The ability of a cell to retain its malignant characteristics in tissue culture has been amply demonstrated, especially by the famous HeLa cell strain isolated by the late George Gey (10). Are these transformations and malignant traits the results of the derepression of an endogenous genome brought about by changes in environmental conditions of the cell or are they the result of exogenous agents transmitted to the cell?

Related to this is the question of whether the transformed cell is maintained by continued expression of the genome(s). Is the transformation a process or an event? The failure to isolate transmissible agents from many tumor cells may reflect the latter.

Another issue that has often confused rather than clarified the picture of cell transformation is one of semantics. For example, in the early forties, Harry Greene (11) of Yale introduced an operational definition of a malignant cell without regard to its morphology. A tumor cell was malignant if it could survive and undergo mitosis in a heterologous host. The experimental model was the successful growth of the tumor in the anterior chamber of the eye of a hamster or other animal. It is obvious that such a definition must take into consideration the role of immunological responses of the host organism to the invasive tumor tissue. The deficiency of this definition of malignant transformation is reflected in the fact that embryonic tissue displays the same degree of autonomy. Nevertheless, this approach does bring into focus the importance of host factors responding to a particular transformed cell. Many of the perplexing questions raised about all transformation may exist because the models used, especially tissue cultures and cell clones, do not include the environment of that cell as it exists in the intact animal. The deficiencies of such models have long been emphasized by Paul Weiss, who gave his Edward Kennaway Memorial Lecture of 1969 the title: "A cell is not an island of itself" (12), a point which molecular biologists are prone to ignore.

The subjects of this two-day symposium find their origins in the brief historical account I have recited. Although many of the questions asked in the first half of this century appear in retrospect to be rather naive, they do so only because today we have the product of thirty years of extremely productive research in molecular biology. That product is a set of models of subcellular systems which can accommodate the phenomenon of cell transformation.

I would close these introductory remarks with a precautionary note. Although the emphasis today, for many obvious reasons, is on the etiology of cancer, we should not lose sight of the fact that transformation of a cell through its genetic apparatus either in the nucleus or possibly in the cytoplasm is a more fundamental issue. The historical lines of battle that have existed since the turn of the century between the virologists and the oncologists have been almost obliterated by the emergence of the new breed, the molecular biologists. There is a commonality of interests and systems in oncogenic and viral processes as well as in genetic engineering or gene therapy. The differences may prove to be more of intent of purposes and goals than in specific mechanisms; hence, the desirability of bringing these fields together at this Symposium.

REFERENCES

1. Borrel, A.: Epithelioses infectienses et epithéliomas. Inst. Pasteur. Ann., 17:81, 1903.
2. Bosc., F. J.: Les epithéliomas parasitaires. La clavalée et l'epithelioma claveleux. Zbl. Bakt. I. Abt. Orig., 34:413, 1903.
3. Rous, P.: Transmission of a malignant new growth by means of a cell free filtrate. JAMA, 56:198, 1911.
4. Iwanowski, von D.: Ueber die mosaikkrankheit der tabakspflanze. Zbl. Bakt. II Abt., 5:250, 1899.
5. Beijerink, von M. W.: Ueber ein contagium vivum fluidum als urache de fleckenkrankheit der tabakblätter. Zbl. Bakt. II. Abt., 5:27, 1899.
6. Loeffler and Frosch: Berichte de Koimmission zur erforschung der Maulund Klanenseuche bei dem Institut für infections krankheiten in Berlin. Zbl. Bakt. I. Abt., 23:371, 1898.
7. Avery, O., Macleod, C., and McCarthy, M.: Chemical nature of a substance inducing transformation of pneumococcal types–induction of transformation by a desoxyribonucleic acid fraction isolated from pneumococcus Type III. J. Exp. Med., 79:137, 1944.
8. Spiegelman, S.: Nuclear and cytoplasmic factors controlling enzymatic constitution. Cold Spring Harbor Symp. Quant. Biol., 11:256, 1936.
9. Earle, W. R., and Nettelship, A.: Production of malignancy in vitro. Part V. Results of injections of cultures into mice. J. Nat. Cancer Inst., 4:213, 1943.
10. Gay, G. O.: Some aspects of the constitution and behavior of normal and malignant cells maintained in continuous culture. Harvey Lecture Series, 50:154, 1954.
11. Greene, H. S. N.: Familial mammary tumors in the rabbit IV. The evolution of autonomy in the course of tumor development as indicated by transplantation experiments. J. Exp. Med., 71:305, 1949.
12. Weiss, P. A.: A cell is not an island entire of itself. Perspect. Biol. Med., 14:182, 1971.

Part I

1. CURRENT WORK ON THE SYNTHESIS OF TRANSFER RNA GENES[1]

Marvin H. Caruthers

The Departments of Biology and Chemistry
Massachusetts Institute of Technology
Cambridge, Massachusetts 02139

INTRODUCTION

The long-range aim of our present work is the development of methods for the total synthesis of biologically specific duplexes. Transfer RNA genes—the double-stranded DNA corresponding in sequence to the transfer RNA—were chosen for these studies because tRNAs are short, biologically active molecules of known function and sequence. The availability of DNAs with defined nucleotide sequence will provide opportunities for studies of several biologically important processes. For example, despite much effort, the relationship of structure to function for tRNA is still an open question. As an approach to this problem, different parts of the tRNA structure could be systematically modified at the gene level and the RNA transcript examined for biological activity. Furthermore, the RNA transcript from tRNA genes could be used to examine the relationship of various base modifications to structure and function. These completely defined DNA duplexes are now being used for studies on transcription of DNA into RNA with RNA polymerase (1,2) and to initiate work on the promoter and terminator sequences of E. coli tyrosine suppressor gene of the transducing bacteriophage $80psu^{+}_{III}$(3). These defined duplexes have also been used for examination of approaches to DNA sequencing (4).

The methodology for synthesis of DNA duplexes of defined sequence was first established with the completion of the structural gene for yeast alanine transfer RNA (5). More recently, we have initiated a program for synthesizing the E. coli tyrosine tRNA suppressor gene. The large amount of data on the molecular biology and genetics of E. coli prompted us to select a gene from this organism. Furthermore, the extensive work at Cambridge on the tyrosine suppressor gene (6) provides a system for analyzing biological activity of any laboratory-produced genes or gene fragments.

The results outlined in this paper will first summarize our progress on the synthesis of the E. coli tyrosine tRNA suppressor gene and then outline a general approach being used to sequence the terminator and initiator regions of this gene.

[1] This work has been supported by grants from the National Cancer Institute of the National Institutes of Health, U.S. Public Health Service (Grants No. 73675, CA05178), The National Science Foundation, Washington, D.C. (Grants No. 73078, G.B.-7474x), and the Life Insurance Medical Research Fund.

ALANINE t-RNA
(NUCLEOTIDES 21-50)

```
                                                                                    END
50 49 48 47 46 45 44 43 42 41 40 39 38 37 36 35 34 33 32 31 30 29 28 27 26 25 24 23 22 21 ↓
G—C—U—C—C—C—U—U—I—G—C—IMε-ψ—G—G—G—A—G—A—G-H2U-C—U—C—C—G—G—T—ψ—C (3')-RIBO
               32P OH
                |   |
      G-A-A-T-C   G-T-A-C-C-C-T-C-T-C-A-G-A-G-G-C-C-A-A-G (5')-DEOXY
                  | | | | | | | | | |
G-C-T-C-C-C-T-T-A-G-C-A-T-G-G-G-A-G-A-G T-C-T-C                (3')-DEOXY
                                      | |
                                     HO P32
```

Fig 1. Chemically synthesized polydeoxynucleotides corresponding to sequence 21 – 50 of yeast alanine tRNA. The icosanucleotide with the 5′-guanosine to the right represents a sequence complementary to nucleotides 21 – 40 of the tRNA and has polarity opposite to that of the tRNA; the pentanucleotide d-C-T-A-A-G contains a sequence complementary to nucleotides 41 – 45 of the tRNA and also has polarity opposite to that of the tRNA. The icosanucleotide with the 5′-guanosine to the left and the tetranucleotide d-T-C-T-C are segments with sequence and polarity identical to the transfer RNA. ^{32}P represents the 5′-phosphate end group wherever shown. In the synthesis, the assumption has been made that the rare bases present in the tRNA arise by subsequent modification of the four standard bases used by the transcribing enzyme. Thus inosine is formed by deamination of adenosine and so comes from an A-T base pair in DNA.

SYNTHESIS OF tRNA SUPPRESSOR GENE

The synthesis of tRNA genes is carried out by a combination of chemical and enzymatic procedures. The general methodology is illustrated in Figure 1. The tetra-, penta-, and two icosanucleotides were first chemically synthesized. When properly aligned, using suitable conditions of temperature and Mg^{2+}, the T4 polynucleotide ligase and ATP bring about joining of the tetra- and pentanucleotides to the appropriate icosanucleotides (7). These results illustrate that oligonucleotides of much longer chain length can be synthesized using a combination of enzymatic and chemical techniques. The strategy for synthesis of bihelical DNA is therefore as follows: 1. chemical synthesis of deoxypolynucleotide segments of chain length in the range 8 to 12 units with free 3′- and 5′-hydroxyl end groups; the segments would represent the entire two strands of the DNA, and those belonging to the complementary strands would have an overlap of four to five nucleotides; 2. the phosphorylation of the 5′-hydroxyl group with ATP carrying a suitable label in the γ-phosphoryl group using the T4 polynucleotide kinase; and 3. the head-to-tail joining with the T4 polynucleotide ligase of the appropriate segments when they are aligned to form bihelical complexes.

The development of this methodology led to the following plan for synthesis of the precursor transfer RNA gene for E. coli su^{+}_{III} tyrosine tRNA. The sequence for this tyrosine precursor transfer RNA as established by Altman and Smith (8) is shown in Figure 2. During maturation the sequence 86–126 is specifically removed. The sequence of the gene corresponding to this tRNA is shown in Figure 3. The numbering begins with the deoxynucleotide corresponding to the 3′-end of the transfer RNA. Nucleotide 126 is the guanosine triphosphate 5′-terminus of the tRNA. This gene sequence assumes that all base modifications such as thiouracil formation take place after transcription. Therefore, at the position of thiouracil, we insert an A-T base pair in the gene sequence. For chemical and enzymatic synthesis, we have divided the DNA duplex into 26 oligonucleotides containing from 4 to 18 nucleotides each. The

E. COLI TYROSINE tRNA PRECURSOR

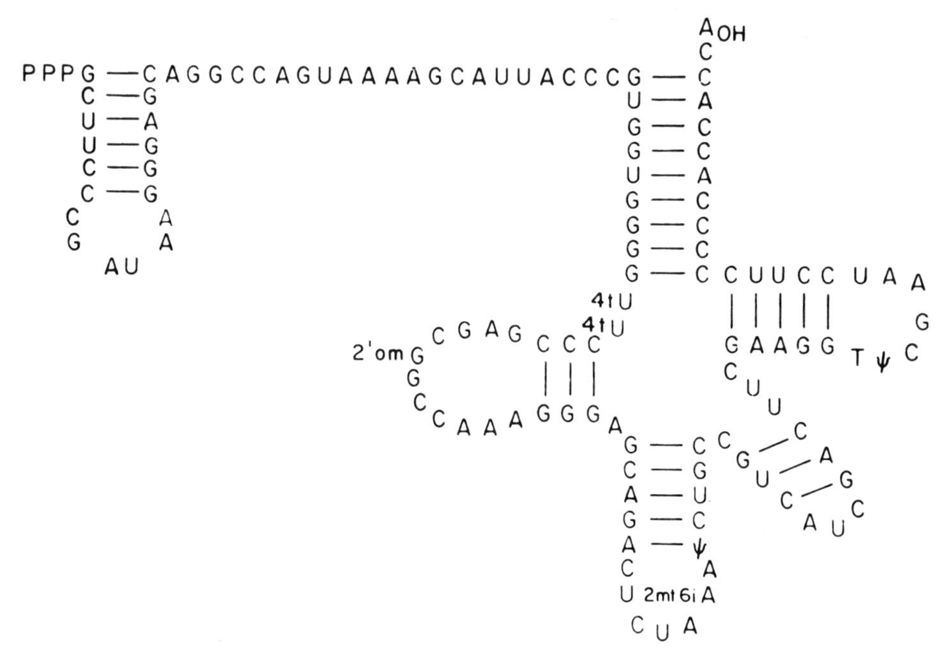

Fig 2. Ribonucleotide sequence of the E. coli tyrosine precursor tRNA. See Altman and Smith for details (8).

brackets, with the serial numbers inserted into the brackets, designate the chemically synthesized oligonucleotides. The segments we have chosen to synthesize overlap one another. As a consequence, a single-stranded region of the growing duplex can be used to guide the next segment into place for the ligase reaction.

The chemical synthesis of each segment shown in brackets requires careful planning with respect to the choice of protected blocks (9). Each segment is usually constructed from preformed di-, tri-, and tetranucleotides. As an illustration of the chemical synthesis, segment 2 is synthesized in the following manner. Our syntheses always start from the 5′-hydroxyl end. Therefore the formation of the first phosphodiester bond in segment 2 involves condensation between N^6-anisoyl-5′-monomethoxytrityldeoxycytidine (d-MMTrC^{An}) and the protected dinucleotide N^6-anisolyl-5′-phosphoryl-deoxycytidylyl-(3′ - 5′)-N^6 - anisoyl-3′ - acetyldeoxycytidine (d-pC^{An}pC^{An} - OAc). Following a simple extraction procedure the trinucleotide (d-MMTrC^{An}pC^{An}) is condensed sequentially with four dinucleotide blocks to form the undecanucleotide. After each condensation, the product is purified from starting materials on DEAE-cellulose columns and then used for the next step. When the chemical synthesis has been completed, all segments are purified, first, at the fully protected stage and, second, after removal of the protecting groups. The final step is purification on DEAE-cellulose in the presence of 7M urea. The chemical synthesis of all but one of the fragments corresponding to the tyrosine structural gene has been achieved. The synthesis of all precursor fragments is well underway and should be completed shortly.

E. COLI TYROSINE t RNA PRECURSOR

```
26 25 24 23 22 21 20 19 18 17 16 15 14 13 12 11 10 9 8 7 6 5 4 3 2 1
              (5)                        (3)                (1)
T-C-C-A-A-G-C-T-T-A-G-G-A-A-G-G-G-G-T-G-G-T-G-G-T-(5')
    T-C-G-A-A-T-C-C-T-T-C-C-C-C-A-C-C-A-C-C-A-(3')
              (4)                        (2)

51 50 49 48 47 46 45 44 43 42 41 40 39 38 37 36 35 34 33 32 31 30 29 28 27 26 25 24 23
              (9)                        (7)
      A-G-A-C-G-G-C-A-G-T-A-G-C-T-G-A-A-G-C-T        -(5')
C-T-A-A-A-T-C-T-G-C-C-G-T-C-A-T-C-G-A-C-T-T-C-G-A-A-G-G-T-(3')
      (10)                    (8)                (6)

70 69 68 67 66 65 64 63 62 61 60 59 58 57 56 55 54 53 52 51 50 49 48 47
              (13)                       (11)
      G-T-T-T-C-C-C-T-C-G-T-C-T-G-A-G-A-T-T-T-(5')
C-G-G-C-C-A-A-A-G-G-G-A-G-C-A-G-A-C-T          -(3')
      (14)                    (12)

94 93 92 91 90 89 88 87 86 85 84 83 82 81 80 79 78 77 76 75 74 73 72 71 70 69 68 67
              (18)                       (15)
      G-G-C-A-C-C-A-C-C-C-C-A-A-G-G-G-C-T-C-G-C-C-G-(5')
A-T-T-A-C-C-C-G-T-G-G-T-G-G-G-G-T-T-C-C-C-G-A-G    -(3')
      (19)                    (17)               (16)

113 112 111 110 109 108 107 106 105 104 103 102 101 100 99 98 97 96 95 94 93 92 91 90
              (22)                       (20)
      T-C-C-G-G-T-C-A-T-T-T-T-C-G-T-A-A-T-G-(5')
G-G-A-G-C-A-G-G-C-C-A-G-T-A-A-A-G-C              -(3')
      (23)                    (21)

126 125 124 123 122 121 120 119 118 117 116 115 114 113 112 111 110 109
                         (24)
C-G-A-A-G-G-G-C-T-A-T-T-C-C-C-T-C-G-(5')
G-C-T-T-C-C-C-G-A-T-A-A-G            -(3')
      (26)                (25)
```

Fig 3. Total plan for the synthesis of the su$^{+}_{III}$ gene of E. coli tyrosine precursor tRNA. The chemically synthesized segments are in brackets, the serial number of the segment being shown within the brackets. The assumption stated in the last sentences of the legend to Figure 1 is valid here for derivation of the gene sequence.

Enzymatic joining of chemically synthesized oligonucleotides corresponding to the first 66 nucleotides (segments (1-13) has been completed (10). One of the ligase-catalyzed joining reactions is shown in Figure 4. This figure illustrates the T4 ligase-catalyzed joining and purification of the duplex from segments 6, 7, 8, 9, and 10. Segments 6, 8, 9, and 10 contain ^{33}P at the 5′ position. These segments are first annealed at high temperature in the presence of $MgCl_2$ and slow-cooled to 5°C. Dithiothreotol, ATP, and T4 ligase are then added. The kinetics of joining (inset of Figure 4) is followed by measuring resistance to bacterial alkaline phosphatase. The general conditions for assay on DEAE-paper strips have been described (7,11). The initial fast reaction is followed by a slow second phase. After two days, the extent of reaction is about 40%. EDTA is added and the joined products are separated from starting materials on an agarose 0.5M column. The first peak from the column contains the desired joined duplex of segments 6, 7, 8, 9, and 10. This duplex is characterized by: 1. Resistance to phosphatase; 2. degradation to 3′ nucleotides, using the micrococcal and the spleen phosphodiesterases; and, 3. hydrolysis to 5′ nucleotides, using pancreatic deoxyribonuclease and venom phosphodiesterase. Results of this characteri-

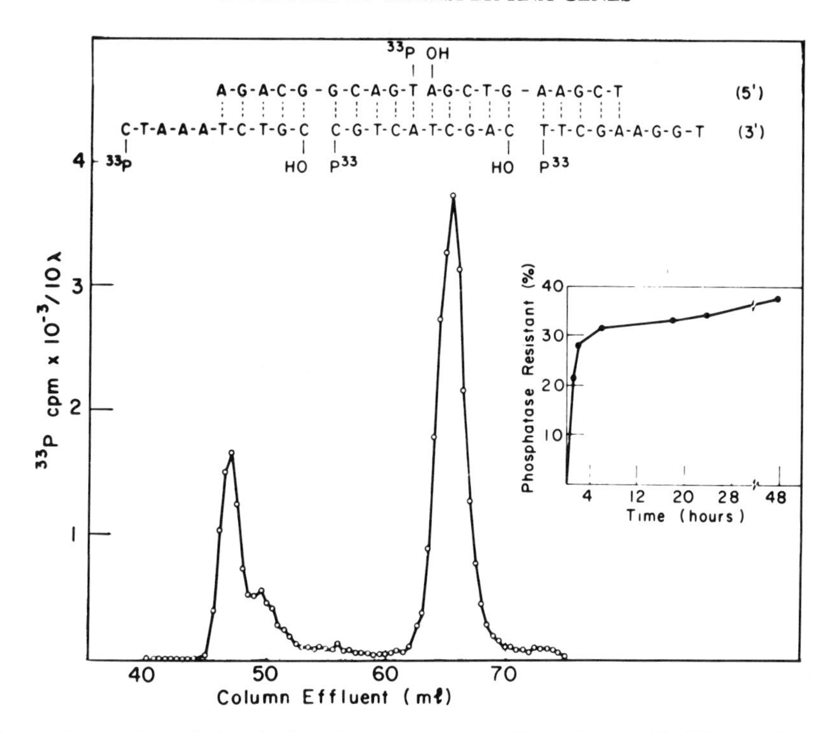

Fig 4. Preparation of the duplex from segments 6, 7, 8, 9, and 10. The reaction mixture contained in 0.23ml: segments 6, 8, 9, and 10, all phosphorylated at the 5′ end using ^{33}P-ATP of the same specific activity, segment 7 (unphosphorylated), 20mMTris-Cl (pH 7.6) and 8mM $MgCl_2$. The final concentration of each segment was 5μM. The mixture was heated to 90°C and slow cooled to 5°C. ATP and dithiothreotol were added and the mixture incubated with T4 ligase at 5°C. At various times of incubation, aliquots were assayed for resistance to bacterial alkaline phosphatase. When the reaction reached a plateau (kinetics in the inset to Fig), EDTA was added (20mM) and the reaction mixture was passed through a 0.5M agarose column (1 times 100cm), the elution buffer being 0.1M TEAB. The first peak, excluding the shoulder, was pooled and characterized as in Table 1.

zation are in Table 1. Degradation to 5′-nucleotides gives significant radioactivity only in thymidylic and cytidylic acid. This is expected, since there are two C and two T 5′ labels in the duplex. The resistance to bacterial alkaline phosphatase (76%) is also as expected (75%). Degradation to 3′-nucleotides after phosphatase should give transfer to cytidylic and adenylic acid. This is observed and the ratio is as anticipated.

For the complete transcription of a piece of synthetic DNA, RNA polymerase must recognize a promoter region on the DNA, transcribe the gene, and terminate the transcription in the terminator region. To do this on a synthetic piece of DNA, we must either construct the promoter and terminator regions onto our structural gene, or discover a method whereby RNA polymerase will initiate synthesis at the beginning of the structural gene. Recent experiments on the latter possibility have shown that RNA polymerase can be made primer-dependent when transcribing synthetic DNA templates (1,2). These experiments used a 7 unit oligoribonucleotide primer with 6 bases complementary to a 29 unit deoxyribopolynucleotide template. Further experiments on synthetic initiation of transcription are in progress, but I will not discuss them at this time. The alternate possibility—construction of the promoter and terminator

TABLE I

Characterization of the Duplex from Segments 6, 7, 8, 9, and 10

		3′-Nucleotide Analysis (cpm)	
Ap	Gp	Tp	Cp
954(1.00)	52	51	1980 (2.08)[a]
		5′-Nucleotide Analysis (cpm)	
pA	pG	pT	pC
236	334	9561(1.01)	9529 (1.00)
		5′-Nucleotide Analysis after Phosphatase (cpm)	
pA	pG	pT	pC
163	154	3557 (2.18)	1619(1.00)

[a]The numbers in parentheses following counts represent experimental molar ratios.

regions—requires that the sequence of nucleotide bases must first be known. Then oligonucleotides corresponding to these sequences could be chemically synthesized and enzymatically joined to the structural gene. The following experiments were carried out (12) in order to lay the foundation for sequencing studies on the tyrosine suppressor tRNA initiator and terminator regions.

SEQUENCE OF THE TERMINATOR AND INITIATOR REGIONS

The approach to sequencing the terminator and promoter regions consists of the following steps: 1. Prepare oligonucleotides by chemical and enzymatic procedures with sequence identical to the known structural gene; 2. separate the DNA of phage $80psu^{+}_{III}$ into its r- and l-strands; 3. hybridize the oligonucleotides to the correct phage DNA strand; 4. carry out repair synthesis using DNA polymerase I and labeled nucleoside triphosphates; 5. sequence the new piece of DNA. The phage $\phi80psu^{+}_{III}$ contains the E. coli suppressor gene su^{+}_{III}. This gene, which normally in E. coli codes for the minor tyrosine transfer RNA, maps near the attachment site of phage $\phi80$ and has been incorporated as an su^{+}_{III} suppressor mutation into $\phi80$-derived transducing phages (12). The r-strand of phage $\phi80$ psu^{+}_{III} is the strand from which the tRNA is made in vivo (14). As shown in Figure 5, by using the r-strand, fragments of the structural gene with sequence identical to the tRNA, and DNA polymerase, repair synthesis into the terminator region of this gene is possible. Sequencing can be carried out in a manner similar to the approach used to sequence the cohesive ends of the bacteriophages λ (15,16) and 186 (17). The same approach can be used to obtain information on the promoter sequence. For this system, the 1-strand with opposite

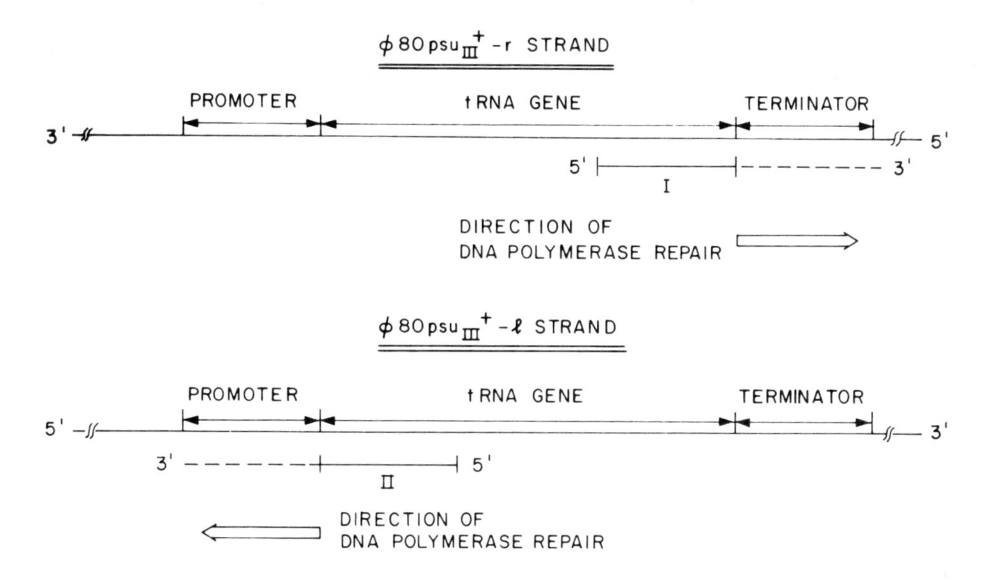

Fig 5. Outline of an approach to sequencing terminator and promoter regions of the E. coli tyrosine t RNA gene from ϕ80psu$^{+}_{III}$ phage. The top part of the slide shows the ϕ80psu$^{+}_{III}$ r-strand containing the promoter, structural, and terminator regions of the tyrosine gene. The 5′ end is to the right. The oligonucleotide I with 5′ end to the left is a chemically and enzymatically synthesized piece of DNA with sequence identical to the sequence of the 3′ end of E. coli tyrosine tRNA. The bottom part of the slide shows the ϕ80 psu$^{+}_{III}$ 1-strand containing the same regions of the tyrosine gene. The 5′ end is to the left. The oligonucleotide II with 5′ end to the right is a chemically and enzymatically synthesized piece of DNA with sequence complementary to the 5′ end of the E. coli tyrosine tRNA. Direction of DNA polymerase repair synthesis is shown by the arrow.

polarity and fragments of the structural gene with sequence complementary to the transfer RNA must be used.

In order to anneal an oligonucleotide to the r-strand of the phage DNA, the oligonucleotide must first be separated from its complement. Figure 6 shows how we separate the strands of a chemically and enzymatically synthesized duplex. The synthesis of this duplex is outlined in Figure 4. The duplex is first denatured at high temperatures and then filtered through a G-75 sephadex column at 65°C. The first peak off the column is the 29 unit oligonucleotide with three ^{33}P-labeled phosphates. The second peak is the 20 unit oligonucleotide with only one ^{33}P-labeled phosphate. The total counts reflect the 3 to 1 ratio of radioactivity in each strand. When the first peak is degraded to 5′-mononucleotides, radioactivity is found only in d-pC and d-pT, the ratio being 1.94. This correlates well with the theoretical ratio of 2 to 1.

The 29 unit oligonucleotide or polynucleotide II can be annealed to the DNA r-strand of phage ϕ80psu$^{+}_{III}$. The annealing conditions are outlined in the legend to Figure 7. After the annealing step, the duplex of the r-strand and polynucleotide II is separated from excess oligonucleotide as shown in Figure 7. The solid line traces the counts observed first in the duplex of high molecular weight DNA and polynucleotide II and next at the position of excess polynucleotide II. The dotted line traces the

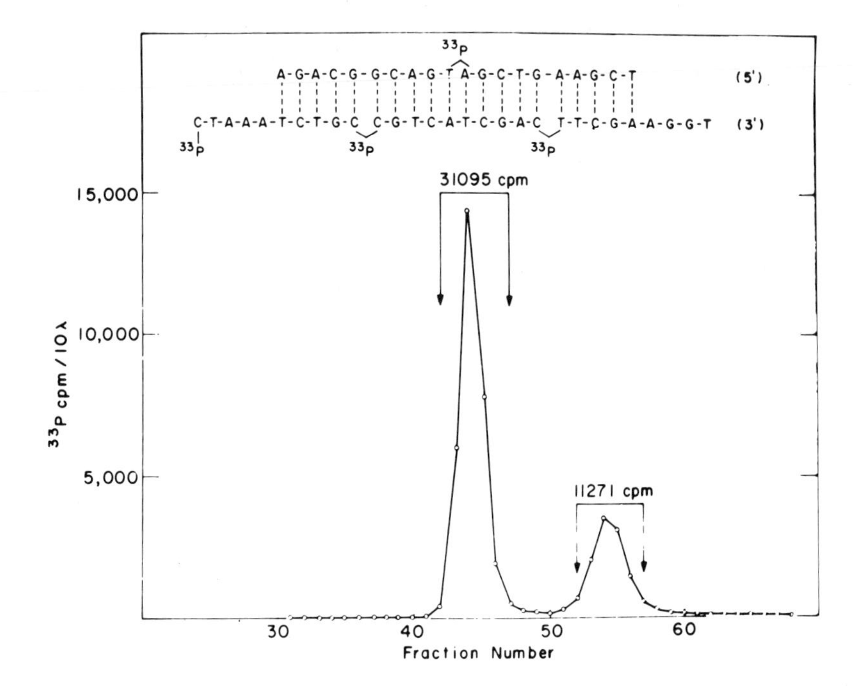

Fig 6. Strand separation of the duplex whose synthesis is outlined in Figure 4. The duplex was heated to 95°C for 2 minutes in 10mM KPO_4 (pH 7.6). The peaks were pooled as shown.

pattern observed when the same experiment is carried out using wild type ϕ80 r-strand DNA which does not contain the tyrosine gene.

In order to show further the specificity of the annealing, polynucleotide II has been joined to the 22 unit deoxypolynucleotide (polynucleotide I) whose sequence corresponds to the 3′-end of the tyrosine tRNA. Deoxypolynucleotide I, therefore, base pairs to the r-strand at a position adjacent to polynucleotide II. Schematically, the system is shown in the top portion of Figure 8. The first step, as before, is annealing of polynucleotide I containing a ^{32}P label and polynucleotide II to the DNA r-strand as outlined in the legend to Figure 8. After separation on an agarose 1.5M column, the annealed duplex is treated with T4 ligase and ATP. The joining reaction is quite fast and complete within an hour. In a control experiment with only $\phi80psu^{+}_{III}$ and polynucleotide I no joining takes place. The product is separated from high molecular weight DNA and starting materials at 76°C on G-100 sephadex column. The pattern for this separation (Figure 8) shows that the joined deoxypolynucleotide I–II nicely separates from $\phi80psu^{+}_{III}$ r-strand and excess deoxypolynucleotide I. The analysis of the joined deoxypolynucleotide indicates transfer of the label completely to d-Tp as expected. Based upon first the specificity of annealing to $\phi80psu^{+}_{III}$ (Figure 7) and second the joining of deoxypolynucleotides I and II by the T4 ligase, the annealing of deoxypolynucleotides I and II appears to be site specific.

The next step is to demonstrate that polynucleotide I can serve as primer for the DNA polymerase repair reaction. Recently in our laboratory, Dr. Robert Miller and Dr. J. H. van de Sande have shown that this deoxypolynucleotide does indeed serve as

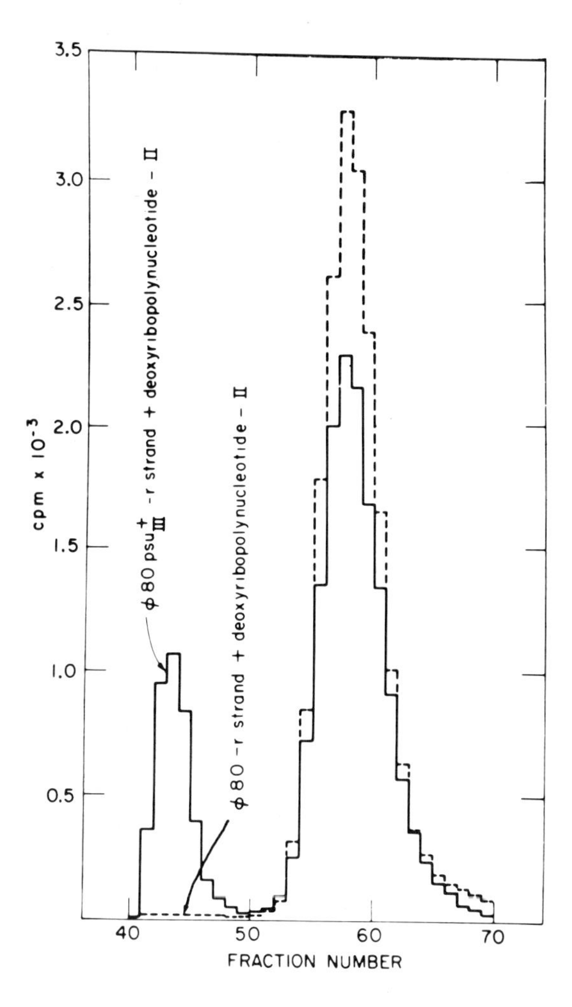

Fig 7. Annealing of polynucleotide II to ϕ 80psu$^{+}_{\rm III}$ r-strand. ϕ 80psu$^{+}_{\rm III}$ r-strand or ϕ80 r-strand (0.5$\mu\mu$ mole) and 3$\mu\mu$ mole of polynucleotide II in 0.1ml of 0.3M NaCl and 0.1M sodium citrate, pH 7.0, are heated to 95°C for 2 minutes and then incubated for 15 hours at 55°C. The mixture was then passed through an agarose 1.5M column (20 times 0.7cm) with a 50mM potassium phosphate, pH 7.6, as eluting buffer and 0.15ml fractions being collected. The fractions were then tested for radioactivity. ———— using ϕ 80psu$^{+}_{\rm III}$ r-strand; -------- using ϕ80 r-strand.

primer for repair synthesis. Experiments such as these are now being carried out to obtain information on the terminator region of this gene.

These experiments demonstrate: 1. our approach to the total synthesis of DNA duplexes of defined sequence; and 2. how we are using these oligonucleotides as probes to obtain information on termination and initiation sequences in DNA. By the combination of these chemical and enzymatic techniques, we hope first to sequence and then synthesize and add these control elements to our structural gene. The control of the transcriptional process using RNA polymerase and synthetic DNAs should then be possible.

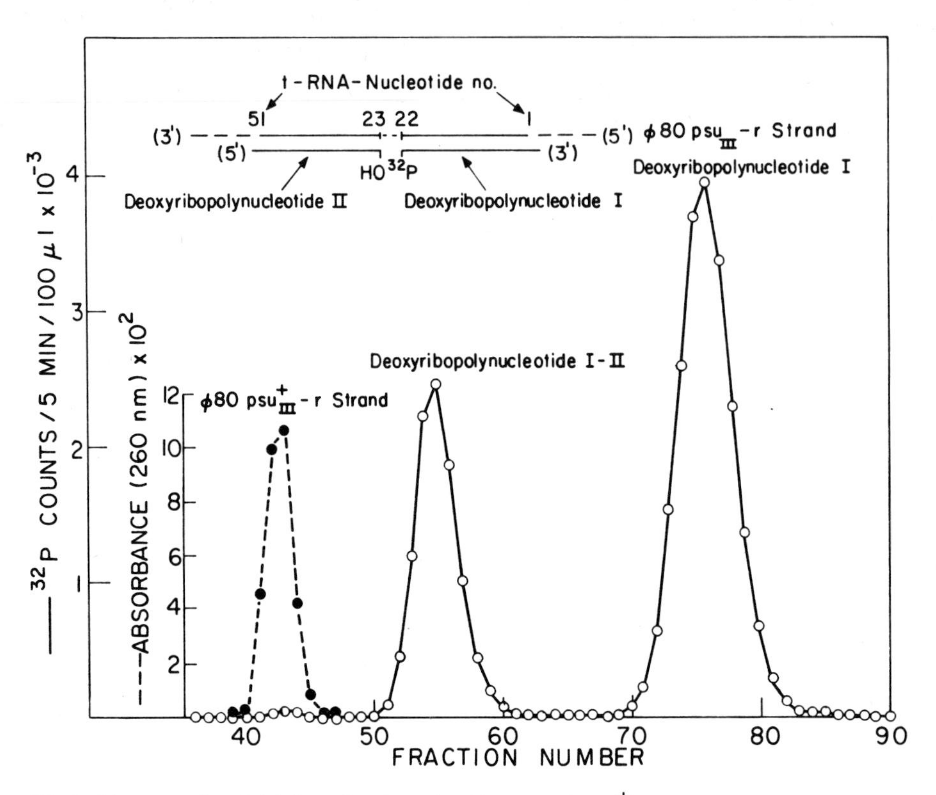

Fig 8. Ligase catalyzed joining of polynucleotides on ϕ 80psu$^{+}_{III}$ r-strand. The top part of this figure shows the ϕ 80psu$^{+}_{III}$ r-strand with 5′ end to the right. Deoxypolynucleotide II is composed of segments 6, 8, and 10 and its isolation was outlined in Figures 4 and 6. Deoxypolynucleotide I is composed of segments 2 and 4 (see Fig 5). Its isolation was similar to deoxypolynucleotide II (see ref 3 for details). These deoxypolynucleotides are annealed in 0.3M NaCl and 0.1M citrate pH 7.0 at 55°C for 15 hours. The duplex of r-strand, deoxypolynucleotides I and II is separated from excess deoxypolynucleotides on a 1.5 M agarose column. The hybridized product was supplemented with ATP (64 M), dithiothreotol (10mM), $MgCl_2$ and T4 ligase. The separation pattern at 76°C on G-100 is shown in the bottom part of the figure.

REFERENCES

1. Kleppe, R., and Khorana, H. G.: J. Biol. Chem., 1972, in press.
2. Terao, T., Dahlberg, J. E., and Khorana, H. G.: J. Biol. Chem., 1972, in press.
3. Bessmer, P., Miller, Jr., R. C., Caruthers, M. H., Kumar, A., Minamoto, M., van de Sande, J. H., Sidarova, N., and Khorana, H. G.: J. Mol. Biol., 1972, in press.
4. van de Sande, J. H., Loewen, P. C., and Khorana, H. G.: J. Mol. Biol., 1972, in press.
5. Agarwal, K. L., Buchi, H., Caruthers, M. H., Gupta, N., Khorana, H. G., Kleppe, K., Kumar, A., Ohtsuka, E., RajBhandary, U. L., van de Sande, J. H., Sgaramella, V., Weber, H., and Yamada, T.: Total synthesis of the gene for an alamine transfer ribonucleic acid from yeast. Nature, 227:27, 1970.
6. Russell, R. L., Abelson, J. W., Landy, A., Gefter, M. L., Brenner, S., and Smith, J. D.: Mutant tyrosine transfer ribonucleic acids. J. Mol. Biol., 47:1, 1970.
7. Gupta, N. K., Ohtsuka, E., Sgaramella, V., Buchi, H., Kumar, A., Weber, H., and Khorana, H. G.: Studies on polynucleotides, XC, DNA polymerase–catalyzed repair of short DNA duplexes with single-stranded ends. Proc. Nat., Acad. Sci., USA 61:215, 1968.
8. Altman, S., and Smith, J. P.: Tyrosine t-RNA precursor molecule polynucleotide sequence. Nature New Biology, 233:35, 1971.

9. Khorana, H. G.: Nucleic acid synthesis. Pure Appl. Chem., 17:349, 1968.
10. Bessmer, P., et al: Process in the total synthesis of the gene for tyrosine suppressor transfer RNA. Fed. Proc., 30:1314, 1971.
11. Gupta, N. K., Ohtsuka, E., Weber, H., Chang, S. H., and Khorana, H. G.: Studies on polynucleotides, LXXXVII–the joinings of short deoxyribonucleotides by DNA-joining enzymes. Proc. Nat. Acad. Sci., Washington, D. C., 60:285, 1968.
12. Miller, R. C. and Khorana, H. G.: Annealing of synthetic deoxypolynucleotides of the tyrosine suppressor t-RNA gene to $80\phi psa_{III}$ r-strand. Fed. Proc., 30:1054, 1971.
13. Smith, J. D., Abelson, J. N., Clark, B. F. C., Goodman, H. M., and Brenner, S.: Studies on amber suppressor tRNA. Sympos. Quant. Biol. (Cold Spring Harbor), 31:479, 1911.
14. Miller, R. C., Besmer, P., Khorana, H. G., Fiandt, M., and Szybalski, W.: Studies on polynucleotides XCVII. Opposing orientations and location of the SU^{+}_{III} and $\phi 80$ dSU^{+}_{III} SU_{III}. J. Mol. Biol., 56:363, 1971.
15. Wu, R.: Nucleotide sequence analysis of DNA. I. Partial sequence of the cohesive ends of bacteriophage and 186 Lena. J. Mol. Biol., 51:501, 1970.
16. Wu, R., and Taylor, E.: Nucleotide sequence analysis of DNA II complete nucleotide sequence of the cohesive ends of bacteriophage DNA. J. Mol. Biol., 57:491, 1972.
17. Padmanabhan, R., and Wu, R.: Nucleotide sequence analysis of DNA. IV. Complete nucleotide sequence of the left-hand cohesive end of coliphage 186 DNA. J. Mol. Biol., 65:447, 1972.

2. REPLICATION OF PLASMID SUPERCOILED DNA[1]

D. R. Helinski, D. G. Blair, D. J. Sherratt,[2]
M. Lovett, Y. Kupersztoch and D. T. Kingsbury

Department of Biology
University of California, San Diego
La Jolla, California 92037

INTRODUCTION

Plasmids are extrachromosomal genetic elements that reside stably in most genera and species of bacteria. In most cases these elements can be transmitted to cells via the process of conjugal mating or viral mediated transduction. When established in a cell, different plasmids determine a wide variety of bacterial traits. The stable existence of these elements and the finding of a relatively constant level of plasmid molecules in a population of cells indicate the operation of a carefully regulated process of plasmid DNA replication and a segregation apparatus that assures at least one copy of the plasmid molecule in each of the two daughter cells after cell division. The nature of genetic and biochemical factors responsible for the duplication of plasmid DNA during the growth of bacterial cells is the subject of this paper.

GENERAL PROPERTIES OF PLASMIDS

The molecular size and genetic content of bacterial plasmids vary considerably. Table I summarizes the properties of the bacterial plasmids that we have examined in our laboratory. These plasmid elements can be grouped into three categories. The colicinogenic (Col) factors of the E-type ($ColE_1$, $ColE_2$, and $ColE_3$) determine the production of the corresponding E colicins, are of relatively low molecular weight, are present almost entirely in the configuration of covalently closed, duplex circles, and are found to the extent of 10 to 15 copies per chromosome in the host cell. The covalently closed duplex, or supercoiled, DNA structure is characteristic of all plasmids isolated to date. A second category of plasmids, the F-type, are distinguished by their possession of genetic determinants of sexuality and, specifically, their determination of an F-type pilus structure. These elements (F_1, Flac, and ColV) are capable of promoting their own transfer or the conjugal transfer of the host chromosome to a suitable recipient cell. The ColV factor determines the production of colicin V, in addition to its sex-factor properties. The third category of plasmid elements, the I-type (ColIb and R64) are sex factors that determine a distinguishable I-type pilus instead of the F-type pilus. ColIb is a colicinogenic factor-sex factor that

[1] The studies described in this presentation were supported by U.S. Public Health Service research grant AI-07194, National Science Foundation research grant 6B-29492, and a U.S. Public Health Service Research Career Development Award (KO4-6M07821).

[2] Dr. Sherratt's present address is: Microbial Genetics Group, School of Biology, University of Sussex, Brighton, England.

TABLE I

Structural Properties of Plasmids[a]

Plasmid	Sex-factor type	Other genetic determinants	Molecular weight	Covalently closed circular DNA	No. copies per chromosome	Reference
$ColE_1$	None	colicin E_1	4.2×10^6	+	10-15	1, 2
$ColE_2$	None	colicin E_2	5.0×10^6	+	10-15	1
$ColE_3$	None	colicin E_3	5.0×10^6	+	10-15	1
F_1	F	–	62×10^6	+	1-2	3, 4, 5
Flac	F	lac operon	135×10^6	+	1-2	Y. Kupersztoch, unpublished data
ColV	F	colicin V	94×10^6	+	1-2	4, & R. Leavitt, unpublished data
ColIb	I	colicin Ib	62×10^6	+	1-2	6
R64	I	Tc^R Sm^R	76×10^6	+	limited	7, & G. Miklos, unpublished data

[a]Parent strains of various plasmids are as follows (in parenthesis): $ColE_1$ (K-30); $ColE_2$ (Shigella sp. P9); $ColE_3$ (CA 38); F_1 (W1485); Flac (C6000); ColV (K94); ColIb (Salmonella typhimurium L2 Cys D36); and R64 (J5-3). Unless otherwise indicated the parent strain is E.coli.

possesses genetic determinants for colicin Ib production, in addition to the determinants of sexuality. The sex-factor plasmid element R64 also determines resistance of the host cell to the antibiotics tetracycline and streptomycin. Both the I-type and F-type plasmids will promote the conjugal transfer of the nonautotransferable E-type plasmid elements, in addition to promoting their own transfer or the transfer of the host chromosome. In contrast to the E-type plasmids, the F-type and I-type sex-factor plasmids are of a relatively high molecular weight and are present to a more limited extent (1 to 2 copies per host chromosome) in the bacterial cell. Other properties of the plasmid elements that determine resistance to antibiotics (R factors) will be considered in the presentations to this symposium of Watanabe and Rownd and their colleagues.

Figure 1 is an electron micrograph of the covalently closed or supercoiled duplex and the open-circular forms of the ColV factor. A break in a single phosphodiester bond in either strand of supercoiled DNA will result in the relaxation of the molecule, or conversion to the open-circular configuration (9). Certain plasmids can also exist as catenated (interlocked circles) or tandemly linked, multiple-circular DNA

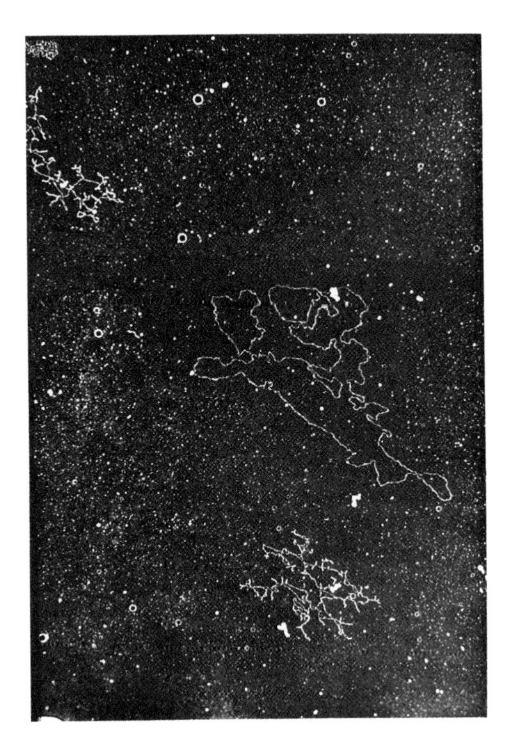

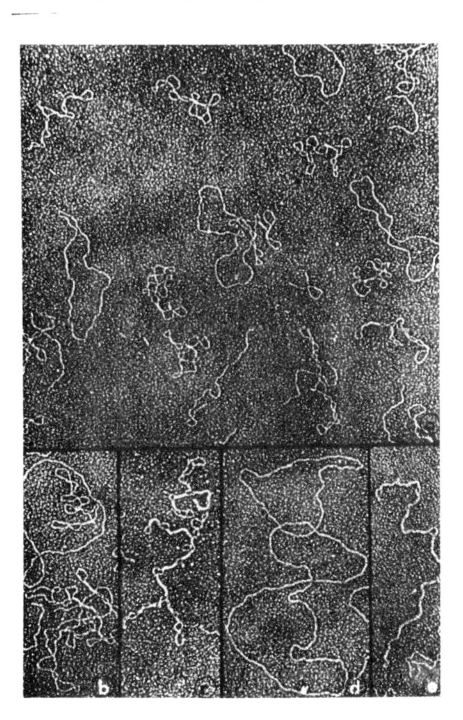

Fig 1. Electron micrograph of the plasmid ColV DNA purified by dye-cesium chloride equilibrium centrifugation. Sample was prepared for electron microscopy by a modification of the Kleinschmidt and Zahn technique (8) as previously described (2). (Ron Leavitt, unpublished data.) (Reduced 38%.)

Fig 2. Electron micrograph of the plasmid $ColE_1$ DNA purified by dye-cesium chloride equilibrium centrifugation from Proteus mirabilis ($ColE_1$) cells. (a) and (b) open and supercoiled circular DNA (2.3 μ contour length). (c) 2.3 μ and 4.7 μ supercoiled DNA forms. (d) 2.3 μ and 4.7 μ open-circular forms. (e) Tightly twisted 4.7 μ supercoiled DNA form. (From Roth and Helinski [2].) (Reduced 38%.)

forms, in addition to their monomer size under certain conditions in their natural host or upon transfer of the plasmid to a foreign host. Figure 2 illustrates the monomer and multiple-circular DNA forms of the $ColE_1$ plasmid extracted from a foreign host of this plasmid, Proteus mirabilis (2). In its natural host $ColE_1$ is found in normal cells almost exclusively as the monomer (4.2×10^6 molecular weight) supercoiled DNA form. While reciprocal recombination events between monomer-circular DNA molecules conceivably could account for the presence of the multiple-circular DNA forms in P. mirabilis, data has been presented in support of the proposition that the oligomers arise largely through errors in replication that occur in the foreign host (10).

GENERAL FEATURES OF PLASMID DNA REPLICATION

The biochemical events in the replication of the plasmid elements are for the most part obscure. A variety of possible replicating structures of $ColE_1$ DNA (double-forked, open, and twisted circles with "tails") have been purified from chromosomeless progeny (minicells) of E. coli (11). Studies with DNA polymerase I mutants of E. coli (12) and temperature-sensitive plasmid replication mutants (D. T. Kingsbury and D. R. Helinski, manuscript submitted for publication) have clearly demonstrated a requirement for DNA polymerase I in the replication of the plasmid $ColE_1$ and to some extent the plasmid $ColE_2$, but not for the replication of the F-type or I-type sex-factor plasmids. The exact nature of this requirement is unknown. The conjugal transfer of the $ColE_1$ plasmid also requires this enzyme (D. T. Kingsbury and D. R. Helinski, manuscript submitted for publication).

It is also clear from topological considerations that a nicking (break in a phosphodiester bond of one of the two DNA strands) event is required at some time during the duplication of a covalently closed, duplex circular structure. Whether a

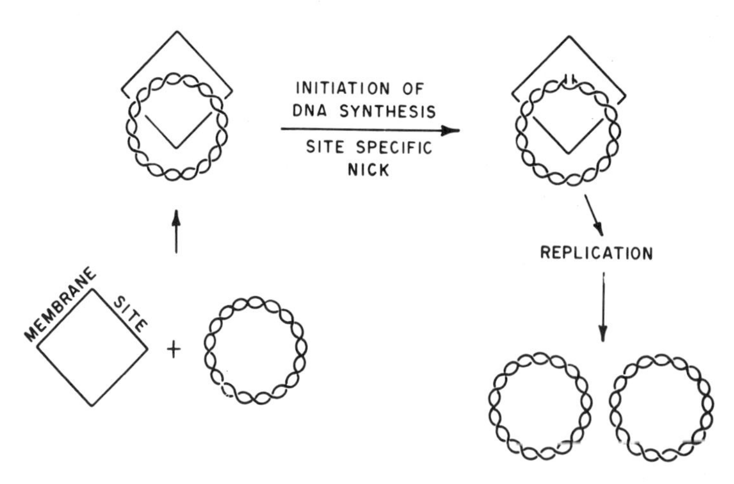

Fig 3. Model for the initiation of DNA synthesis requiring a site-specific endonucleolytic break in one of the two polynucleotide strands of covalently closed circular DNA. A membrane site is proposed for the nicking event.

single nicking event or multiple nicking and sealing events occur, or at what particular stage of the duplication process the endonuclease event(s) occurs, remains to be determined. While very little is known regarding the biochemical events responsible for the initiation of the duplication of the covalently closed circular DNA molecules of bacterial plasmids, mitochondrial DNA or the genome of oncogenic viruses, it is not unlikely that a nicking event is an early regulatory step in the duplication process, as shown in Figure 3. According to this model, a single nicking event that is site-specific with respect to the polynucleotide strand is an integral part of the initiation process for the duplication of covalently closed circular DNA. This model also proposes that the nicking even occurs at a membrane site in the host cell. As in the case of a variety of other DNA replication systems, evidence has been obtained for the replication of the $ColE_1$ plasmid at a membrane site (D. Sherratt and D. R. Helinski, unpublished data). A potential candidate for the nicking endonuclease in this model for the initiation of plasmid DNA duplication is the relaxation complex of supercoiled DNA and protein characterized for a variety of plasmids of E.coli.

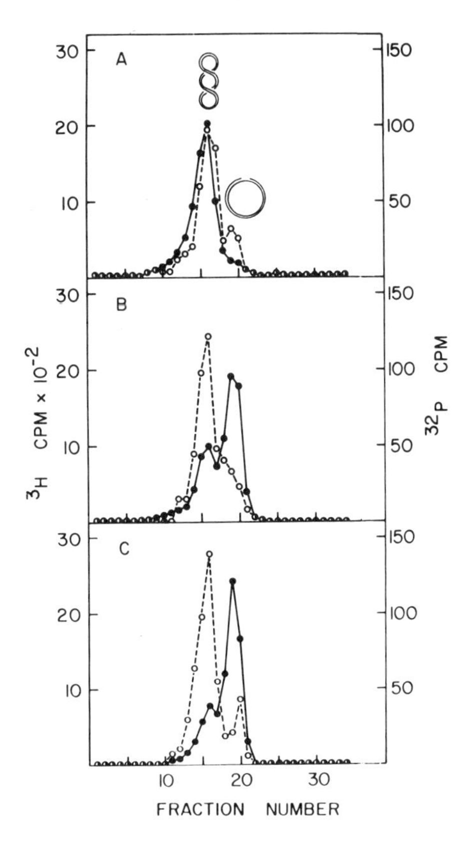

Fig 4. Sucrose gradient analysis of $ColE_2$ relaxation complex. A mixture of [^{3}H]-thymine-labeled complexed $ColE_2$ DNA and [^{32}P]-labeled noncomplexed $ColE_2$ DNA was centrifuged in a 5 - 20% neutral sucrose gradient after no treatment (A), treatment with 1.25 mg/ml pronase (B), and treatment with 0.25% SDS (C). The positions of the supercoiled and open-circular DNA forms of $ColE_2$ DNA in the gradients are indicated. The top of the gradient is on the right in each case. (From Blair et al. [14].)

RELAXATION COMPLEXES OF PLASMID DNA AND PROTEIN

Lysis of plasmid containing E.coli cells with the aid of nonionic detergents yields a substantial proportion of the plasmid DNA (the level depends on the plasmid, bacterial strain, and growth conditions) in the form of a complex of supercoiled DNA and protein (13). The remainder is found as supercoiled DNA without associated protein. This complex, purified by preparative sucrose density gradient centrifugation, when treated with an ionic detergent (sodium dodecyl sulfate [SDS]), protease (pronase), or alkali (pH 12.5) undergoes a unique conversion. As shown in Figure 4 treatment of a differentially labeled mixture of complexed and noncomplexed supercoiled ColE$_2$ DNA with SDS results in the conversion of the supercoiled DNA of a substantial proportion of the complexed ColE$_2$ DNA to a slower sedimenting form (14). The noncomplexed ColE$_2$ DNA is unaffected by this treatment. Electron microscopy and alkaline sucrose gradient analyses of the product of this conversion have demonstrated that it involves a transition from the supercoiled form to an open circular DNA molecule possessing a single nick (13, 14, 15). Prior heat treatment (60°C, 20 min) of the complex results in the removal of the protein and prevents the SDS or the pronase-induced relaxation of the ColE$_2$ DNA (14). Heat inactivation of the complex is also found for the relaxation complex of the F$_1$ sex factor (3). In the

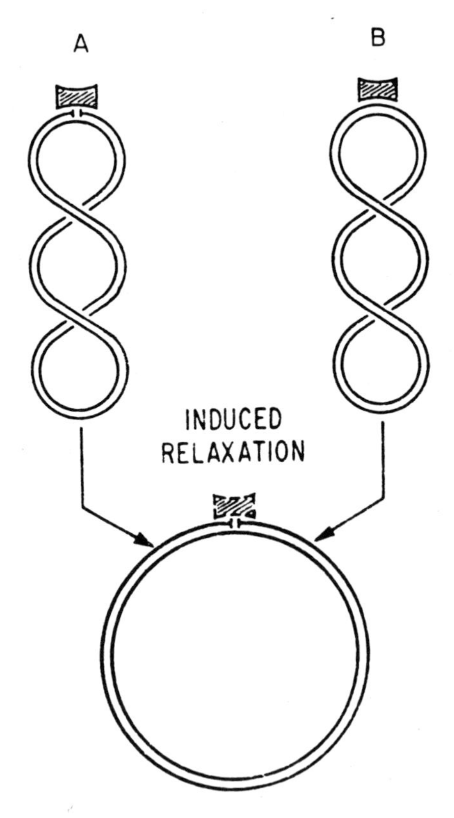

Fig 5. Two possible models for the relaxation complex of DNA and protein. (A) Binding protein associated at a nick or gap at a specific site on one of the two plasmid DNA strands. (B) Latent nickase at a specific site on the covalently closed circular DNA form of the plasmid. (From Clewell and Helinski [15].)

case of the $ColE_1$ relaxation complex, however, heat treatment induces the relaxation of the supercoiled DNA in the complex (13).

Two models, shown in Figure 5, have been considered for the structure of the relaxation complex (15). One (Fig 5A) proposes that the relaxation complex consists of a nicked-circular DNA molecule and a binding protein that is associated with the DNA at the nicked site. The binding protein maintains the DNA molecule in a supercoiled DNA configuration. Denaturation or destruction of the protein results in a release of the configurational restraints and the transition to an open-circular molecule. The second model (Fig 5B) proposes that the complex consists of a latent endonuclease associated with a covalently closed circular DNA molecule. Protein denaturing agents or a protease release the activity of the endonuclease and permit an endonucleolytic cleavage of one of the DNA strands before inactivating the enzyme itself. The heat-induced resistance of the complex to the agents SDS, or pronase in the case of the $ColE_2$ and F_1 complexes, favors the latent endonuclease model, at least for these complexes (3,14).

The strand specificity of the relaxation event was examined in the case of the $ColE_1$, $ColE_2$, and F_1 complexes. On the basis of DNA-DNA hybridization studies and results with the strand separation technique of Szybalski and his co-workers (16)

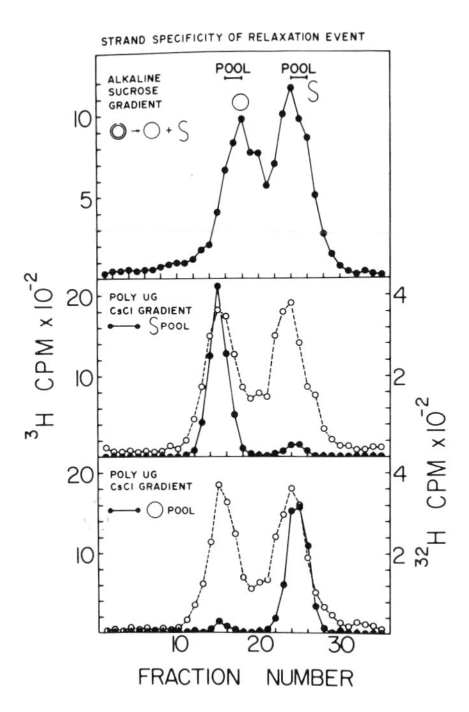

Fig 6. Test of strand-specificity of the nick in relaxed $ColE_2$ DNA. Relaxation was induced in the DNA of the $ColE_2$ relaxation complex by treatment with pronase. The circular and linear single strands of the relaxed $ColE_2$ DNA were separated by centrifugation in an alkaline sucrose density gradient. Fractions of the single-stranded circular and linear peaks were pooled as shown and the pooled material centrifuged to equilibrium in the presence of an excess of poly- (U,G) and a differentially labeled marker containing equal amounts of both strands of $ColE_2$ DNA. (●----●) ^{3}H-labeled complexed $ColE_2$ DNA. (○----○) ^{32}P-labeled noncomplexed $ColE_2$ DNA marker. (From Blair et al. [14].)

involving the centrifugation of denatured DNA to equilibrium in a CsCl gradient in the presence of poly-(U,G), the strand specificity of the induced relaxation event was established for the $ColE_1$ (14,15), $ColE_2$ (14) and F_1 (3) plasmids. In each case the heavy strand of the DNA possessed the break as defined by CsCl-poly-(U,G) technique (3,14). A typical result with this technique is shown in Figure 6. It is of interest that in the case of several sex-factor plasmids (including F_1), the unique strand transferred to the recipient cell during conjugal mating also is the heavy strand as defined by the CsCl-poly-(U,G) technique (7,17). Table II lists the plasmid elements found to date in the form of a relaxation complex.

TABLE II

Relaxation Complexes of Supercoiled DNA and Protein

Plasmid	% of molecules complexed	Strand-specificity of relaxation event	References
$ColE_1$	15-90[a]	heavy strand (poly-U,G)	13, 14
$ColE_2$	20-90[a]	heavy strand (poly-U,G)	14, 18
$ColE_3$	70-80	not known	18
F_1	~ 50	heavy strand (poly-U,G)	3
Flac	> 50	not known	Y. Kupersztoch, unpublished data
ColV	~ 50	not known	R. Leavitt, unpublished data
ColIb	> 80	not known	6
R64	> 50	not known	G. Miklos, unpublished data

[a]Extent of $ColE_1$ or $ColE_2$ plasmid molecules in the form of relaxation complex varies with the host strain, growth medium, and incubation temperature ([19] and L. Katz, D. T. Kingsbury, and D. R. Helinski, manuscript submitted for publication).

In an attempt to determine the fate of the protein in the relaxation complex after the induction of relaxation, the protein of the complex was labeled with ^{3}H-leucine by the addition of this radioactive amino acid to $ColE_1$ containing E.coli cells at the time of active synthesis of the relaxation complex. $ColE_1$ complex labeled with ^{14}C-thymidine in the DNA and ^{3}H-leucine in the protein was purified and analyzed by sucrose density gradient centrifugation before and after treatment with SDS and pronase (M. Lovett and D. R. Helinski, unpublished data). As shown in Figure 7 treatment with SDS results in the induced relaxation of a substantial portion of the supercoiled DNA. ^{3}H-leucine labeled protein, associated with both the supercoiled and spontaneously relaxed DNA before treatment with SDS, is found associated with the open circular DNA and at the top of the gradient after SDS treatment. Pronase also induces relaxation of the DNA and removes all of the protein

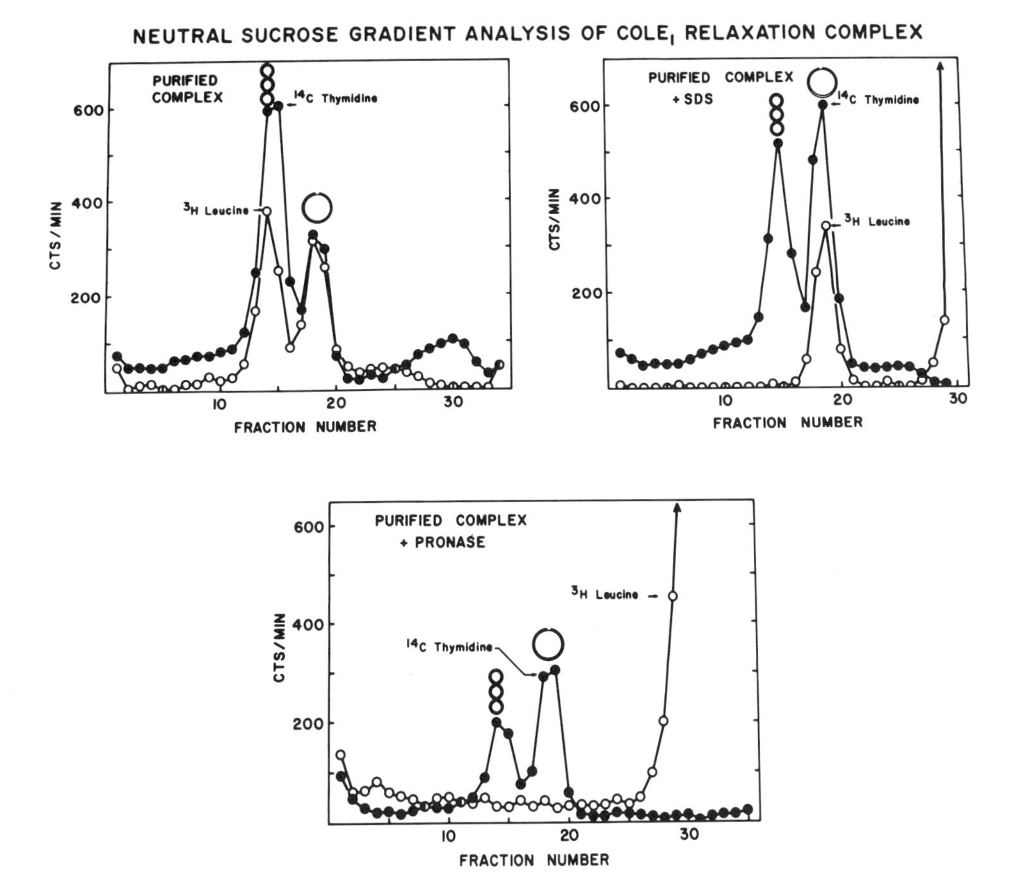

Fig 7. SDS and pronase-induced relaxation of ^{3}H-leucine and ^{14}C-thymidine-labeled ColE$_1$ relaxation complex. E. coli CR34 (ColE$_1$) cells were grown in M9 glycerol medium (20) containing ^{14}C-thymidine. During a period of active synthesis of ColE$_1$ relaxation complex, ^{3}H-leucine was added to label the protein. The ColE$_1$ relaxation complex was purified by column chromatography and adsorption to and elution from hydroxylapatite. Neutral sucrose density gradient analysis was carried out on the purified relaxation complex after (a) no treatment; (b) treatment with 0.5% SDS (final concentration) for 15 min at 25°C; and (c) treatment with 1.25 mg/ml of pronase for 15 min at 25°C. (M. Lovett and D. R. Helinski, unpublished data.)

associated with the DNA. The high degree of spontaneous relaxation in this preparation is due to the additional purification steps that were required to obtain a relaxation complex free of contaminating protein. The association of protein with the open-circular DNA product of the relaxation event is unusual in that it resists treatment with hot SDS (60°C, 60 min) (M. Lovett and D. R. Helinski, unpublished data) and is specific for the nicked strand as shown in Figure 8. In this experiment the open-circular and linear-single strand components of the open-circular DNA product of the relaxation event were separated by centrifugation in an alkaline sucrose gradient. The ^{3}H-leucine labeled protein is found associated exclusively with the nicked strand, which is the heavy strand as defined by CsCl gradient centrifugation in the presence of (poly-(U,G) (as shown previously in Fig 6). The strand-specific DNA-protein association also resists centrifugation to equilibrium in alkaline (pH 12.5) cesium chloride gradient or treatment of the nicked strand-protein complex (purified by alkaline cesium

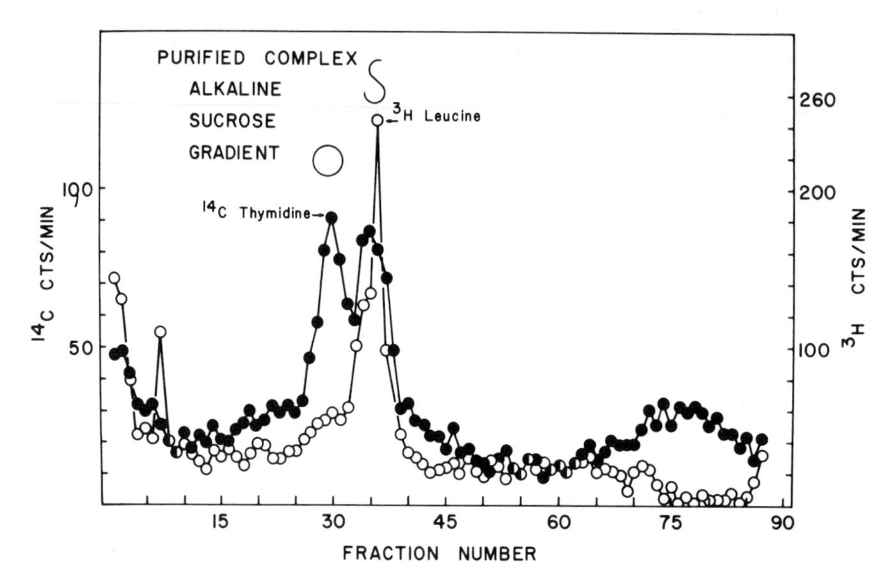

Fig 8. Alkaline sucrose density gradient analysis of ^{3}H-leucine and ^{14}C-thymidine-labeled $ColE_1$ relaxation complex. $ColE_1$ relaxation complex, prepared as described in Figure 7, was centrifuged in a 5 – 20% alkaline sucrose density gradient in an SW65 rotor, as previously described (14). The peaks of circular and linear single strands are indicated. (M. Lovett and D. R. Helinski, unpublished data.)

chloride centrifugation) with 8M urea, 2M sodium thiocyanate, or 2M lithium chloride (D. G. Blair and D. R. Helinski, unpublished data). These properties in general are exhibited by the open circular DNA product of the relaxation of either the complexed $ColE_1$ or $ColE_2$.

The unusual resistance of this strand-specific DNA-protein association to a variety of agents and conditions that normally disrupt hydrophobic or ionic interactions suggests a covalent linkage between the protein and the nicked strand. Since, at least for the $ColE_2$ complex, there is evidence indicating that the DNA in the complex prior to the induction of relaxation is covalently closed, it is reasonable to consider the formation of the linkage between DNA and protein during or after the relaxation event. A scheme is presented in Figure 9 that proposes the formation of a covalent linkage between the endonuclease and one of the termini of the nicked strand after the induction of relaxation. The protein-DNA linkage may function to conserve the high-energy character of the nicked phosphodiester bond to facilitate the sealing reaction after replication or conjugal transfer of the plasmid DNA. Conceivably, the protein-DNA linkage could also play a role in transferring the terminus of the DNA strand to a membrane receptor site, as was proposed by Gilbert and Dressler (21) in the rolling-circle model of DNA replication, or, possibly, may be a component in a swivel apparatus that may facilitate unwinding of the DNA strands during replication of the plasmid DNA. Finally, the linkage of a specific protein to the terminus of the nicked strand of plasmid DNA may serve as a protective and facilitative device for the conjugal transfer of a specific strand of the plasmid DNA to a recipient cell. While this scheme proposes several important physiological roles for the relaxation complex, it should be emphasized at this point that evidence for the role of the relaxation complex in either the replication of plasmid DNA or its conjugal transfer has not been

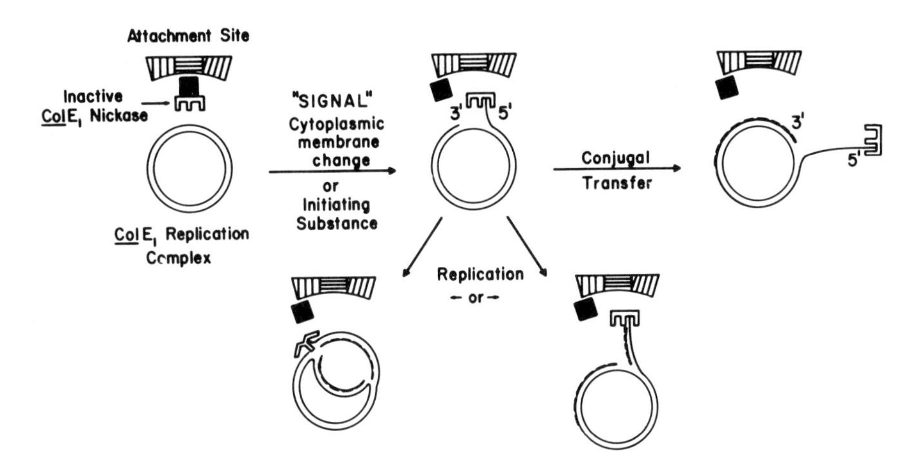

Fig 9. A possible role for the relaxation complex in the replication of plasmid DNA. In this model two functions are considered for the relaxation complex. (a) Catalysis of a strand-specific nick in the covalently closed plasmid DNA in response to an initiating signal (cytoplasmic substance or membrane change). (b) Covalent linkage with the 5′ terminus of the nicked strand. The covalently linked protein mediates the subsequent steps of replication and possibly the conjugal transfer of the DNA. Two types of replication cyclic intermediates are considered: Cairns forked circle and rolling circle. (Light lines represent parental DNA; heavy lines represent newly synthesized DNA.)

obtained and, therefore, the physiological meaning of these interesting properties of the relaxation complexes remains to be elucidated.

RNA-CONTAINING SUPERCOILED COLE$_1$ DNA

Unlike chromosomal DNA, the initiation of ColE$_1$ DNA synthesis continues in E.coli cells in the presence of the protein synthesis inhibitor chloramphenicol (19,22). Replication of ColE$_1$ DNA under these conditions will continue for 10 to 15 hours (Fig 10). After several hours of synthesis of ColE$_1$ DNA in the presence of chloramphenicol, only a small percentage of the supercoiled DNA molecules are in the form of relaxation complex. The supercoiled ColE$_1$ DNA synthesized in the presence of chloramphenicol (CM-ColE$_1$) possesses the same sedimentation properties and buoyant density as supercoiled ColE$_1$ DNA generated in the absence of chloramphenicol (non-CM-ColE$_1$) (23). The synthesis of supercoiled ColE$_1$ DNA in the presence of chloramphenicol, however, exhibited two unusual properties that called for the further investigation of the CM-ColE$_1$ DNA. In the first place, Clewell and his co-workers (22) reported that the RNA polymerase inhibitor rifampicin blocks ColE$_1$ synthesis in E.coli cells growing in the presence of chloramphenicol. Second, it was observed that supercoiled CM-ColE$_1$ DNA is unusually sensitive to exposure to alkaline pH conditions (23). As shown in Figure 11 most of the molecules in a preparation of supercoiled CM-ColE$_1$ DNA lose their covalently closed circular structure after exposure to pH 13 conditions. Under these conditions supercoiled non-CM ColE$_1$ DNA is relatively unaffected by this treatment. The conversion in alkali is time-dependent in that the rapid denaturation and neutralization of covalently closed CM-ColE$_1$ DNA does not result in a loss of the supercoiled property. Similarly, when the CM-ColE$_1$ DNA is heat-denatured there is no loss of the covalently closed circular DNA structure.

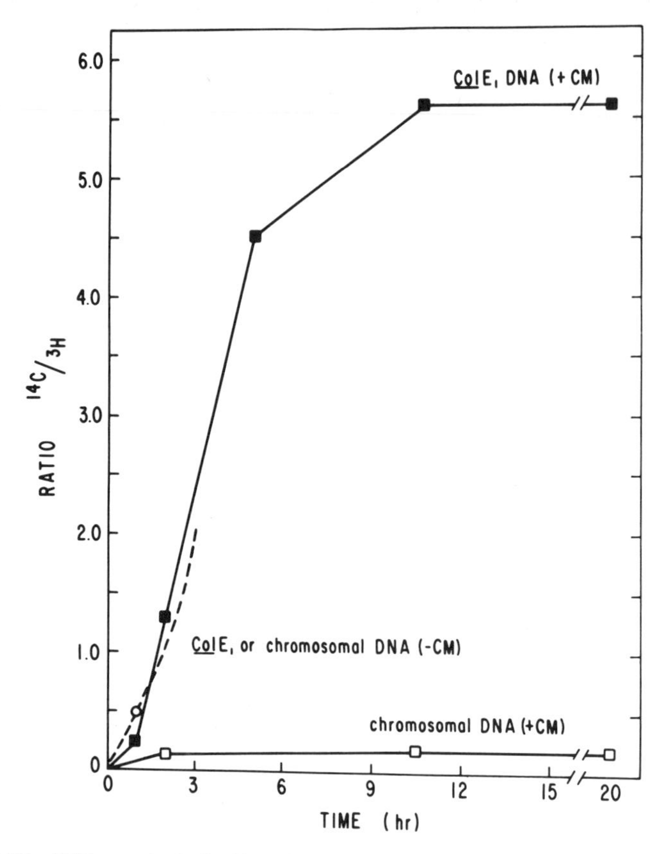

Fig 10. $ColE_1$ DNA synthesis in E. coli cells after the addition of chloramphenicol. JC411 ($ColE_1$) cells were labeled with ^3H thymine prior to the addition of CM and with ^{14}C thymine after the addition of CM (150 μg/ml). The single point for the relative DNA synthesis of both plasmid and chromosomal DNA in the absence of CM was determined after one generation after ^{14}C-thymine addition. The broken line was drawn assuming an exponential rate of synthesis. synthesis. (D. G. Blair and D. R. Helinski, unpublished data.)

The observed requirement for an RNA primer in the in vitro synthesis of M13 double-stranded circular DNA from the single-stranded phage DNA (24), the unusual alkaline lability of supercoiled CM-$ColE_1$ DNA, and the rifampicin sensitivity of the synthesis of CM-$ColE_1$ DNA (22) suggested the presence of ribonucleotides in the structure of supercoiled CM-$ColE_1$ DNA. This was tested by examining the sensitivity of CM-$ColE_1$ DNA to pancreatic ribonuclease A (RNase A) (23). Mixtures of differentially labeled CM- and non-CM $ColE_1$ DNA were treated with RNase A at a concentration of 1 mg/ml in a low-salt buffer for 1 hour at 37°C. As shown in Figure 12, approximately one-half of the supercoiled DNA molecules in the CM-$ColE_1$ DNA preparation are converted to a slower sedimenting form in 10 minutes. Essentially no conversion of the non-CM $ColE_1$ DNA occurs, even after 2 hours of incubation. The RNase A-resistant fraction of CM-$ColE_1$ DNA is no more sensitive to alkali than non-CM $ColE_1$ DNA.

To determine the structure of the slower sedimenting product of the ribonuclease treatment, RNase A-treated CM-$ColE_1$ DNA was examined in the electron microscope (D. G. Blair and D. R. Helinski, unpublished data). In this experiment a

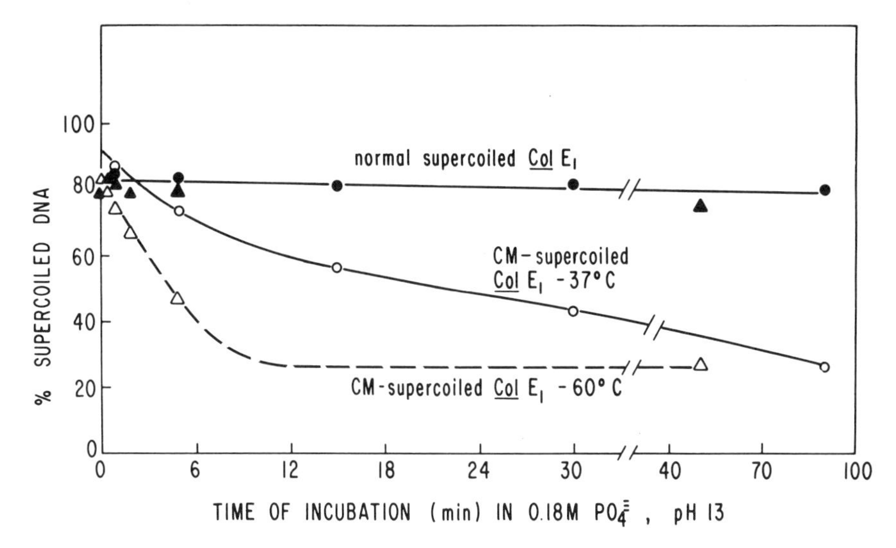

Fig 11. Alkaline lability of covalently closed CM-$ColE_1$ DNA. Samples of supercoiled non-CM-$ColE_1$ and CM-$ColE_1$ DNA were incubated at pH 13 in 0.18M PO_4 at the temperatures and for the times indicated, then neutralized and analyzed on 5 – 20% sucrose density gradients. Supercoiled DNA is expressed as a percentage of total $ColE_1$ DNA recovered from the gradients. (From Blair et al. [23].)

mixture of supercoiled $ColE_1$ DNA from chloramphenicol treated and untreated cultures of the E.coli strain JC411 ($ColE_1$) (Pool A of the dye-cesium chloride gradient in Fig 13) was treated with ribonuclease A and the mixture centrifuged to equilibrium in a dye-cesium chloride gradient to separate the product of the reaction from the unconverted supercoiled DNA. The DNA of various fractions of the two dye-cesium chloride gradients was examined by electron microscopy. As shown in Figure 13, the RNase A treatment results in a conversion of the CM-$ColE_1$ from the supercoiled to the open-circular DNA form. Less than 3% of over 300 randomly selected molecules converted by the RNase A treatment were linear in structure. RNase H, specific for the RNA portion of an RNA-DNA hybrid and ineffective against double-stranded or single-stranded DNA or RNA (25), also converts CM-$ColE_1$ DNA to the open-circular DNA form (23). High concentrations of RNase T_1, however, showed no effect on the CM-$ColE_1$ DNA under the reaction conditions employed (23).

Additional evidence for the open-circular nature of the product of the RNase A treatment came from an alkaline sucrose density gradient analysis of RNase A-treated supercoiled CM-$ColE_1$ DNA (23). Purified supercoiled CM-$ColE_1$ DNA was treated with RNase A and the open-circular DNA product of the reaction was isolated by dye-cesium chloride equilibrium centrifugation. This material was then centrifuged in an alkaline sucrose density gradient under conditions that separate the circular and linear single strands. The ratio of circular and linear strands observed (Fig 14a) indicates that at least 80% of the RNase A-nicked open circles possessed only one nicked strand. In other experiments the ratio of linear single strands indicates that approximately 100% of the molecules are of this type. The fact that the slower sedimenting peak of linear material exhibits little trailing indicates that most of the nicked DNA strands contained only one RNase A-sensitive site.

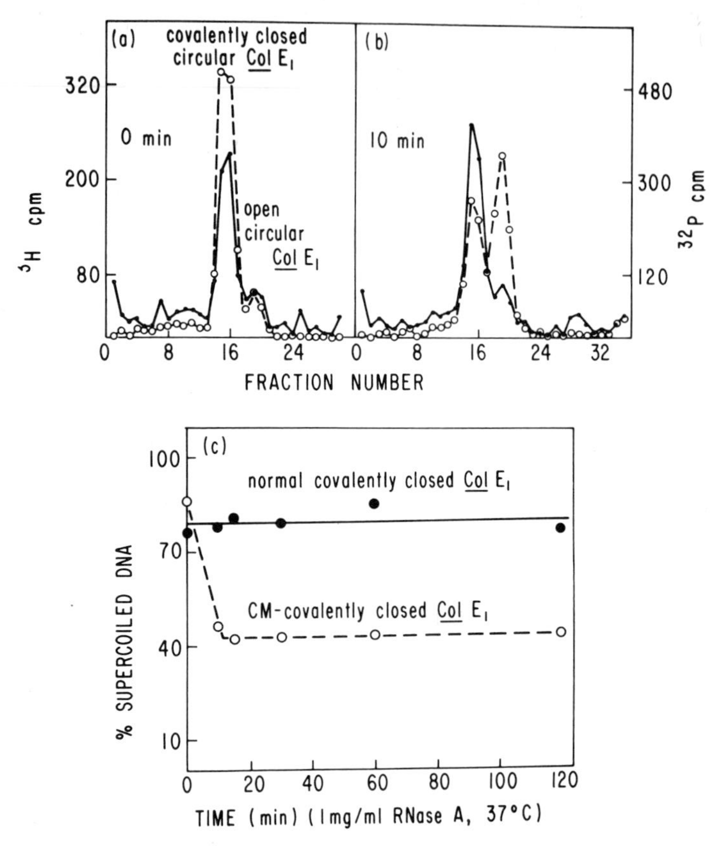

Fig 12. Sensitivity of covalently closed CM-ColE$_1$DNA to RNase A. Mixtures of supercoiled non-CM ColE$_1$ and CM-ColE$_1$ were incubated with RNase A for various time periods. Following RNase A treatment, the reaction mixture was analyzed on sucrose density gradients. (a) Incubation for 10 min at 37°C in the absence of RNase A; (b) incubation for 10 min at 37°C in the presence of RNase A; (c) relative amounts of the supercoiled DNA form of ColE$_1$ DNA purified from chloramphenicol treated and untreated E. coli (ColE$_1$) cells after incubation for various lengths of time with RNase A. The top of the sucrose density gradient in (a) or (b) is on the right in each case. (●----●) non-CM ColE$_1$; (○---○) CM-ColE$_1$. (From Blair et al [23].)

When the separated strands of RNase-nicked CM-ColE$_1$ DNA were centrifuged to equilibrium in a cesium chloride gradient in the presence of poly-(U,G), the results shown in Figure 14b and c were obtained (23). It is clear from this result that both the circular and linear strands consist of equal numbers of each of the complementary ColE$_1$ DNA strands. The RNase-sensitive site, therefore, is found with equal probability in each of the two strands in CM-ColE$_1$ DNA.

RIFAMPICIN EFFECT ON THE GENERATION OF RNASE A-SENSITIVE COLE$_1$ DNA

The appearance of RNase A-sensitive ColE$_1$ DNA is detected approximately 1 to 2 hours after the addition of chloramphenicol to a logarithmically growing culture of colicinogenic E.coli cells. The proportion of molecules of ColE$_1$ DNA that are

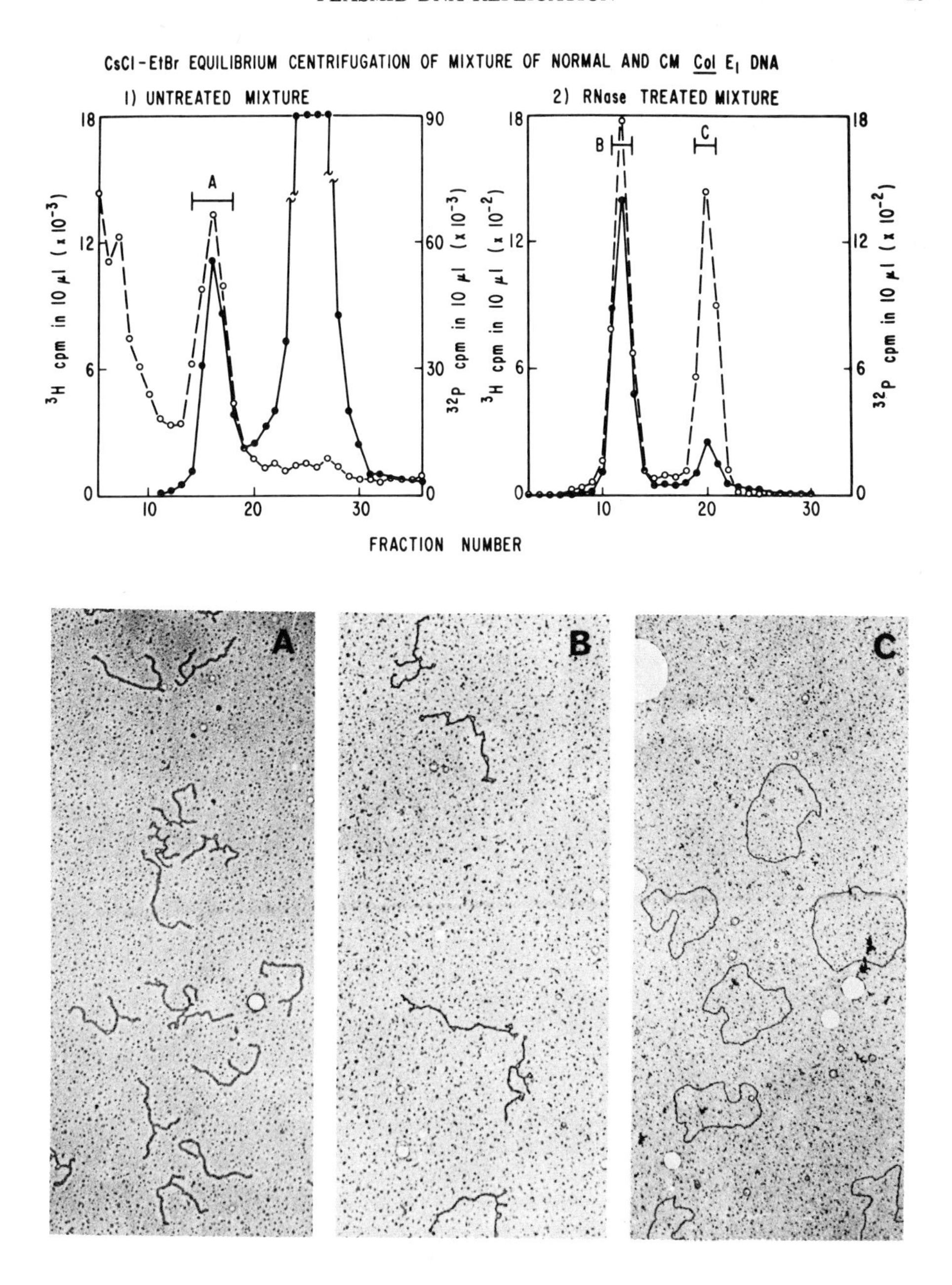

Fig 13. RNase A-induced conversion of supercoiled CM-ColE$_1$ DNA to the open-circular form. Upper left panel represents a preparative dye-cesium chloride equilibrium gradient of a mixed lysate of CM treated and untreated JC411 (ColE$_1$) cells. The supercoiled ColE$_1$ DNA from this mixed lysate (pool A of upper left panel) was further purified by sucrose density gradient centrifugation, RNase treated, and rebanded in a dye-cesium chloride gradient (upper right panel). (•——•) ^{3}H-labeled non-CM ColE$_1$ DNA; (o---o) ^{32}P-labeled CM-ColE$_1$ DNA. The ratio of CM treated and untreated cells was adjusted so that more than 90% of the ColE$_1$ DNA molecules in the gradients were from the CM treated culture. (A), (B), and (C) represent electron micrographs of material from pools A, B, and C of the gradients, respectively. (Reduced 74%.) (D. G. Blair, unpublished data.)

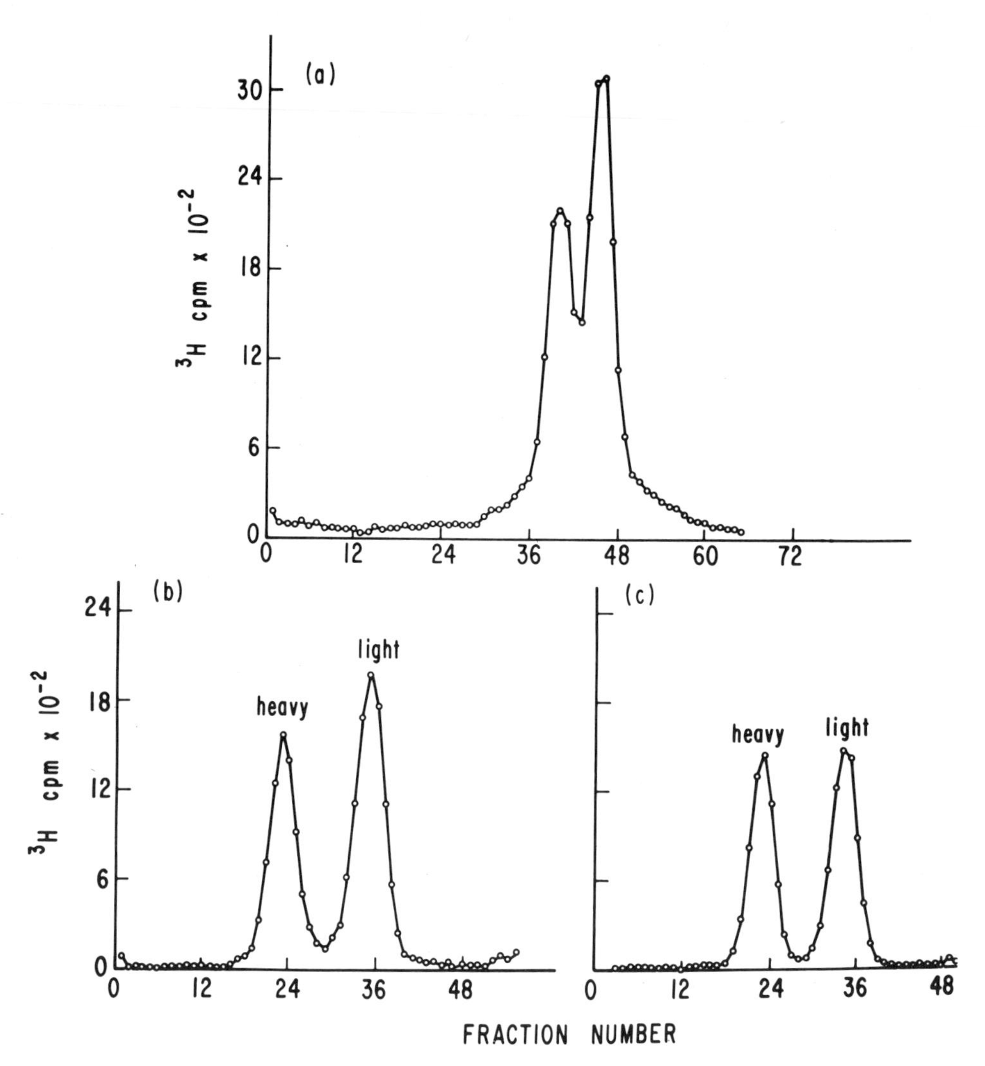

Fig 14. Strand-specificty of the RNase A-sensitive site. After treatment of CM-ColE$_1$ DNA with RNase A, the open-circular DNA product of the reaction was purified by dye-cesium chloride equilibrium centrifugation. The purified open-circular ColE$_1$ DNA was layered on a 5 –20% alkaline sucrose density gradient and centrifuged under conditions that separate the circular and linear single strands. The top of the alkaline sucrose density gradient (a) is on the right. Fractions of the faster (circular) peak and the slower (linear) peak were pooled separately and analyzed on cesium chloride gradients containing poly-(U,G). (b) Poly-(U,G) cesium chloride equilibrium centrifugation of the material from the single-stranded circular pool. (c) Poly-(U,G) cesium chloride equilibrium centrifugation of the material from the single-stranded linear pool. (From Blair et al [23].)

RNase A-sensitive continues in the presence of chloramphenicol to increase at a linear rate (Fig 15). On the basis of several experiments, it is clear that the increase in the relative percentage of RNase A-sensitive ColE$_1$ DNA ceases when plasmid DNA synthesis stops after long periods of incubation of the cells in the presence of chloramphenicol (23). This result is consistent with a requirement for active ColE$_1$ DNA synthesis for the generation of the RNase A-sensitive form of ColE$_1$ DNA. In addition, as shown

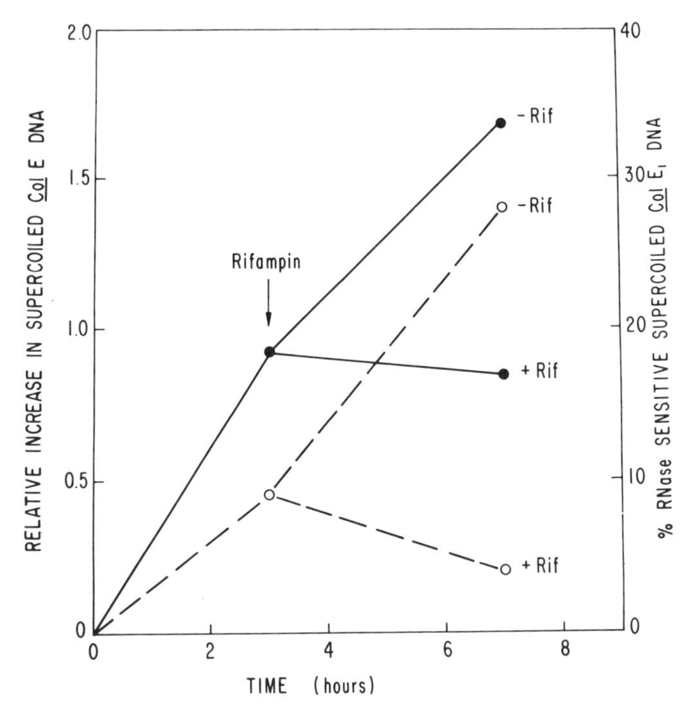

Fig 15. Effect of rifampicin on synthesis of supercoiled ColE$_1$ DNA and the generation of RNase-sensitive ColE$_1$. (+ Rif) refers to a culture of JC411 (ColE$_1$) cells grown in the presence of rifampicin (3 μg/ml). (- Rif) represents the nontreated control culture. The arrow indicates the time of addition of rifampicin. (D. G. Blair, unpublished data.)

in Figure 15, rifampicin prevents both the synthesis of ColE$_1$ DNA and the generation of RNase A-sensitive ColE$_1$ DNA molecules (23).

RNA AS A PRIMER OF COLE$_1$ DNA SYNTHESIS

The above results indicate that when protein synthesis is inhibited by chloramphenicol, supercoiled ColE$_1$ DNA containing one or more ribonucleotides at a single site in either one or the other complementary strands accumulates in those cells that are actively synthesizing this plasmid. While it is possible that these molecules arise as a result of the erroneous insertion of ribonucleotides during ColE$_1$ DNA synthesis in the presence of chloramphenicol, the finding of a single RNase-sensitive site and the reported sensitivity of ColE$_1$ DNA synthesis to rifampicin (22) suggests a physiological role for this RNA in ColE$_1$ DNA synthesis. Figure 16 illustrates a possible model for the formation of RNA-containing supercoiled ColE$_1$ DNA. In this scheme the RNA serves as a primer in the initiation of DNA synthesis similar to the role of RNA in the in vitro synthesis of the replicative form of M13 DNA (26). The finding that the RNase A-sensitive site is present in either strand requires, according to this model, that the primer segment by synthesized randomly off either strand and, therefore, initiation of DNA synthesis is random with respect to the DNA strand, or synthesis is bidirectional and initiated with short RNA segments from both strands of the replicating molecule. An important test of this proposed physiological role for the RNA in

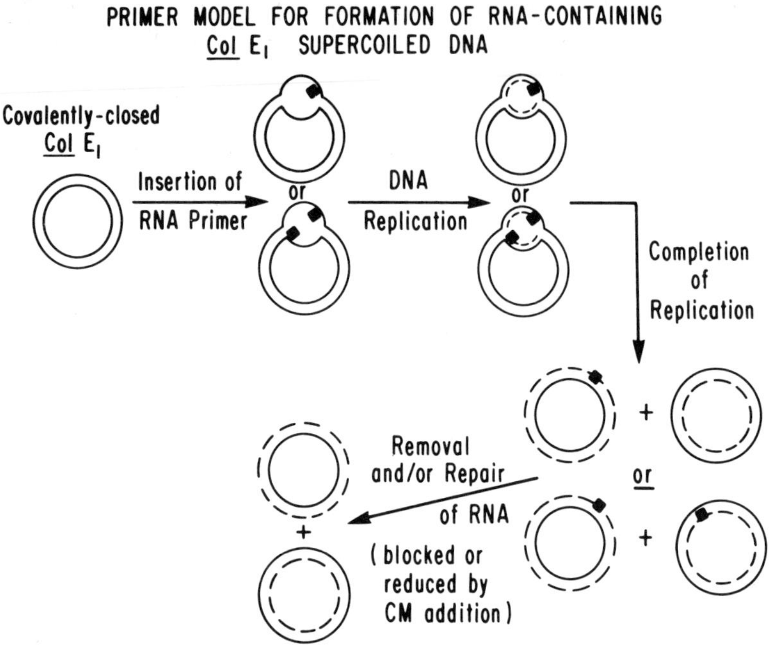

Fig 16. Primer model for the formation of RNA-containing $ColE_1$ DNA. The initial steps of DNA synthesis involve the addition of deoxyribonucleotides to primer RNA that is complementary to a single site on either DNA strand. Either a single RNA primer molecule, complementary to one or the other strand is involved in the duplication of any one circular DNA molecule, or synthesis proceeds bidirectionally from two primer RNA molecules present in a single replicating DNA molecule. Normally, the RNA primer is removed immediately before or after completion of replication of the circular DNA, however, in the presence of chloramphenicol RNA-containing, supercoiled $ColE_1$ DNA molecules accumulate. This model is fundamentally different from the rolling circle model of DNA duplication that is considered in part in Figure 9. (From Blair et al [23].)

CM-$ColE_1$ DNA is an examination of the site-specificity of the RNA in the two strands.

POSSIBLE EVENTS IN THE REPLICATION OF PLASMID DNA

The majority of plasmid DNA molecules in each case examined is in the form of covalently closed circular DNA. At least one nicking event is required at some stage in the duplication or conjugal transfer of these molecules. The relaxation complexes of supercoiled plasmid DNA and protein may play an important role in the duplication and the transfer of plasmid DNA due to the activateable strand-specific nicking associated with these complexes. It is conceivable that the replication of supercoiled plasmid DNA initially requires the formation of a relaxation complex between specific protein components and the DNA. In addition to providing endonuclease activity the relaxation complex may facilitate the attachment of the plasmid DNA to a "replicator" site. Activation of the relaxation complex by a cytoplasmic initiating substance or a membrane associated event may be a rate-limiting step in the duplication process. The strand-specific nick of the supercoiled DNA and covalent attachment of protein to one of the termini of the nicked DNA strand could facilitate the unwinding of the DNA strands during the duplication process. It is also possible that the proposed covalent linkage of DNA and protein facilitates the sealing of the nicked region at some stage in the duplication process. A primer RNA fragment may be synthesized and associate with the plasmid DNA at the initiator site. After bidirectional synthesis of plasmid DNA extending from the RNA primer site, covalently closed circular DNA daughter molecules containing the primer RNA may be generated. The RNA could then be removed from the supercoiled DNA by means of a system similar to that carrying out DNA repair. Under normal growth conditions the RNA may be removed prior to closure of the circular plasmid DNA. Following this sequence of events the plasmid DNA would be released from its membrane-located replicator site.

To date the evidence for this sequence of events in the replication of $ColE_1$ DNA is circumstantial at best. There is even less information on the possible events responsible for the duplication of the larger plasmid DNA molecules that are present in the cell to a more limited extent. We are just now beginning to approach the question of the degree of relatedness between the biochemical events in the replication of different plasmid elements and the host chromosome. The answer to this question is critical not only to our understanding of plasmid DNA replication, but it may be a necessary step toward the understanding of the duplication of extrachromosomal DNA and the chromosomes of higher organisms.

REFERENCES

1. Bazaral, M., and Helinski, D. R.: Circular DNA forms of colicinogenic factors E1, E2 and E3 from E.coli. J. Mol. Biol., 36: 185, 1968.
2. Roth, T. F., and Helinski, D. R.: Evidence for circular DNA forms of a bacterial plasmid. Proc. Nat. Acad. Sci. USA, 58:650, 1967.
3. Kline, B. C., and Helinski, D. R.: F_1 sex factor of Escherichia coli: Size and purification in the form of a strand-specific relaxation complex of supercoiled deoxyribonucleic acid and protein. Biochemistry, 10:4975, 1971.
4. Clowes, R. C.: Molecular weight and number of copies of various sex factor DNA molecules in Escherichia coli. X Int. Congr. Microbiol., Mexico, p. 58, 1970.
5. Sharp, P., Hsu, M., Ohtsubo, E., and Davidson, N.: J. Mol. Biol., in press.
6. Clewell, D. B., and Helinski, D. R.: Existence of the colicinogenic factor-sex factor coli–pq as a supercoiled circular DNA–protein relaxation complex. Biochem. Biophys. Res. Commun., 41:150, 1970.
7. Vapnek, D., Lipman, M., and Rupp, W. D.: Physical properties and mechanisms of transfer of R factors in Escherichia coli. J. Bact., 108:508, 1971.
8. Kleinschmidt, Von A., and Zahn, R. K.: Über desoxyribonuclein säuremolekein in proteinmischfilmen. Z. Naturforsch., 146:770, 1959.
9. Vinograd, J., Lebowitz, J., Radloff, R., Watson, R., and Laipis, P.: The twisted circular form of polyma viral DNA. Proc. Nat. Acad. Sci. USA, 53:1104, 1965.
10. Goebel, W., and Helinski, D. R.: Generation of higher multiple circular DNA forms in bacteria. Proc. Nat. Acad. Sci. USA, 61:1406, 1968.
11. Inselburg, J., and Fuke, M.: Isolation of catenated and replicating DNA molecules of colicin factor El from minicells. Proc. Nat. Acad. Sci. USA, 68:2839, 1971.
12. Kingsbury, D. T., and Helinski, D. R.: DNA polymerase as a requirement for the maintenance of the bacterial plasmid colicinogenic factor E_1. Biochem. Biophys. Res. Commun., 41:1538, 1970.

13. Clewell, D. B., and Helinski, D. R.: Supercoiled circular DNA protein complex in Escherichia coli: purification and induced conversion to an open circular DNA form. Proc. Nat. Acad. Sci. USA, 62:1159, 1969.
14. Blair, D. G., Clewell, D. B., Sherratt, D. J., and Helinski, D. R.: Strand-specific supercoiled DNA-protein relaxation complexes: comparison of the complexes of bacterial plasmids col E_1 and col E_2. Proc. Nat. Acad. Sci. USA, 68:210, 1971.
15. Clewell, D. B., and Helinski, D. R.: Properties of a supercoiled deoxyribonucleic acid-protein relaxation complex and strand specificity of the relaxation event. Biochemistry, 9:4428, 1970.
16. Szybalski, W., Kubinski, H., Hradecna, Z., and Summers, W. C.: Analytical and preparative separation of the complementary DNA strands. Methods in Enzymology. Edited by L. Grossman and K. Moldave. Academic Press, New York, XXI:383, 1971.
17. Vapnek, D., and Rupp, W. D.: Asymmetric segregation of the complementary sex-factor DNA strands during conjugation in Escherichia coli. J. Mol. Biol., 53:287, 1970.
18. Clewell, D. B. and Helinski, D. R.: Evidence for the existence of the colicinogenic factors E_2 and E_3 as supercoiled circular DNA-protein relaxation complexes. Biochem. Biophys. Res. Commun., 40:608, 1970.
19. Clewell, D. B., and Helinski, D. R.: The effect of growth conditions on the formation of the relaxation complex of supercoiled Col E_1 deoxyribonucleic acid and protein in Escherichia coli. J. Bact., 110:1135, 1972.
20. Herschman, H. R., and Helinski, D. R.: Comparative study of the events associated with colicin induction. J. Bact., 94:691, 1967.
21. Gilbert, W., and Dressler, D.: DNA replication: the rolling circle model. Sympos. Quant. Biol. (Cold Spring Harbor), 33:473, 1972.
22. Clewell, D. B., Evenchick, B., and Cranston, J. W.: Direct inhibition of col E_1 plasmid DNA replication in Escherichia coli by rifampicin. Nature New Biology, 237:29, 1972.
23. Blair, D. G., Sherratt, D. J., Clewell, D. B., and Helinski, D. R.: Isolation of supercoiled colicinogenic factor E_1 DNA sensitive to ribonuclease and alkali. Proc. Nat. Acad. Sci. USA, 69:2518, 1972.
24. Brutlag, D. R., Schekman, R., and Kornberg, A.: A possible role for RNA polymerase in the initiation of M13 DNA synthesis. Proc. Nat. Acad. Sci. USA, 68:2826, 1971.
25. Hausen, P., and Stein, H.: Ribonuclease H. An enzyme degrading the RNA moiety of DNA-RNA hybrids. Eur. J. Biochem., 14:278, 1970.
26. Wickner, W., Brutlag, D., Schekman, R., and Kornberg, A.: RNA synthesis initiates in vitro conversion of M_{13} DNA to its replicative form. Proc. Nat. Acad. Sci. USA, 69:965, 1972.

3. THE POLYOMA PSEUDOVIRION–ITS PROPERTIES, PRODUCTION, AND USE IN TRANSFERRING DNA TO MOUSE AND HUMAN CELLS[1]

H. Vasken Aposhian, Pradman K. Qasba, David B. Yelton, Quentin A. Pletsch and V. Sagar Sethi

Department of Cell Biology and Pharmacology
University of Maryland School of Medicine
Baltimore, Maryland 21201

INTRODUCTION

For the past few years we have been interested in developing a DNA or gene delivery system for use in whole animals as well as in cultured animal cells (1,2,3). For these studies, we decided to use animal virus-like particles called polyoma pseudovirions. This paper describes: 1). the intracellular events related to polyoma pseudovirus production; 2). the use of these pseudovirions to deliver fragments of mouse DNA to the nuclei of mouse embryo and human embryo cells; and 3). some preliminary experiments dealing with our attempts to reassemble polyoma virus from its individual components.

POLYOMA PSEUDOVIRUS PRODUCTION

The infection of mouse cells by polyoma virus may produce four polyoma-related particles: infectious polyoma virions, defective polyoma virions, polyoma pseudovirions, and empty polyoma capsids. The *infectious polyoma virions* contain the complete polyoma genome which consists of supercoiled, double-stranded DNA with a molecular weight of approximately 3 X 10^6 daltons (4,5). The *defective polyoma virions* contain only 50 to 75% of the polyoma genome, as reported by Blackstein, Stanners, and Farmilo (6). The DNA of the defectives, however, is also double-stranded and supercoiled. These defective particles appear to be produced when the virus used for infection has undergone numerous passages in the laboratory at a high multiplicity of infection. Weil's group in Switzerland (7) and Winocour (8) in Israel, independently have reported the production of polyoma pseudovirions. These *pseudovirions* contain linear double-stranded fragments of mouse DNA encapsidated by polyoma coats. The DNA has a molecular weight estimated to be between 1 X 10^6 and 3 X 10^6 daltons. Finally, *empty capsids* which contain no detectable DNA are also found.

Pseudovirions appear to be relatively unusual particles in animal virology. They have been found thus far only after infection of mouse cells with polyoma (2,7,8,9) or of monkey cells (10,11) with SV40. Generally, about 10% of the full

[1] This work was supported in part by Contract 70-2082 of the National Institute of General Medical Sciences and grants from The John A. Hartford Foundation, Inc., the National Cancer Institute (CA10497), and the National Cystic Fibrosis Research Foundation.

particles produced after polyoma infection of mouse kidney cells are pseudovirions. We have known for some time that a strikingly different result occurs in our laboratory. The infection of primary mouse embryo cells grown in one-gallon glass roller bottles yields full particles of which approximately 80% or more are pseudovirions (2,12).

Polyoma pseudovirions are produced in our laboratory (2) as follows. To 24-hour-old primary mouse embryo cells in glass roller bottles, radioactive thymidine is added. Five days later the medium is removed and the radioactivity is chased for two hours with medium containing nonradioactive deoxythymidine and deoxycytidine. The medium is then removed and the cells are infected with polyoma virus at a multiplicity of 0.1 PFU/cell. Fresh medium containing a high level of nonradioactive deoxythymidine and deoxycytidine is added. Six days postinfection the cells are harvested and disrupted by cycles of freezing and thawing. The virus is released by treatment with receptor destroying enzyme and purified by differential centrifugation, two equilibrium centrifugations in CsCl gradients, followed by sedimentation through a neutral sucrose gradient.

We have examined some of the intracellular events that occur during the production of polyoma pseudovirions in primary mouse embryo cells grown in glass roller bottles (13).

Synthesis of polyoma DNA was detected 18 hours after infection (Fig 1). This was followed by an increase of the empty particles which began 30 hours after infection. Fragmentation of host cell DNA to a molecular weight of about 3 X 10^6 daltons (14S) started between 30 and 36 hours postinfection. Assembly of polyoma virus and pseudovirions began between 36–42 hours after infection and by 96 hours replication was complete.

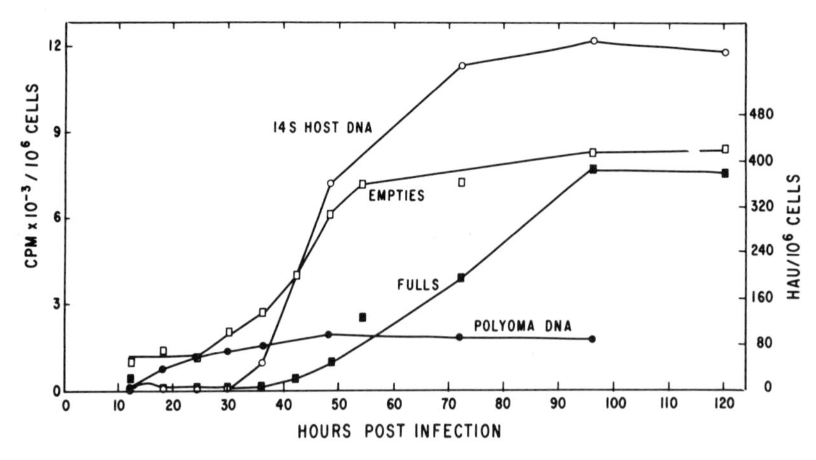

Fig 1. Relationship of the intracellular events leading to the production of pseudovirions. Primary mouse embryo cells were grown in glass roller bottles and infected with polyoma virus. At the indicated times, infected cells were analyzed for polyoma DNA, 14S host cell DNA fragments, full particles, and empty particles. Polyma DNA synthesis and production of 14S host cell DNA fragments were measured by counting the radioactivity in the 53S and 16S peaks after sedimentation through alkaline sucrose. HAU of the peaks in CsCl equilibrium centrifugation having a density of 1.32 g/cc and 1.28 g/cc determined the full and empty particles, respectively (13).

It should be emphasized that significantly more radioactivity was associated with the 14S host DNA fragments than was associated with the polyoma DNA molecules. Since the specific activities of the 14S DNA and the polyoma DNA were found to be about the same, the ratio of their radioactivities gives an indication of the relative quantity of each produced. When full particles were first found (42 hours), the ratio of 14S DNA to polyoma DNA was about 3 to 1; by 96 hours, it was 7 to 1. The fragmented host DNA sediments at 14S in neutral sucrose and 16S in alkaline sucrose.

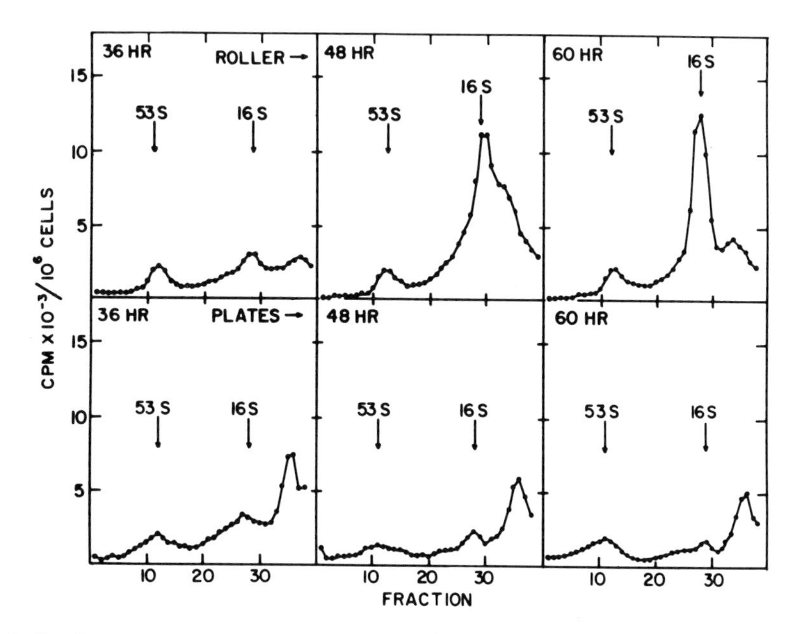

Fig 2. Synthesis of polyoma DNA and fragmentation of host cell DNA. Primary mouse embryo cells, grown in glass roller bottles or plastic plates, were infected with polyoma virus. The cells were labeled with ^{3}H-thymidine after infection. At the indicated times, the low molecular weight DNA was extracted from the cells by the method of Hirt (19). The supernatant was sedimented for 2 hours at 50,000 rev/min in the Spinco SW 50.1 rotor at 4 C. The total gradient volume was 4.6 ml. The gradient was prepared using 5 – 20% sucrose dissolved in 0.5M NaCl and 0.001M EDTA at pH 12.5.

The production of 14S DNA in amounts much larger than polyoma DNA appears to be unique with primary mouse embryo cells grown in roller bottles. This large amount of 14S host DNA relative to polyoma DNA was not found when primary mouse embryo cells were grown on plates (Fig 2) or when baby mouse kidney cells or 3T3 cells were grown on plates (Fig 3). Apparently, the mouse embryo-roller bottle system produces more pseudovirions than polyoma virions because there is more 14S host DNA than polyoma DNA competing for encapsidation.

We have been indeed fortunate to have this primary mouse embryo-roller bottle system available to us, since it has enabled us to have preparations containing large amounts of pseudovirions which can be separated from the small amounts of infectious polyoma virions present.

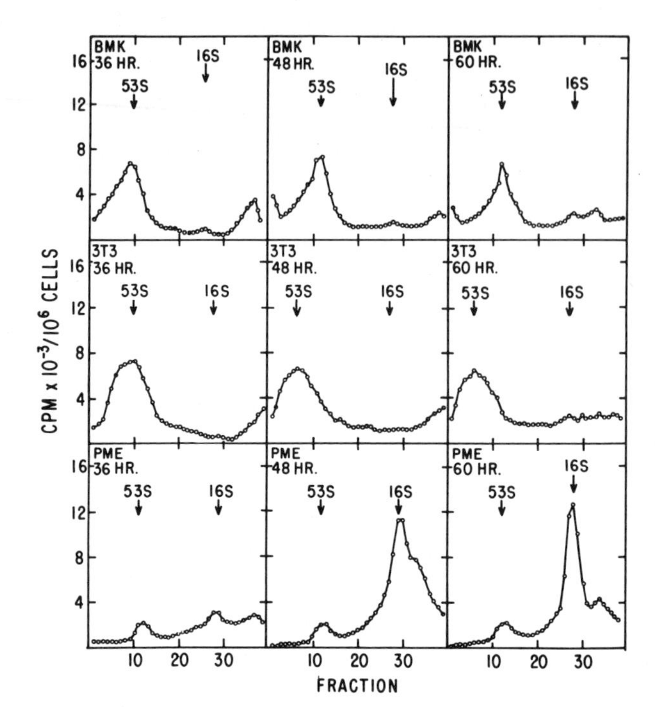

Fig 3. Synthesis of polyoma DNA and fragmentation of host cell DNA. BMK (primary baby mouse kidney cells), 3T3 mouse cells, and PME (primary mouse embryo cells). BMK and 3T3 grown on plastic petri dishes, PME in glass roller bottles. Procedure as outlined in legend to Figure 2.

TRANSFER OF DNA FROM ONE CELL TO ANOTHER

Can pseudovirions be used as a delivery system to transfer host cell DNA from one cell to another? We shall present evidence that pseudovirions can be adsorbed to and uncoated by cells, and that the naked pseudoviral DNA is found associated with the nuclear fraction of these cells.

For such experiments, our polyoma pseudovirions must be free of any radioactive polyoma virions or polyoma DNA. We have used over seven critera to characterize our pseudovirions (2). The most important of these are plaque formation and DNA-DNA hybridization. Polyoma virus forms plaques on mouse embryo monolayers. Pseudovirions do not. Using plaque assays, we have found that our most purified polyoma pseudovirion preparations contain less than one polyoma particle per 10^6 pseudovirus particles (2). The DNA of these highly purified pseudovirions hybridizes with mouse embryo DNA but does not hybridize with polyoma DNA (3).

For uncoating evidence we used the following approach: radioactively labeled DNA that has been uncoated is susceptible to pancreatic DNase; that is, the radioactivity will be converted from an acid-insoluble to an acid-soluble form. Encapsidated radioactive DNA, however, is protected from DNase and is *not* converted from an acid-insoluble to an acid-soluble form. The data of Table I indicate that pseudovirions are adsorbed to and uncoated by secondary mouse embryo cells.

TABLE I

The Uncoating of Pseudovirus in Secondary Mouse Embryo Cells as Measured by Susceptibility to Pancreatic DNase (2)

	Time cells exposed to virus, min				
	0	60		90	
	(cpm)	(cpm)		(cpm)	
Experiment 1					
Pseudovirus adsorbed	1,470	22,300		26,500	
Pseudovirus uncoated[a]					
- DNase	5	30		75	
+ DNase	85	915		1,560	
Experiment 2					
Pseudovirus adsorbed	3,750		25,200		25,600
Pseudovirus uncoated					
- DNase	0		55		70
+ DNase	140		985		930

To each plate of secondary mouse embryo cells, purified pseudovirus (36,200 cpm/plate and 33,000 cpm/plate for Experiments 1 and 2, respectively) was added at 0 time.

[a]Measured as a conversion by pancreatic DNase of acid-insoluble radioactivity to acid-soluble radioactivity.

Does pseudovirus enter the nuclei? Pseudovirions labeled with H^3-thymidine were added to secondary mouse embryo cells; at various times after infection the nuclei were isolated by the method of Penman (14). By 24 hours postinfection, approximately 23% of the radioactivity was associated with the nuclear fraction of the cells (Fig 4). Sixty-eight percent of the radioactivity in the nuclear fraction was sensitive to pancreatic DNase (Table II). The nuclei were gently homogenized and separated by centrifugation into supernatant and pellet fractions. After DNase treatment, 53% of the radioactivity in the supernatant fraction and 78% of the radioactivity in the pellet fraction became acid soluble. Since the intact viral particles are not sensitive to DNase treatment, these results further indicate that pseudovirions are uncoated and their naked DNA is present in the nuclear fraction.

There are over 100 single gene defects known in the human which have been proved to result in the synthesis of an altered enzyme (15). Cells from humans with some of these genetic disorders can be propagated in culture. These single-gene defects provide a wide variety of potential genetic markers for transduction experiments. For this reason, the interaction of polyoma pseudovirions with human cells was studied.

Transfer of mouse DNA to human cells was studied by infecting human embryonic cells with radioactive pseudovirions. Twenty-four hours after infection, cells were harvested by trypsinization and the cytoplasmic and nuclear fractions were prepared. Twenty-two percent of the input radioactivity remained firmly bound to or within the cells (Table III). Of the total radioactivity found in the cells, 7.3% was

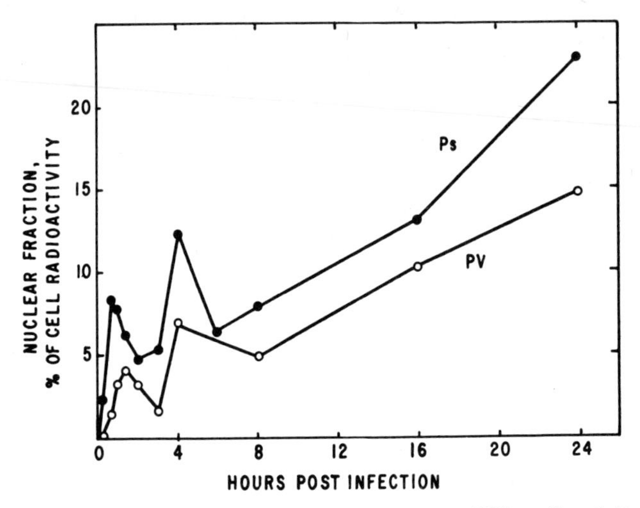

Fig 4. Appearance of pseudovirus (Ps) or polyoma virus (PV) radioactivity in the nuclear fractions of secondary mouse embryo cells (3).

TABLE II

DNase Sensitivity of Nuclear Radioactivity after Infection of ME-Cells [^{3}H]-Thymidine Pseudovirus (3)

Fraction	% radioactivity after treatment with DNase	
	TCA soluble	TCA insoluble
Whole nuclei	68	32
Homogenized nuclei		
Supernatant	53	47
Pellet	78	22

TABLE III

Distribution of [^{3}H]-Pseudovirus Radioactivity in Human-Embryonic Cells after Infection with [^{3}H]-Thymidine Pseudovirus (3)

Time (hr)	Input cpm bound to or in cells (%)	Percent of total cellular radioactivity	
		Cytoplasmic fraction	Nuclear fraction
0	0.8	99.3	0.7
24	21.9	92.7	7.3

associated with the nuclear fraction. After homogenizing the nuclei from infected cells and pelleting the debris, the supernatant fluid was analyzed by sedimentation through a neutral sucrose gradient (Fig 5). The radioactivity in the nuclear fraction was shown to be representative of uncoated pseudoviral DNA. Some tritiated material cosedimented with [^{14}C]-labeled pseudovirus used as a marker. Most of the tritiated material, however, sedimented in a more heterogeneous manner than the pseudovirus used for infection. About 50% of the radioactivity was found near the top of the gradient in the position where free DNA would sediment under such conditions. These results indicate that polyoma pseudovirions entered and were uncoated by human embryonic cells.

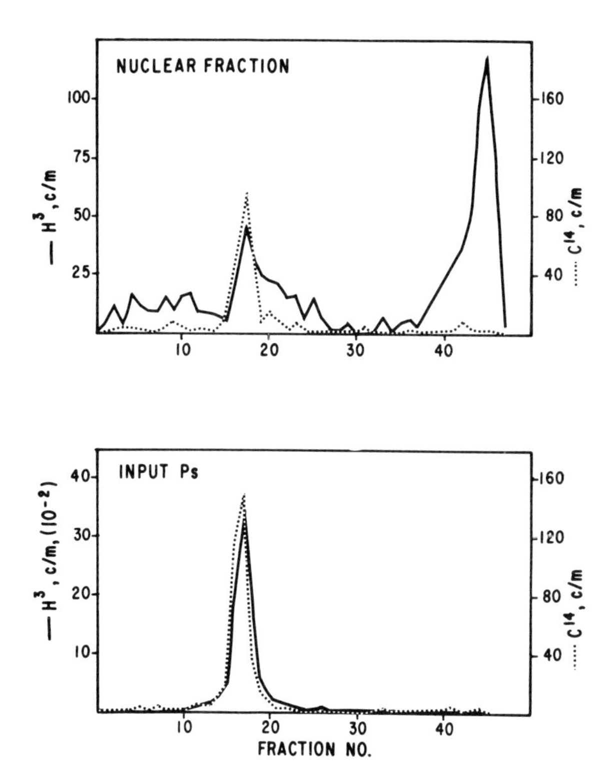

Fig 5. Sedimentation in neutral sucrose of the nuclear fraction of human embryonic cells infected with ^{3}H-thymidine pseudovirus (3).

REASSEMBLY OF POLYOMA VIRUS

Grady, Trilling, and Axelrod have shown that SV40 pseudovirions contain no unique sequences of monkey DNA (10,16). C_0t analysis supports their proposal that SV-40 pseudovirions contain random fragments of monkey DNA. Hirt's group (17) has tried to transfer information for the synthesis of thymidine kinase to mouse cells lacking the thymidine kinase gene by using polyoma pseudovirions. Their results were negative. However, the results of our experiments demonstrate that pseudovirions can be used to transfer DNA to the nuclei of mammalian cells. Therefore, we have decided to attempt the reassembly of polyoma pseudovirions. In vitro assembly of pseudovirions would allow us to encapsidate a specific gene. Certainly in the next few years specific isolated genes will become available either by chemical synthesis à la Khorana or by the use of reverse transcriptase, for example, the work on the hemoglobin gene.

We have begun some experiments designed to find out whether there are specific binding sites for DNA within the polyoma capsids. Our assay consists of incubating purified polyoma empty particles with radioactive type I polyoma DNA in the presence of $MgCl_2$ and pH 7.5 Tris buffer. After a 60-minute incubation at 37°, a portion of the reaction mixture was sedimented through a 10 - 40% neutral sucrose gradient. C^{14}-thymidine-labeled pseudovirions were used as a marker. The top panel of Figure 6 shows the sedimentation analysis in sucrose of polyoma DNA incubated in the absence of empty particles. One peak of radioactivity representing DNA was found at the top of the gradient. However, when the polyoma DNA was incubated with empty particles, three discrete peaks of radioactivity were found. The slowest sedimenting peak represented unreacted polyoma DNA, the second peak had an S value of about 115 S and the third peak had an S value of about 180 S. When the polyoma DNA was incubated with full virions, the formation of these complexes did not occur.

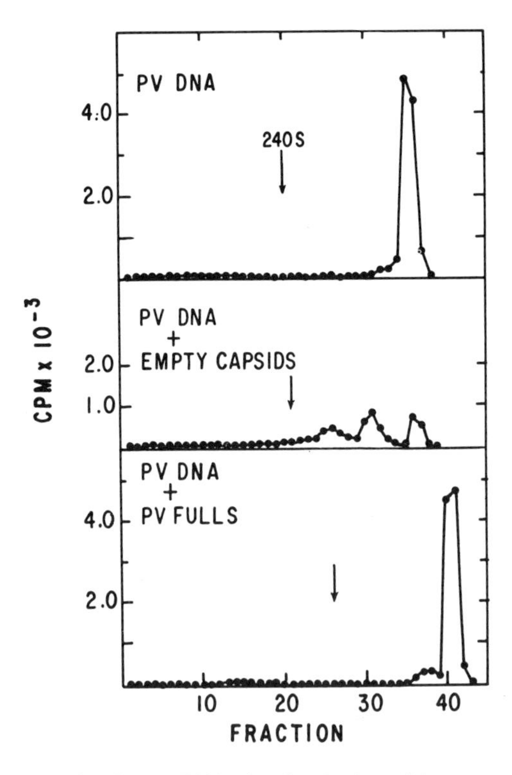

Fig 6. Sedimentation of polyoma DNA after incubation with empty polyoma capsids.

These results show that polyoma DNA can associate with empty polyoma particles and that this association does not occur if full virions are used instead of empties. Mixing experiments using empty and full particles showed that the association of polyoma DNA with empty particles is not inhibited in the presence of full particles.

Is this binding of DNA to empty particles due to a viral protein or specific site which is within the complete virus particle?

Figure 7 shows the association of polyoma DNA with the empty particles produced from intact virions by disruption. The top panel shows the sedimentation of radioactive polyoma DNA incubated with full particles. Again, the DNA does not bind to full particles. The next panel is the sedimentation of radioactive polyoma DNA incubated with disrupted full particles. These disrupted virus preparations contain empty capsids plus the endogenous DNA. This mixture also does not bind to the added radioactive DNA. The endogenous DNA is still present in these disrupted fulls. The bottom panel of Figure 7 shows the sedimentation of radioactive polyoma DNA following incubation with empty particles purified from the preparation of disrupted full particles. Notice that the DNA binds to these empty particles and the sedimentation pattern is similar to the one obtained when naturally occurring empty particles were used for the interaction.

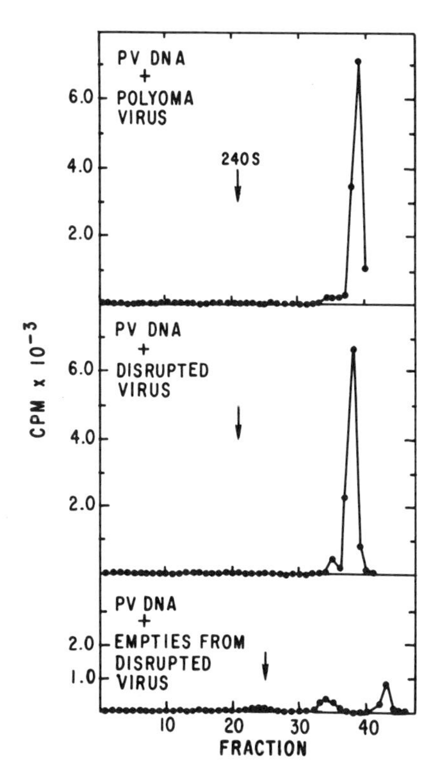

Fig 7. Sedimentation of polyoma DNA after incubation with empty polyoma capsids prepared by disruption of intact virus.

Our interpretation of these results is that a viral protein(s) or a specific binding site(s) for DNA is present within polyoma virions and becomes available only after disruption of full particles into empty particles and removal of the endogenous DNA from this binding site(s). These capsid-DNA complexes are nucleoproteins, since the complex is destroyed by incubation with pronase and dissociates at high salt concentrations. The binding of polyoma DNA to a specific site in the capsid may be the first step in the assembly of this virus. Obviously we have a long way to go. Friedmann (18) is also studying polyoma reassembly.

SUMMARY

Large amounts of polyoma pseudovirions can be produced by infecting primary mouse embryo cells grown in glass roller bottles with polyoma virus. This appears to be due to the large amount of 14 S host DNA fragments produced, which compete with polyoma DNA for the virus assembly process. Extensive fragmentation of host cell DNA is not found when baby mouse kidney cells, 3T3 cells, or primary mouse embryo cells grown in plates are used.

Polyoma pseudovirions can transfer fragments of mouse DNA to the nuclear fraction of mouse embryo and human embryo cells.

There appear to be specific binding sites within polyoma capsids for polyoma DNA which may be of help in reassembly experiments.

REFERENCES

1. Aposhian, H. V.: The use of DNA for gene therapy–the need, experimental approach, and implications. Perspect. Biol. Med., 14:98, 1970.
2. Osterman, J. V., Waddell, A., and Aposhian, H. V.: DNA and gene therapy: uncoating of polyoma pseudovirions in mouse embryo cells. Proc. Nat. Acad. Sci., USA, 67:37, 1970.
3. Qasba, P. K., and Aposhian, H. V.: Transfer of mouse DNA to human and mouse embryonic cells by polyoma pseudovirions. Proc. Nat. Acad. Sci. USA, 68:2345, 1971.
4. Dulbecco, R., and Vogt., M.: Evidence for a ring structure of polyoma virus DNA. Proc. Nat. Acad. Sci. USA, 50:236, 1963.
5. Weil, R., and Vinograd, J.: The cyclic helix and cyclic coil forms of polyoma viral DNA. Proc. Nat. Acad. Sci. USA, 50:730, 1963.
6. Blackstein, M. E., Stanners, C. P., and Farmilo, A. J.: Heterogeneity of polyoma virus DNA: isolation and characterization of non-infectious small supercoiled molecules. J. Mol. Biol., 42:301, 1969.
7. Michel, M. R., Hirt, B., and Weil, R.: Mouse cellular DNA enclosed in polyoma viral capsids. Proc. Nat. Acad. Sci. USA, 58:1381, 1967.
8. Winocour, E.: Further studies on the incorporation of cell DNA into polyoma-related particles. Virology, 34:571, 1968.
9. Basilico, C., Maksuya, Y., and Green, H.: The interaction of polyoma virus with mouse-hamster somatic hybrid cells. Virology, 41:295, 1970.
10. Trilling, D. M., and Axelrod, D.: Encapsidation of free host DNA by simian virus 40: a simian virus 40 pseudovirus. Science, 168:268, 1970.
11. Levine, A. J., and Teresky, A. K.: Deoxyribonucleic acid replication in simian virus 40-infected cells. II. Detection and characterization of SV40 pseudovirions. J. Virol., 5:451, 1970.
12. Qasba, P. K., Yelton, D. B., Pletsch, Q. A., and Aposhian, H. V.: Properties of polyoma pseudovirions in mouse and human cells. In: Molecular Studies in Viral Neoplasia. M. D. Anderson Symposium, Houston, Texas, 1972.
13. Yelton, D. B., and Aposhian, H. V.: Polyoma pseudovirions. I. Sequence of events in primary mouse embryo cells leading to pseudovirus production. J. Virol., 10:340, 1972.
14. Penman, S.: Preparation of purified nuclei and nucleoli from mammalian cells. In: Fundamental Techniques in Virology. Edited by K. Habel and N. P. Salzman. Academic Press, New York, p. 35, 1969.

15. McKusick, V. A.: Mendelian Inheritance in Man. The Johns Hopkins Press, 3rd edition. Baltimore and London, 1971.
16. Grady, L., Axelrod, D., and Trilling, D.: The SV40 pseudovirus: its potential for general transduction in animal cells. Proc. Nat. Acad. Sci. USA, 67:1886, 1970.
17. Hirt, B.: Inclusion of host DNA in viral capsids. Presented at Conference on the Prospects of Gene Therapy, Fogarty Center, NIH, 1971.
18. Friedmann, T.: In vitro reassembly of shell-like particles from disrupted polyoma virus. Proc. Nat. Acad. Sci. USA, 68:2574, 1971.
19. Hirt, B.: Selective extraction of polyoma DNA from infected mouse cell cultures. J. Mol. Biol., 26:365, 1967.

4. CELL SURFACE MODIFICATIONS IN VIRUS-TRANSFORMED CELLS –GLYCOLIPID CHANGES IN THE NIL2 HAMSTER CELLS

I. MacPherson, D. R. Critchley,
K. A. Chandrabose and Diane Humphrey

Imperial Cancer Research Fund Laboratories
London WC2A 3PX

INTRODUCTION

Animal cells transformed in culture by cancer viruses provide simple model systems for the study of cancer cell physiology. An understanding of the key modifications affecting a malignant cell's response to homeostatic controls is more likely to be obtained from the study of these systems than from comparisons between normal and cancer tissues per se. The latter type of study has often been undertaken without any clear indication that the cells compared had a common origin. This important requirement can be fulfilled with cultured cells, particularly, with cloned lines of cells and their transformed derivatives.

Transformation in vitro is a "package deal" in which a variety of cell modifications are expressed within a few cell generations of virus genes integrating with the cell. Cell variants in which transformed cell characteristics are segregated give some indication as to how the various properties are linked. Revertants with partial or complete recovery of normal cell properties may result from chromosome loss or suppression of virus genes by cellular factors (1). As yet there is no clear indication which cell modifications are most closely associated with malignancy and which should therefore be studied in most detail to give a clue to the nature of the change in chemical terms. It is, however, now widely believed that the crucial change occurs at the cell surface, a belief which is supported by evidence from a wide variety of sources. The plasma membrane has become a very fashionable organelle to study and currently is being coaxed to reveal its secrets to electronmicroscopists, immunologists, chemists, physicists, and others.

From these wide-ranging studies the idea has emerged that glycosylated macromolecules at the surface play a role in the regulation of cellular physiology and the social behavior of cells. Of particular interest are the glycolipids which, although a minor component of the cellular lipid, are the most characteristic of a particular tissue and are affected by the genetic strain, sex, and age of the animal (2,3). They are exposed at the surface of the cell and may be concentrated in the plasma membrane (4,5).

Perhaps the earliest indication that glycolipid changes occurred in malignant tissue came from the work of Rapport and his coworkers (6). They found that an antigen frequently expressed in tumor tissue was a dihexosyl ceramide with lactose as the carbohydrate moiety. Later Tal (7) showed that the sera from a wide range of cancer patients and pregnant women agglutinated suspensions of HeLa cells and that agglutination was specifically inhibited by lactosyl ceramide or lactose. Interest in the role of glycolipids in cultured cells was stimulated by the work of Hakomori and his colleagues (8,9). They found that in BHK21 hamster cells transformed by polyoma or Rous sarcoma virus there was a reduction in the amount of hematoside which is normally synthesized from dihexosyl ceramide by the addition of N-acetylneuraminic acid. This reduction in hematoside was sometimes accompanied by an accumulation of dihexosyl ceramide. Subsequent studies have also shown that carbohydrate chain elongation of glycolipids is affected in other virus-transformed cell systems (10). The glycolipids affected were different in different cells, but the basis of these changes seems to be a reduction in the activity of the appropriate sugar transferase. In mouse cells transformed by SV40 virus the enzyme catalyzing the transfer of N-acetylgalactosamine to hematoside is much reduced and consequently causes a considerable diminution in the amount of Tay-Sachs ganglioside, the product of the reaction (11).

Our own studies have been made with a cloned line of NIL2 hamster cells (12,13) and its transformed derivatives. When these cells are transformed by cancer viruses they show a reduction or loss of neutral glycolipids with carbohydrate residues larger than dihexoside (14,15,16). In this line the enzyme block affects the addition of another galactosyl residue to form trihexosyl ceramide from lactosyl ceramide. In transformed cells this results in the disappearance of the trihexoside and the higher glycolipids derived from it, that is, ceramide tetra and penta hexoside (Fig 1). Of particular interest is the observation that these same three compounds are reduced in sparse noncontacted NIL2 cultures and that they only reach maximal amounts in

Fig 1. Possible Synthetic Pathway of Glycolipid Synthesis in NIL2 Cells

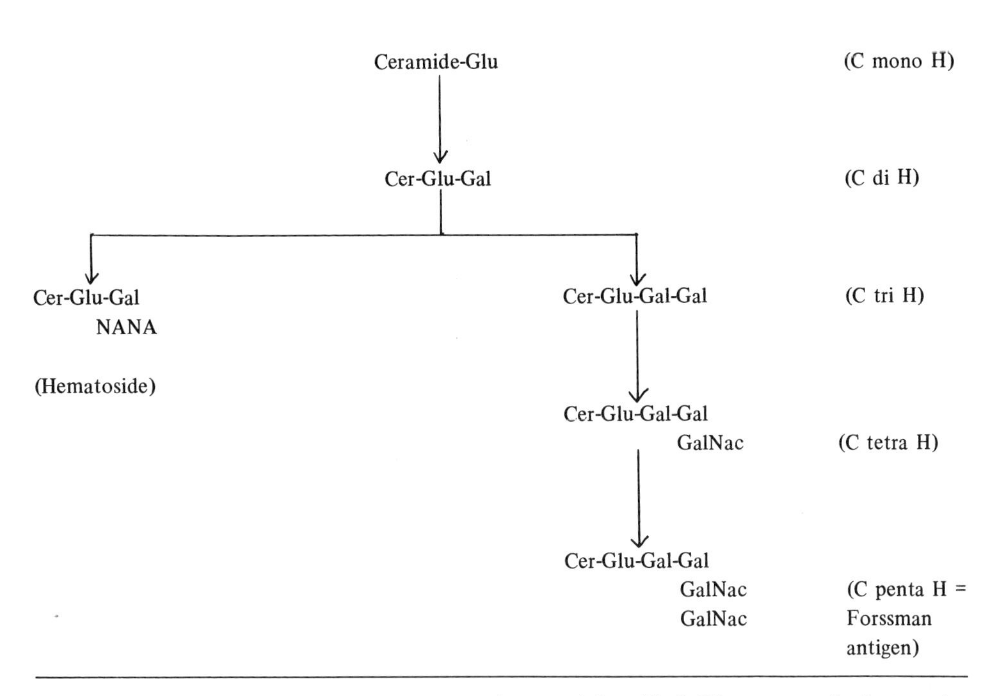

Glu = glucose; Gal = galactose; NANA = nacetyl neuraminic acid; GalNac = n-acetyl-galactosamine
Enzyme block in transformed cells

confluent cultures (14,15), suggesting that cell-to-cell contact may influence synthesis. Similar results have been obtained by Hakomori (17).

Here we describe further studies on the control of glycolipid synthesis in the NIL2 cell and some of its variants under different growth conditions.

MATERIALS AND METHODS

Cell Cultures

Clones of NIL2 hamster cells were derived from single cells isolated by micromanipulation (18). They were studied within a limited number of generations after isolation or in cultures recently derived from low passage cells stored in liquid nitrogen. Conventional methods of culture were used. Clones with low saturation densities, that is, 4 - 15 times 10^6 cells/9cm diameter polystyrene dish, were used in most experiments. The medium was Dulbecco's modification of Eagle's with 10% calf serum (heated at 56°C for 30 min). Cultures of cells transformed by polyoma virus or hamster sarcoma virus (HSV) were derived from colonies growing in soft agar medium (19). These formed ten days after seeding with infected cells. Transformed cells were cloned in the same way as normal cells. Tumors were produced by injecting $>10^7$ NIL2 cells subcutaneously into adult female hamsters. Tumor tissue was trypsinized and cultured. In all cases the cells were karyologically male like NIL2, showing that the tumors were derived from the implanted cells.

Glycolipid Analysis

Cellular lipids were labeled with ^{14}C-palmitate for periods of up to 72 hours by growing 5 X 10^5 X 1.5 X 10^7 cells per 9 cm dish in the presence of 1 μc/ml palmitate. Label was taken up rapidly by the cells–up to 60% being incorporated over a 12-hour period. After the labeling period the cell sheet was washed with medium, then scraped into methanol, sonicated, and samples taken for scintillation counting and protein estimation. The methanol suspension was dried under nitrogen and the lipid extracted with chloroform:methanol 2:1 then 1:2 at 4° overnight. Lipids were then separated by two dimensional thin-layer chromatography (TLC) on silica gel precoated plates (Merck). The first solvent mixture was chloroform:methanol:water 65:25:4 and in the second dimension tetrahydrofuran:methylal:methanol:water, 10:6:4:1. Autoradiography was used to determine the position of the labeled lipid spots. These were subsequently assayed by scintillation counting of the silica gel scraped from the radioactive areas. The variation in incorporation per unit of sphingosine into individual glycolipids is <1.5 fold after labeling periods of >48 hours. Since there is one molecule of sphingosine per glycolipid molecule this shows that the method gives a quantitative as well as qualitative measure of the glycolipids.

Characterization of Individual Glycolipids

The labeled lipids were characterized in the following way:

1. ^{14}C-palmitate:^{32}P incorporation ratios to identify phospholipids.
2. Co-chromatography with known glycolipid standards.
3. Stability in methanol:water (9:1) 0.1 N with respect to NaOH. Glycerol based lipids are hydrolyzed under these conditions whereas sphingosine-based lipids are stable.
4. ^{14}C-glucosamine labeling to identify glycolipids containing amino sugars or sialic acid residues.
5. Partial acid hydrolysis with 0.1 N H_2SO_4 at 80°C for 1 hour to cleave sialic residues followed by TLC product analysis.
6. Partial acid hydrolysis with 0.1 N HCl at 100°C for 1 hour to remove terminal neutral and amino sugar residues followed by TLC product analysis.

RESULTS

1. NIL2 Cell Lipids

Seven of the lipids of NIL2 cells were unaffected by treatment with mild alkali and were therefore presumably sphingosine based. One of these compounds incorporated ^{32}P and behaved chromatographically like a sphingomyelin standard. Partial acid hydrolysis of the sphingosine-based lipids followed by the characterization of the products showed that five were neutral glycolipids ranging from ceramide monohexoside (C mono H) to ceramide pentahexoside (C penta H) (Table I). Since the tetra and penta hexoside could also be labeled with ^{14}C-glucosamine they probably contain N-acetylhexosamine residues. One glycolipid was hydrolyzed by H_2SO_4 and was probably sialyl-dihexosyl ceramide (hematoside). From the results of these tests it

TABLE I

Characterization of Glycolipids by Partial Acid Hydrolysis

Compounds	Hydrolysis products	
	0.1 N HCl 100°C 1 hr	0.1 N H_2SO_4 80°C 1 hr
C mono H	C mono H	C mono H
C di H	C mono H	C di H
C tri H	C di H, C mono H	C tri H
C tetra H	C tri H, C di H	C tetra H
C penta H	C tetra H, C tri H C di H	C penta H
Hematoside	C di H	C di H

C mono H etc. = ceramide monohexoside (see Fig 1)

was possible to propose chemical structures for the glycolipids in NIL2 cells and a pathway for their synthesis (Fig 1). This agreed with recently published work (16).

2. Glycolipids of Virus-transformed NIL2 Cells

Clones of NIL2 cells transformed by polyoma virus or hamster sarcoma virus (HSV) consistently lacked ceramide tri, tetra, and pentahexosides. These results confirmed previous observations by Robbins and Macpherson (14). Representative results are shown in Table II. The amount of radioactivity incorporated by the glycolipids has been calculated per mg of cell protein. In Table II it can also be seen that there are no major changes in the phospholipids of NIL2 cells following transformation or under different cell culture densities. A possible exception is phosphatidyl inositol which may be density dependent.

TABLE II

Incorporation of ^{14}C-palmitate into the Lipids of NIL2 and Transformed NIL2/HSV Cells

Lipid	NIL2		NIL2/HSV	
	Sparse	Dense	Sparse	Dense
SM	62[a]	58	58	66
PC	130	157	155	212
PE	43	72	33	61
PI	15	19	5	23
C mono H	5	2	8	2
C di H	2	2	5	3
C tri H	0.8	6	0	0
C tetra H	0.6	3	0	0
C penta H	2	4	0	0
Hematoside	14	11	14	14
Hem/C tri H	17.5	1.8	14	14

[a] cpm/10 μg/protein
SM = sphingomyelin
PC = phosphatidyl choline
PE = phosphatidyl ethanolamine
PI = phosphatidylinositol
Hem = hematoside
C mono H etc. = ceramide monohexoside (see Fig 1)
HSV = hamster sarcoma virus

3. Effect of Cell Density on NIL2 Glycolipid Synthesis

Studies on the ^{14}C-palmitate labeling pattern in NIL2 cells in different culture conditions had previously (14) shown that the glycolipids missing in virus-transformed cells (C tri H, C tetra H, and C penta H) were reduced in normal cells when in sparse culture. These observations were confirmed (Table II) and extended to include clones with different saturation densities. Essentially similar results were obtained in each case. As the cells became dense and made contact with each other there was a concomitant increase of the density-dependent glycolipids. The same differences were obtained in single cultures with areas of low and high cell densities in which the two areas of cells shared the same medium. This showed that (a) low pH of the medium does not trigger C tri H synthesis or (b) cells in dense culture deplete the medium of an inhibitor which blocks C tri H synthesis. Experiments in which normal cells were grown in the same culture as an area of dense transformed cells showed that soluble inhibitors of C tri H synthesis were not produced by the latter, since the NIL2 glycolipid synthesis was unaffected.

These results can best be explained on the basis that the enzyme catalyzing the transfer of UDP-galactose to ceramide dihexoside is affected by cell density and also by virus transformation. As mentioned in the introduction, the diminution of certain glycolipids in transformed cells has been shown to be associated with a loss of the appropriate sugar transferase activity. This also seems to be the case with NIL2 cells. Dense cultures synthesized 1560 $\mu\mu$moles of C tri H/hour/mg protein under optimal conditions in a total incubation volume of 0.25 ml, using a microsomal preparation as an enzyme source. These results are the average of five experiments. When a crude homogenate is used as an enzyme source the activity is 685 $\mu\mu$moles/hour/mg protein under the same conditions of incubation. NIL/HSV cells had no detectable activity in the same conditions either as a crude homogenate or as a microsomal fraction.

To gain a better understanding of how this enzyme's activity is affected by cell density, we studied the kinetics of change in glycolipid synthesis in cells undergoing a change in their culture environment from sparse to dense and vice versa. We also studied the synthesis of "contact sensitive" glycolipids in sparse NIL2 cells in which cell growth and division were inhibited in various ways.

4. The Kinetics of Glycolipid Synthesis in NIL2 Cells

(a) Sparse → Dense Culture

The change in synthesis of glycolipids in NIL2 cells growing from sparse to dense culture was studied in 9 cm Petri dishes with an initial population density of 2 X 10^5 cells. The cells were allowed to grow to a level at which cell contacts were still minimal, then 4 - 8 dishes were each pulsed for 6 hours with 10 μc of ^{14}C-palmitate. Replicates were used for cell counts or ^{3}H-thymidine labeling for 1 - 6 hours. As the cultures increased in density similar pulse labelings and cell counts were made. The incorporation of ^{14}C-palmitate was variable. The amount incorporated per unit of protein tended to decrease as the cultures increased in density. Accordingly, the ratio of ^{14}C incorporation into the hematoside:ceramide trihexoside was used to reflect the pathway preferred by the cell under different growth conditions (Table III and Fig 1). This ratio was compared to cell density and incorporation of ^{3}H thymidine per 10^6 cell (Fig 2). Increase in cell number from 2 times 10^6 per dish (in which cell contact was minimal) to 5 times 10^6 per dish resulted in a drop in the hematoside/ceramide trihexoside ratio even though the cells had the capacity for further cell

TABLE III

Glycolipid Synthesis in NIL2 Cells Growing from a Sparse Inoculum

Time of ^{14}C-palmitate pulse	0-6	21-27	25-31	29-35	44-50 hr.
CPM/MG protein x 10^6	3.2	2.7	2.2	2.1	0.86
$\frac{\text{Phospholipid CPM}}{\text{Glycolipid CPM}}$	7.9	11.0	10.3	19.0	38.6
PC + SM[a]	740[b]	745	712	760	780
PE	51.2	54.1	53.0	54.0	51.0
PI	26.2	42.5	64.0	66.0	79.0
C mono H	50.0	30.4	33.2	10.8	6.1
C di H	49.0	36.9	38.7	21.1	1.5
C tri H	1.1	2.4	3.5	4.4	4.1
C tetra H	0.6	0.6	0.9	1.4	3.2
C penta H	1.5	2.0	2.2	2.9	5.2
Hem	11.6	11.5	9.4	8.7	5.0

[a] See Table II for contractions.

[b] $\frac{\text{CPM} \times 10^3}{\text{Total phospholipid \& glycolipid CPM}}$

division. The decrease continued as the cells became more dense, finally becoming constant at about 10^7 per dish. Thymidine incorporation was used to detect the cessation of DNA synthesis resulting from cell contacts, but as can be seen in Figure 2 incorporation per cell was not constant even in cells growing logarithmically. From these observations it is clear that the synthesis of glycololipids in these cells favors the ceramide trihexoside pathway when cell contacts are being made rather than at a later stage when the cells' growth has become dense and quiescent.

(b) Dense → Sparse Culture

In this experiment, the reverse of that described in (a), cells were first grown to confluence and maintained until ^{3}H-thymidine incorporation was minimal and there was stability of cell counts over a two-day period. At this time four dishes were pulsed with ^{14}C-palmitate and other dishes harvested with trypsin and transferred at 5 X 10^5 cells per dish. As a control, dense cultures were trypsinized and allowed to settle and attach in the same dish without cell dilution. Cell counts, ^{3}H-thymidine, and ^{14}C-palmitate incorporation were examined at times before and after the first cell division. From Figure 3 it can be seen that the hematoside:ceramide trihexoside ratio remained unaltered for the first 10 hours, and ^{3}H-thymidine incorporation remained low. After 18 hours DNA synthesis had started and there was a marked increase in the Hem/C tri H ratio, even though the cells had not divided. After 35 hours the cells had divided and the ratio increased still further. In the dense culture trypsinized and recultured at high density there was a small transient increase in the Hem/C tri H ratio

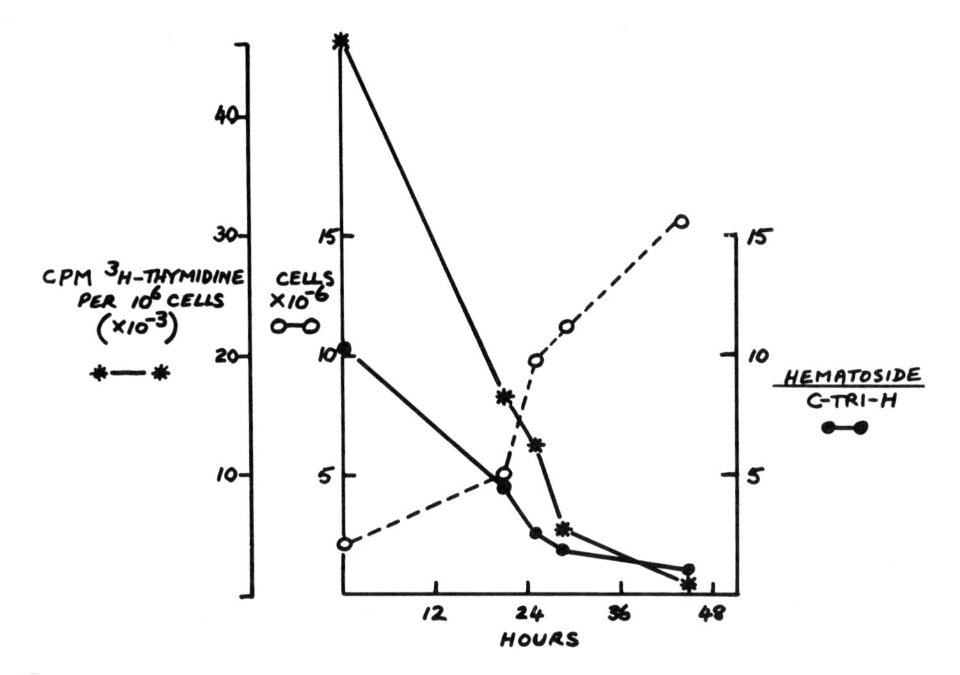

Fig 2. Synthesis of glycolipids in NIL2 cells growing from sparse to dense culture. Cells were pulsed with ^{14}C-palmitate (1 μc/ml) or ^{3}H-thymidine (0.1 μc/ml) for 6 hrs at various stages in growth from sparse to dense culture. A ratio of incorporation of ^{14}C-palmitate into Hematoside:C tri H is used to reflect the preferred route of glycolipid synthesis.

which could not be accounted for by the preferential loss of C tri H, since trypsin does not remove glycolipids from the cell.

Cells prelabeled with ^{14}C-palmitate were used in a similar dense-to-sparse experiment to test whether C tri H is actively broken down before the first cell division. The results are shown in Table IV. It can be seen that over a 46 hour period in which the sparse cells divided twice there was no major change in the Hem/C tri H ratio. Cells which were trypsinized and maintained at high cell density showed a similar pattern. The constancy of the ratio suggests that C tri H is not actively degraded by dividing cells but is probably diluted by cell division.

Glycolipids of Noncontacted Nondividing NIL2 Cells

To elucidate further the role of growth and cell contacts in glycolipid synthesis experiments were carried out in which the growth of the NIL2 cells was inhibited in various ways while maintaining the cultures in a sparse condition. This can be achieved in the following ways:

a) by maintaining the cells in medium containing only 0.25% serum. This results in a block in the Gl phase. Addition of serum up to the usual levels (10%) results in cells entering the S phase about 10 hours later (20).

b) by growing cells in the absence of glutamine. This causes the accumulation of cells in Gl (21).

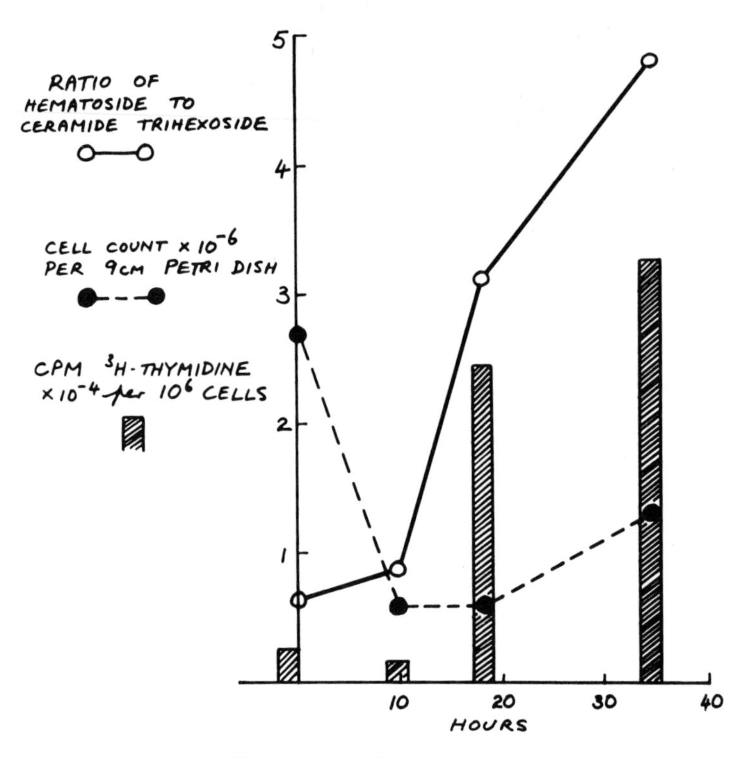

Fig 3. Glycolipid synthesis in NIL2 cells seeded from dense-to-sparse culture. Dense cultures harvested with 0.25% trypsin in tris/saline were seeded to low density and pulsed with ^{14}C palmitate (10 μc/ml, 6 hrs) or ^{3}H-thymidine (0.1 μc/ml, 1 hr) at time intervals before and after the first cell doubling.

TABLE IV

Turnover of Ceramide Trihexoside in Cells Seeded from Dense (D) to Sparse (S) Culture

Time of labeling with ^{14}C-palmitate	Hem / C Tri H		CPM ^{3}H thymidine x 10^{-3} per 10^{6} cells		Cell number /cm^{2} x 10^{3}	
hrs	S	D	S	D	S	D
0-6	–	1.2	–	2.4	–	80
6-12	0.8	0.6	3.2	0.4	6	100
13-19	0.8	0.8	34.0	3.6	6	100
28-34	0.9	1.2	48.0	2.8	14	160
36-42	1.3	1.2	27.0	0.8	28	180

c) Addition of 2 mM thymidine. Excess thymidine causes an inhibition of CDP reductase with a consequent reduction in the level of de CTP which is insufficient to maintain DNA synthesis at the normal rate. Cells are therefore blocked at the beginning of S phase or in S phase (22).

In each case the efficiency of the inhibitor was tested by a reduced incorporation of ^{3}H-thymidine in the case of methods (a) and (b) and stationary cell counts in the case of (c). About two days after seeding the cells, ^{14}C-palmitate was added and the cells cultured for a further 48 hours. The viability of the cells at the end of the experiment was determined by removing the inhibitor and estimating ^{3}H-thymidine incorporation and cell counts.

NIL/HSV cells failed to survive in the absence of serum or in the presence of excess thymidine, but glutamine deprivation produced the required inhibition of DNA synthesis and cell division. The results are shown in Table V. Increased levels of ceramide trihexoside typical of dense culture were only found in NIL2 cells blocked in low serum medium. Cells in excess thymidine showed some tendency toward increased levels, but cells without glutamine synthesized very little of the trihexoside. Nondividing NIL/HSV cells also failed to synthesize any of the density-dependent glycolipids.

TABLE V

Distribution of ^{14}C-palmitate in the Glycolipids of Growth-inhibited NIL2 Cells

Lipid	Cells in sparse culture	Cells inhibited by: High density	Low serum	No glutamine	Excess thymidine
C mono H	16[a]	5	24	23	19
C di H	25	6	6	13	9
C tri H	6	34	20	8	10
C tetra H	2	17	8	3	6
C penta H	6	c.10	8	11	17
Hematoside	44	<28	33	41	37
Hematoside / C tri H	7.3	0.8	1.6	5.0	3.7

[a]% total glycolipids.

Glycolipids in NIL2 Variants

The NIL2 cell line appears to be genetically unstable in that some clones lose their original growth characteristics and social behavior quite rapidly and give rise to variants with altered properties. Other clones, such as those used for the kinetic studies, are relatively stable.

In an attempt to correlate some of these properties with changes in the glycolipids 12 subclones of the original NIL2 clone were studied. These clones had a range of cell morphologies and saturation densities. Some were tending to epithelial growth, others were fibroblastic. All clones synthesized the "density-dependent" glycolipids, although there was some variation in the proportions between the three components (Table VI). Transformation of these clones with hamster sarcoma virus leads

TABLE VI

Incorporation of ^{14}C-palmitate in the Cell Density-dependent Glycolipids of NIL2 Cells and Variants

Cell type	C tri H	C tetra H	C penta H
NIL2 clones no. 1-6	23.5[a]	26.5	21.6
NIL2 clones no. 7-12	15.0	12.5	30.5
NIL2/HSV	0.9	1.6	7.0
Tumor lines no. 1-10	0	0	20.5
Tumor line no. 11	22.7	17.5	27.2
Tumor line no. 11 (Passaged in hamsters)	19.5	5.4	32.5
Tumor line no. 11 (Transformed by HSV)	1.2	1.5	13.0

[a]% total glycolipid

to complete loss of the tri and tetrahexoside, although small amounts of the pentahexoside remained.

Of 11 tumor lines 10 had only trace amounts of the tri and tetrahexoside, but all contained some pentahexoside (Forssman antigen). The pentahexoside was, however, not density dependent and, interestingly, this was also the least density-dependent glycolipid of many of the normal NIL2 cell clones. One tumor line was exceptional in that it retained all three components at about normal levels, and they were still density dependent. Cells from this line were passaged twice in hamsters and the tumors obtained on each occasion were cultured and their glycolipids studied. Passage did not lead to any major modification. Transformation of this tumor line by HSV resulted in the loss of the tri and tetrahexosides and a considerable reduction in the amount of pentahexoside.

DISCUSSION

Recent studies on the glycolipids of normal, virus-transformed, and tumor cells in culture have revealed some general characteristics of the biosynthesis of these compounds. With few exceptions virus-transformation results in a reduction in the carbohydrate chain length of the larger glycolipids made by the cell. This change is associated with, and is almost certainly the result of, a considerable diminution in a particular sugar transferase activity. The glycolipids of different species and even cells from different tissues have their own characteristic glycolipid complement. However, when a particular cell type is transformed the same glycolipids are usually affected, regardless of whether the transforming virus is a small DNA virus like polyoma or a large RNA virus like Rous sarcoma virus. The change in the glycolipids is transformation-specific rather than virus-specific, suggesting that the alteration results from some property common to transformed cells. Thus the mechanism is not a specific modification inserted at a particular point in the cell's biochemical machinery, like that imposed on *Salmonella* O-antigen polymerase by the ϵ series of temperate bacteriophages (23).

There is now evidence from several systems suggesting that the glycolipids missing in virus-transformed cells may also be depleted in sparse cultures of their normal precursors. Maximal synthesis is only obtained when normal cells make contact. As has been previously pointed out (14), this suggests that synthesis of these glycolipids responds to a positive control activated by cell contact and that transformed cells are unresponsive to this signal. Roseman and Roth (24,25) have suggested that cell recognition and adhesion may be mediated by the binding of sugar transferases on one cell to the target molecule of substrate on another cell in a process of transglycosylation. One explanation for the failure of transformed cells to "complete" their glycolipids may be that on contact with other cells this interaction does not take place because the appropriate transferase is absent. The absence of the transferase may also account for the loss of contact inhibition of movement in transformed cells and their loss of density-dependent inhibition of growth.

Our kinetic studies support the idea that cell contact is the prime effector for the synthesis of the "density-dependent" glycolipids, rather than the cessation of cell division. However, the fact that cells in sparse culture in low serum are also able to synthesize these compounds suggests that other effectors may operate. Experiments in progress in which sparse cultures were blocked by glutamine deprivation and held in low serum medium show that this combination does not permit the synthesis of the density-dependent glycolipids. This indicates that inhibitors of chain elongation are not present in serum.

The belief that a similar abbreviation of glycolipid synthesis may be characteristic of malignant cells received good support from the observations of Siddiqui and Hakomori (26). They found that Morris rat hepatomas all had depleted glycolipids compared with rapidly dividing neonatal liver. The greatest effect being present in the most rapidly growing hepatomas. However, Brady and Mora (27) have found that certain mouse epithelial lines which had spontaneously become highly tumorigenic had the same glycolipid distribution as the nontumorigenic cells they were derived from. Our own observations show that the cells from NIL2 tumor no. 11 have glycolipids like the control cells they were derived from and, moreover, they increase in amount as cell density increases.

Correlative studies such as those we have described can, of course, only be signposts to possible areas in which to discover the fundamental changes that influence a cell's behavior. To implicate glycolipid changes as key events in cell membrane physiology requires a more direct approach. Perhaps by simulating the transformed cell glycolipid structure in normal cells with specific glycosidases it may be possible to induce some transformed cell properties. Reciprocally, it may be possible to block the "reduced" glycolipids of transformed cells with specific antibodies or lectins and recover some normal cell property.

REFERENCES

1. MacPherson, I.: Reversion in cells transformed by tumour viruses. Proc. Roy. Soc. London (Biol.), 177:41, 1971.

2. Hay, J. B., and Gray, G. M.: The effect of testosterone on the glycosphingolipid composition of mouse kidney. Biochim. Biophys. Acta, 202:566 – 68, 1970.
3. Hay, J. B., and Gray, G. M.: Glycosphingolipid biosynthesis in kidneys of normal C3H/Hc mice and those with BP8 ascites tumours. Biochem. Biophys. Res. Commun. 38:527, 1970.
4. Dod, B. J., and Gray, G. M.: The lipid composition of rat liver plasma membrane. Biochim. Biophys. Acta (Amst.) 150:397, 1968.
5. Renkonen, O., Gahmberg, C. G., Simons, K., and Kääriäinen, L.: The lipids of the plasma membranes and endoplasmic reticulum from cultured baby hamster kidney cells (BHK21). Biochim. Biophys. Acta, 255:66, 1972.
6. Rapport, M. M., Graf, L., Skipski, V. P., and Alonzo, N. F.: Cytolipin H, a pure lipid hapten isolated from human carcinoma. Nature (London), 181:1803, 1958.
7. Tal, C., Dishon, T., and Gross, J.: The agglutination of tumour cells *in vitro* by sera from tumour patients and pregnant women. Brit. J. Cancer, 18:111, 1964.
8. Hakomori, S., and Murakami, W. T.: Glycolipids of hamster fibroblasts and derived malignant-transformed cell lines. Proc. Nat. Acad. Sci. USA, 59:254, 1968.
9. Hakomori, S., Teather, C., and Andrews, H.: Organizational differences of cell surface "hematoside" in normal and virally transformed cells. Biochem. Biophys. Res. Commun., 33:563, 1968.
10. Mora, P. T., Brady, R. O., Bradley, R. M., and McFarland, V. W.: Gangliosides in DNA virus-transformed and spontaneously transformed tumorigenic mouse cell lines. Proc. Nat. Acad. Sci. USA, 63:1290, 1969.
11. Cumar, F. A., Brady, R. O., Kolodny, E. H., McFarland, V. W., and Mora, P. T.: Enzyme block in the synthesis of gangliosides in DNA virus-transformed tumorigenic mouse cell lines. Proc. Nat. Acad. Sci. USA, 67:757, 1970.
12. Diamond, L.: Two spontaneously transformed cell lines derived from the same hamster embryo culture. Int. J. Cancer, 2:143, 1967.
13. McAllister, R. M., and Macpherson, I.: Transformation of a hamster cell line by adenovirus type 12. J. Gen. Virol., 2:99, 1968.
14. Robbins, P. W., and Macpherson, I.: Control of glycolipid synthesis in a cultured hamster cell line. Nature (London), 229:569, 1971.
15. Robbins, P. W., and Macpherson, I.: Glycolipid synthesis in normal and transformed animal cells. Proc. Roy. Soc. London (Biol.), 177:49, 1971.
16. Sakiyama, H., Gross, S. K., and Robbins, P. W.: Glycolipid synthesis in normal and virus-transformed hamster cell lines. Proc. Nat. Acad. Sci. USA, 69:872, 1972.
17. Hakomori, S.: Cell density dependent changes of glycolipids in fibroblasts and derived malignant-transformed cells. Proc. Nat. Acad. Sci. USA, 67:1741, 1970.
18. Macpherson, I.: Cell cloning in microdrops. In: Fundamental Techniques in Virology. Edited by Habel and Salzman. Academic Press, New York, p 17, 1969.
19. Macpherson, I., and Montagnier, L.: Agar suspension culture for the selective assay of cells transformed by polyoma virus. Virology, 23:291, 1969.
20. Bürk, R. R.: Growth inhibitor of hamster fibroblast cells. Nature (London), 212:1261, 1966.
21. Ley, K. D., and Tobey, R. A.: Regulation of the initiation of DNA synthesis and cell division by isoleucine and glutamine in Gl arrested cells in suspension culture. J. Cell Biol., 47:453, 1970.
22. Bootsma, D. L., Budke, L., and Vos, O.: Studies on synchronous division of tissue culture cells initiated by excess thymidine. Exp. Cell Res., 33:301, 1964.
23. Robbins, P. W., and Uchida, T.: Determinants of specificity in *Salmonella*: changes in antigenic structure mediated by bacteriophage. Fed. Proc., 21:702, 1962.
24. Roseman, S.: The synthesis of complex carbohydrates by multiglycosyltransferase systems and their potential function in intercellular adhesion. Chem. Phys. Lipids, 5:270, 1970.
25. Roth, S., and White, D.: Intercellular contact and cell surface galactosyl transferase activity. Proc. Nat. Acad. Sci. USA, 69:485, 1972.
26. Siddiqui, B., and Hakomori, S.: Change of glycolipid pattern in Morris Hepatomas 5123 and 7800. Cancer Res., 30:2930, 1970.
27. Brady, R. O., and Mora, P. T.: Alteration in ganglioside pattern and synthesis in SV40 and polyoma virus transformed mouse cell lines. Biochim. Biophys. Acta, 218:308, 1970.

5. DISCUSSION

Dr. Helinski:
(La Jolla, Calif.)

I would like to direct a question to Dr. Aposhian with respect to his pseudovirions, and ask whether he found any circular DNA in the 14-S DNA that may have arisen from the mouse chromosome.

Dr. Aposhian:
(Baltimore, Md.)

From the mitochondria?

Dr. Helinski:

Yes.

Dr. Aposhian:

We used both neutral and alkaline gradients. The S value of a circular, supercoiled kind of DNA, especially in the alkaline gradient, would be drastically different.

Dr. Helinski:

Circular rather than co-valently closed? I guess I am bringing out what the larger issue is in terms of encapsulation of DNA. Does its circularity play any kind of role? The fact that you get pseudovirions suggests that it may not.

Dr. Aposhian:

To answer your question, we have not carefully looked for this with the electron microscope. However, Roger Wild's group has. Electron micrographs have been published of from 200 to 400 molecules. No circular DNA was found at that time. In our case, we haven't looked that carefully. We have based our work primarily on sedimentation and density values and DNA-DNA hybridizations.

Dr. Caruthers:
(Cambridge, Mass.)

Dr. Helinski, have you carried out any procedures or attempted to isolate this nickase that you elaborated upon?

Dr. Helinski:

One of the disappointments in our study of this proposed nickase is that we have been unable to dissociate it from the DNA molecule in an active form. We could determine the daltons, but the procedures we have to use to dissociate it from the DNA inactivate the protein.

Dr. Beers:
(Baltimore, Md.)

Dr. Caruthers, in the synthesis for genomes by chemical means is the approach going to be through the synthesis of the DNA genome directly, through the synthesis of the messenger, itself, followed by reverse transcription to synthesize the genome?

Dr. Caruthers:

I feel now, based upon the enzymes and the methodology available, it would be much easier to carry out the synthesis of the DNA genome both chemically and enzymatically and then go into the transcription to the messenger RNA rather than to go in the reverse direction. The chemical synthesis, for example, of any type of long or high molecular weight RNA is rather difficult. Some enzymatic procedures in this field are being explored and in some cases results are encouraging.

Dr. Beers:

Dr. Macpherson, it wasn't quite clear to me in your presentation what the implications were with respect to the modification of cell surfaces by viruses and transformation processes in systems other than, perhaps, those involved in oncogenic transformation.

Is this a phenomenon which you would anticipate to have much wider implication than simply in growth or contact phenomena?

Dr. MacPherson:
(London, England)

In spontaneous transformation the change doesn't seem to occur unless the cells are malignant. One bit of information which I didn't present in my talk suggests that there is a good correlation between the changes I described and malignancy work done by Siddiqui Hakomori. They studied the glycolipid pattern in Morris hepatoma cells from three different hepatomas. One was a very slow-growing hepatoma that proliferated slower than newborn rat liver and was used for a comparison with the more rapidly growing hepatomas. They found a very good correlation between the rapidity with which these hepatomas developed and the amount of carbohydrate loss in the tumors.

Newborn rat liver had the same glycolipid pattern as fully differentiated liver. That is probably one of the best examples in which this change is, in fact, associated with malignancy. There are exceptions as illustrated in my last slide.

Dr. Beers:

Dr. Aposhian, do you feel that a rather universal requirement in the use of specific genomes for somatic changes is always an association of the genome with either a protein coat or something comparable to the transfer factors to be discussed later by Dr. Rownd?

Dr. Aposhian:

I don't know. Our calculations are based on the fact that many people for many years have tried to use naked DNA with a number of animal virus systems. Some have been successful, some have been unsuccessful. We have come to the conclusion that the protein coat was involved not only in allowing the virus to absorb to the cell, but also in protecting it.

Now, as far as viral infection is concerned, there is a much poorer chance of getting naked animal virus DNA into a cell and having it expressed than with encapsulated DNA. For example, with polyoma virus, it takes approximately 75 particles to produce one plaque. It takes 10^8 naked DNA molecules to produce one plaque. Because of the difference in efficiency of infection between a naked DNA and a coated DNA, we have elected to use the whole particle. However, I am certain we and other people will also try naked DNA if a successful transduction in systems like this can be obtained.

Dr. Beers: Is there any possibility that the relative efficiency of these various naked polynucleotides is a function of either their conformation or their molecular weight?

Dr. Aposhian: Yes. I think a number of people have shown that the size of the polymer determines how much of it gets into an animal cell or in tissue culture. There is a cutoff point. We really don't know in many cases how many viruses do get into a cell. There is the idea of pinocytosis with lysosomal uptake or of specific pores.

Dr. Caruthers: Dr. Aposhian, have you looked at the nature of the DNA in pseudovirions that is encapsulated compared with the host GC content or AT content? Have you looked for a specificity for certain types of host cell DNA?

Dr. Aposhian: We have not as yet. The C_0t analysis has been done with the SV-40 pseudovirions.

Dr. Helinski: Dr. Aposhian described defective polyoma which has less DNA than is normally found. What about the virion that has more DNA than is normally found? Is it possible to package a piece of DNA that is longer?

Dr. Aposhian: One of the young men who worked with me, Dr. Kosbar, has seen this. We found heavier material sedimenting, as Paul Berg and his collaborators did in the Salk Institute, that would appear to be dimers or trimers of polyoma DNA. We have never really gone into that area because we have been interested in other things.

But let me turn the question around and add something else, if I may. In addition to defective virions the group in Canada first reported for polyoma, Vinograd and others have reported that pseudopolyoma virus can be isolated with fairly good evidence that some host DNA is integrated into the polyoma genome. This brings up the whole lambda story again. The evidence from a number of groups indicates that even during a lytic infection with polyoma the polyoma genome must be integrated and then released.

Dr. Helinski: But these are defective virions that presumably carry a bit of the host DNA. Are they defective virions, then?

Dr. Aposhian: There have been three papers within the last two or three months in which plaque assays were not performed.

Dr. Helinski: Considering the conceptual basis of the approach, it would be valuable now to have a viable virus particle that can get the DNA in and can sustain itself in the host cell for a reasonable period of time. One approach, of course, is to have all of the genes that the SV-40 requires for its duplication and maintenance in addition to extra DNA. But, clearly, there is much activity involved in chemical synthesis of such viruses. I guess I am questioning whether you do come across these viable viral particles in vivo that are not defective but carry some of the host genetic material.

Dr. Aposhian: We have not come across those.

Dr. Helinski: I would just ask another question of Dr. Aposhian. You said that there is no indication that the DNA in pseudovirions is circular. There is the possibility that this DNA may have cohesive ends which you can use to cyclicize that DNA. This may be important for packaging the DNA. Has that ever been looked for?

Dr. Aposhian: No one has looked for that. I think one of the problems to be realized in animal virology is the very great difficulty in getting large amounts of material. You set certain priorities because of the scarcity of material available to you. Consequently, we and other people have not gotten around to doing this kind of study requiring large amounts of material.

Dr. Helinski: Even those systems where you have as much as 80% of the particles, pseudovirions, will not provide you with enough material?

Dr. Aposhian: Our primary objective has been to obtain genetic information. Our main efforts are designed to develop genetic transduction experiments with these particles, both in the whole animal and in tissue culture. There are many, many problems. Come on over into animal virology. It is great.

Dr. Helinski: Depending upon the criticism I get from today's talk I may take you up on it.

Dr. Aposhian: Your talk was excellent; don't worry about criticism.

Dr. Meltz: (San Antonio, Texas) Dr. Helinski, you were just picking on Dr. Aposhian; I propose some experiments for you, possibly. Have you tried physical breaking, possibly through X-radiation, of the supercoiled strands to see what effect this might have on releasing the protein and/or exposing its activity toward other supercoiled molecules in random locations?

Dr. Helinski: We have not used X-ray but we have used pancreatic DNase to induce a limited number of nicks in the supercoiled DNA protein complex. Releasing super helical conformation does not in any way facilitate the removal of that protein.

Dr. Hamori: (Newark, Del.) I would also like to ask Dr. Helinski about his interesting plasmid work. Is there a physical necessity for a nick when you want to uncoil supercoiled circular DNA into a regular ring? Can you do it without a nick or will you end up with a heavily strained circle?

Dr. Helinski: Well, you clearly can unwind one of the strands around the other, that is, unwind the double helical structure. There is a limit as to the extent DNA can be unwound. At some point a nick must occur to permit its further unwinding.

Also, it is quite clear that even if the system were able to unwind completely, the two strands will still be interlocked. On duplication of those two strands there is the problem of releasing the two daughter duplexes, circles, from each other. Clearly, one would expect a chopping event or a nicking of these to be required for the release of the two daughter molecules.

Dr. Hamori: I also felt that there must be a topological reason for that nick, because every time you unwind a supercoil you generate a twist in the original helix which has to be dissipated somewhere.

Dr. Helinski: You can get partial unwinding without any nicking.

Dr. Hamori: Yes.

Dr. Helinski: You release the super helical turns, then you can actually turn in the other direction to a partial extent, but there is a limit as to how far you can go without inducing a nick.

Dr. Oleinick: (Durham, N.C.) Dr. Aposhian, I wonder if the binding areas or binding sites on the capsid of the polyoma virus are specific for the polyoma DNA or, for example, would the polyoma DNA bind to disrupt an SV-40 capsid.

Dr. Aposhian: Type One, SV-40 DNA, will also bind. If we fragment mouse DNA to a size of about 2,000,000, the results are slightly different. We don't get as many peaks, as much formation of the complex, and under our conditions the results have not been very reproducible.

We were excited about these first, because it was only type One polyoma DNA that worked, whereas the host DNA fragments would not bind at all. We were off to the races. Then as we checked every now and then things began to change. The SV-40 One DNA always bound. But we have obtained variable results with the mouse DNA fragments that we don't understand.

Dr. Hompshire: (Norfolk, Va.) Dr. Macpherson, could you briefly give us some information about the effects of the viruses on the glycoproteins?

Dr. MacPherson: The glycoproteins, of course, are a much more complex set of molecules on the surface of the cell. They haven't been characterized nearly in the same way as the glycolipids.

Leonard Warren, Clayton Bach, and Suzie Blick in Philadelphia have done an interesting series of investigations. They dilute the incorporation of the fucose label into normal and transformed cells. In general, following transformation a higher molecule weight glycoprotein containing the fucose appears.

This is also found in normal mitotic cells. Another example of something that is coming up repeatedly is the fact that many of the fixed transformed cell properties are also found in rapidly growing normal cells.

Leonard Warren, who worked with me last year, and I looked at fucose incorporation into chicken cells transformed by a temperature-sensitive mutant of the Rous sarcoma virus. We found that the synthesis of this fucose-containing glycoprotein was temperature sensitive. So this was more directly correlated with transformation.

There have been a lot of studies on the salic acid composition of glycoproteins. It has generally been supposed that the higher negative charge in transformed cells is a function of higher content of salic acid. But the best studies, to my mind, done by Phil Robbins and Bill Grimes, indicate that there is very little difference in negative change. So it must be a topographical phenomenon that causes the higher electrophoretic mobility.

So, in summary, to answer your question, there is not a great deal known; it is so much more complex a problem than that of the glycolipids.

Dr. Henderson: (Albuquerque, N. Mex.) Dr. Helinski, I would like to know whether or not there is any evidence that supercoiled DNA may also be a repressed form of DNA?

Dr. Helinski: I think that the in vitro evidence is to the contrary. Supercoiled DNA molecules are, in fact, even more effective in an in vitro transcription system than the open-circular DNA form of these molecules. It also seems quite clear, at least in the bacterial systems and probably also in the mitochondrial systems, that the great majority of this DNA is indeed in the co-valently closed circular form.

Now, one might expect that this DNA is actively transcribing during the life of that cell. If the supercoiled form represses transcription in any way, when the cells are active metabolically, then you might expect that the majority of the DNA molecules would not be in supercoiled form but in some nicked or open-circular state. This doesn't seem to be the case.

Dr. Aloysius: (Syracuse, N.Y.) Dr. Helinski, your data suggest that the protein bound to the complex binds at a specific site. Since it binds to a supercoiled DNA, I presume there is also a topological enhancement of this binding process. Did you try to get that protein complex out by removing supercoils through titration with ethidium bromide?

Dr. Helinski: No, we have not used that approach, but it is something that we have thought of. We think it is conceivably a very good idea. I didn't have a chance to show some recent studies we have done using electron microscopy techniques. They seem to convince us at least that the supercoiled DNA which is complexed is associated with membrane cell wall preparations. We can actually visualize these supercoiled DNA molecules attached to this membrane preparation. We can activate these complexes in this membrane-associated state.

Then we can actually denature the DNA while it is still on the membrane. When we do this the single-strand circle comes off the membrane and the linear single strand stays attached to the membrane. This indicates to us that, indeed, there is some association between this complex and the membrane site and that the complex itself may be in some way facilitating this attachment of the DNA to the membrane, conceivably through this hypothetical DNA protein bond that I propose.

Dr. Aloysius: How did you get the DNA off the membrane? Just what technique did you use?

Dr. Helinski: Just heat to denature the DNA strands.

Dr. Abu Sabi: (New Brunswick, N.J.) We have recently been interested in regulation of growth in bacteria. Cultures were grown under conditions which will support rapid growth or slow growth. We have studied relationships between the rate of growth and the ability of the organism to synthesize inducible enzymes, the adenylcyclase and cyclic AMP contents. The rapidly growing culture has lower synthetic ability for inducible enzymes and much lower content of AMP, compared with the slower growing culture.

Dr. Macpherson, I wonder if your results are due to repression of the particular enzyme because of the fact that they are rapidly growing. There is no cyclic AMP; there is no transcription for this enzyme; you don't have a complete system.

Is the difference in types or kinds of glycolipids between transformed versus nontransformed cells the cause of transformation?

Dr. MacPherson: There is a correlation between the failure to synthesize trihexicide and the amount of enzyme present. One of the features of the study of Moore and Hakomori is the direct correlation between a considerable drop in the synthesis of the appropriate enzyme and transformation.

With respect to your other point about cessation of growth being relevant, we had done some studies in which sparse cultures of cells were blocked in various ways. Two minimoles of thymidine will block cells regardless of the density of the culture. In the S phase one can reduce the serum to quarter percent and again this prevents growth. Another way is to remove glutamine from the medium. These cells stop somewhere in the G-1 phase.

In the case of the glutamine block and the excess thymidine block, we found that cells that were held as sparse cells for several days still failed to make these contact-sensitive glycolipids. Rather surprisingly, in the case of cells blocked by low serum, some of the density-dependent glycolipids were formed. So again this is another anomalous result to go along with the results of the tumors.

One explanation was an inhibitor of the glycosylation in calf serum. By reducing the calf serum down to a quarter percent we allowed glycosylation to occur. If you do a double block and grow sparse cells in the absence of glutamine in low serum, again you fail to get the glycolipids synthesized.

So it is not clear. There certainly doesn't seem to be just one pathway that affects these cells.

Dr. Ng: (Bloomington, Ind.) Dr. Aposhian, have you done any experiments to characterize what happens to the mouse DNA that is transduced by the pseudovirion?

Dr. Aposhian: I wouldn't use the word "transduce" in this case, but we have left pseudovirus with mouse embryo and human embryo cells for as long as 24 hours. We have then examined the DNA in the virus under a number of conditions. In most cases by 24 hours about 50% of the naked DNA, or the uncoated DNA, has had nicks put into it. In other words, it has gotten smaller, as shown by our alkaline gradients. The amount of acid-soluble material is relatively small. We don't pick it up on our gradients. We have not attempted yet to see whether there is integration. That is about all we know at the present time.

Dr. Nussbaum: (Nutley, N.J.) Dr. Caruthers, just how efficient is the annealing of the 29-mer to the Phi-80 PSU?

The reason that I ask the question is that in a somewhat similar attempt we have synthesized one of the sticky ends of phage lambda. Then we tried to put it back into denatured lambda and found that the annealing process is relatively inefficient. Of course, this is a much smaller piece, only 12 nucleotides rather than 29. There may be other reasons. So could you comment on this?

Dr. Caruthers: Our best results are approximately 85% on a molar basis of the monomer oligo nucleotide bound to the R strand using a five-fold excess of the 29 oligomer. I might add, that we also find that the 22 oligomer, that I previously mentioned, binds to the extent of two- to three-fold greater than the 29 oligomer. In other words, between two and three moles of 22 oligomer will bind to the R strand, compared with only one mole of the 29 oligomer. One interpretation of this is that during the incorporation of the E.coli portion of the chromosome, there could be a portion of the doublet still within our singlet phage. So this is one explanation as to why the 22 oligomer binds to a much greater extent than the 29 oligomer.

I might add that Ray Wu at Cornell has done experiments similar to the ones you are talking about with the sticky end of lambda. He has found a limit of approximately eight nucleotides bound to the GC-rich sticky end of lambda. I don't think he has done any studies with the other sticky end of lambda.

I might add in our studies we have a C-rich 8-oligomer at the three-prime end of C-C-C-C-C-A-C-C-A with which one can show marginal binding. But one gets good binding with something at least 20 oligomer. The region is somewhere between 8-mer and 20-mer. I think it is more an effect of the conformation of the individual polynucleotide as a single strand.

Dr. Grafstrom: (Hersey, Pa.) Dr. Helinski, have you found the RNA moiety in your supercoiled DNA in other conditions of protein synthesis inhibition other than with chloramphenicol?

Dr. Helinski: No, we have not. We have not looked for other means of protein synthesis inhibition, such as amino acid starvation, to see if this would result in the accumulation of the RNA-sensitive supercoils.

Clearly, the supercoiled DNA synthesis of the col-1 plasmid does continue when you inhibit protein synthesis by amino acid starvation just as it does upon chloramphenicol addition. But we haven't looked at the material there. If you look at plasmid elements such as col-1 , they generally are synthesized in a fairly accurate way when they are in their natural hosts, col-1 and E.coli. If you now transfer this plasmid by bacterial mating to a foreign host, or an unnatural host, such as the related enteric organism, Proteus mirabilis, things don't quite happen so normally. Instead of just getting nice supercoiled DNA molecules which are of the order of 4,000,000 in molecular weight, you pile up multiple forms of the circle DNA elements as dimers, trimers, tetramers, and so on.

We propose that the reason for this generation of the

multiple circular DNA forms is through errors in the synthesis, that is, the foreign host didn't have all the machinery to carry out all the synthesis with good fidelity. For this reason we are now looking at the plasmid col-1 in Proteus mirabilis with the possibility that under normal growth conditions of this organism the supercoiled DNA has the RNA in it. Just as a replication process may be inefficient in generating multiple circular forms, there may be inefficiency in repairing or removing the RNA in the supercoiled DNA.

Dr. Mills: (Salt Lake City, Utah) Dr. Macpherson, did you culture the cells from the tumor which had the higher chain length carbohydrate? Did these show accumulations in culture, and did they maintain their ability to synthesize the longer chain?

Dr. MacPherson: Yes. These experiments with induction of tumors were done, of course, with the clone line of NIL cells which is male and has the large marker X-chromosome. The cells were inoculated into female hamsters so that we could check on the carrier type of the cultured cells. In all cases I have described, they only had one X-chromosome.

Tumor number eleven was abnormal. The first ten tumors looked like transformed cells in their glycolipids. There was a variation in the morphological pattern from something that looked rather like the untransformed NILs to cells which were quite rounded and refractile. In the case of tumor number eleven, which retained the higher glycolipids, there wasn't really too much difference in the appearance of these cells. If anything, they were rather more refractile. In the second pass, again they were slightly more refractile, but they didn't have such a striking change as one sees in virus transformation.

Dr. Rome: (Washington, D.C.) Dr. Macpherson, have you observed these transferases in C-type virus particles?

Dr. MacPherson: We haven't looked. I wouldn't expect to see them there.

Dr. Rome: Dr. Aposhian, do you know whether the pseudovirions are antigenically identical to the normal polyoma virion or the infectious polyoma virion?

Dr. Aposhian: Antiserum–I have forgotten whether it is rabbit or rat antiserum–to polyoma virus will precipitate polyoma pseudovirions also. That's all the information I have.

Dr. Meisler: (Rochester, N.Y.) Dr. Macpherson, I wonder to what extent the hexose portion of the glycolipid reflects the composition of the media? Is there any evidence for a preferential sequestering of hexoses to the membrane, for example, in the case of limiting glucose concentrations? Does the reduction in growth rate and reduction in membrane simply follow hand in hand or does the synthesis of glycolipid continue?

Dr. MacPherson: We haven't done any experiments along these lines, but since there is a very high degree of specificity in the glycosyl transferase, I wouldn't expect changes in the composition of the medium to make too much difference.

Dr. Hackel: (Brooklyn, N.Y.) Dr. Aposhian, you say that about 20% of the DNA in the pseudovirions winds up in the nuclear fraction. Have you been able to determine whether or not this DNA is associated, perhaps, with particular chromosomes in the nuclear fraction, or could you, perhaps, use autoradiography to determine if certain segments of this pseudovirion DNA wind up in particular chromosomes?

Dr. Aposhian: We haven't done that yet. David Yelton has gotten some cells from Howard Green. These cells are hybrids of mouse and human strains. They contain primarily mouse chromosomes, the number 17 human chromosome for thymidine kinase or the chromosome that contains thymidine kinase.

Not only is there the number 17 chromosome but there are anywhere from four to six copies of that particular chromosome. David Yelton is infecting these cells with R strain of polyoma virus to see whether we can get a polyoma pseudovirion that contains the thymidine kinase gene. There should be at least an enrichment of some kind. These experiments are in progress.

Dr. Hackel: Dr. Helinski, with regard to supercoiling in plasmid DNA, how does this degree of supercoiling in plasmid DNA of E. col-1 compare with the degree of supercoiling in a mitochondrial DNA? How does this degree of supercoiling compare with DNA eukaryotic chromosomes in general?

Dr. Helinski: The superhelical density of the plasmid coli in DNA is very similar to that of the mitochondrial DNA.

The mitochondrial DNA has about the same superhelical density as the plasmid DNA. The plasmid DNA's seem to be quite similar in the superhelical density.

Dr. Beers: I can recall some criticism leveled at Dr. Aposhian once when he referred to DNA as a potential drug, for the simple reason that DNA can replicate itself or can be replicated, whereas most drugs disappear when therapy is stopped. I would like to pose a question both to Dr. Aposhian and the others with respect to the long-term picture. Having gotten to this stage, where are we going now, and how close are we to getting there?

Dr. Aposhian: Well, one always has to stick his neck out, as the saying goes. I believe that within five years a monogenic defect in an animal, if not a human being, will be treated successfully by a piece of DNA, a gene, used as a drug. I think Dr. Merril will tell you this afternoon about some very interesting experiments that he has done. He might want to go on and tell you about what someone else has done.

I think that it is a matter now of applying what we know about molecular biology, genetics, and cellular pharmacology–it is going to sound as though I were waving the American flag–to help people who are ill. For the young people who often feel that biology is no longer relevant, there are a lot of problems in biology, especially in gene therapy, that you can really help people with; DNA has a great deal of potential for use in human beings for the treatment of genetic disorders. Where people get hung up is when they confuse treatment of a somatic cell mutation with treatment of the germinal cells.

The treatment of somatic cell mutation usually does not create a stir. It is reasonable in many ways. One is dealing with the phenotypic expression of DNA.

But when people start talking about dealing with the genotypic expression of DNA, dealing with the germinal cells, I think there we are talking about 50 to 100 years.

We usually include in our talks something about the responsibility of all biologists to educate people, laymen, about the whole problem of genetics and gene therapy. I don't use the term "genetic engineering." I find the term quite repugnant.

Dr. Beers: If you were to pick a particular clinical syndrome that would be most amenable to this kind of approach, would you suggest something such as sickle cell anemia, which is fairly simple as far as its molecular biological mechanism is concerned, or would you pick something in which there is an actual enzymatic defect?

Dr. Aposhian: I believe in enzymes. I think that the better-defined biochemical system you have the better off you are going to be in trying to do this. However, I think there is a tremendous advantage now to the reverse transcriptase story and the isolation, I guess it was from Sol Spiegelman's lab, of the product of the reverse transcriptase using the globin RNA messenger from human reticulocytes.

If we could do genetic surgery and prevent the synthesis of the sickle cell, I think the greatest chance of something happening within the next five years will be in the sickle cell area. That is how I see the current experiments going. The globin gene work that a number of people are doing looks very, very good.

Part II

6. EFFECTS OF BACTERIOPHAGE ON EUKARYOTES

Carl R. Merril

Laboratory of General and Comparative Biochemistry
National Institute of Mental Health
Bethesda, Maryland 20014

INTRODUCTION

Discussions of bacteriophage effects in human cells, transformation of eukaryotic cells with exogenous DNA, or the possible dangers of nonprimate tumor viruses are often met with considerable resistance. Why is resistance generated by such observations and possibilities? Perhaps the answer lies in the model most biologists have formulated regarding these topics. From a review of the current literature most biological models portray prokaryotic and eukaryotic cells as isolated genetic entities which rarely engage in the uptake of foreign genes (gene transfer). Furthermore, it is assumed that the more distant organisms are taxonomically the less the chance of a successful gene transfer.

Cellular membranes are considered to act as boundaries, preventing introduction or loss of genetic information which each cell uses exclusively until the time of cell division. This genetic information is then duplicated and passed on to the direct progeny.

Intercellular transfers of genetic information are generally divided into three main groups. The first involves uptake and genetic utilization of soluble DNA by cells and is known as transformation. The second, conjugation and cell fusion, requires cell to cell contact, sometimes requiring specialized cell appendages to facilitate genetic communication. The third involves viral transmission of genetic information. Such gene transfers may either cause stable or transient alterations in the host.

MECHANISMS OF GENE TRANSFER

Uptake of Soluble Genetic Material

It is generally assumed that the closer organisms are taxonomically the more frequent are the events of successful gene transfer (1). This assumption has empirical support in transformation studies utilizing soluble DNA isolated from microorganisms. It has been demonstrated that the ratio of transformants obtained following treatment of an organism with its own DNA compared to the transformants obtained with DNA from another strain correlates well with the taxonomic similarities between the strains. As an example, exchanges within the same genus have a ratio of 10^{-3} while strains in different genera have a ratio of 10^{-6} or less (1). Mammalian DNA has been reported to be capable of "helping" transformation of streptococci with streptococcal DNA (2), but this appears to be a special case in which the "helper" DNA functions by saturating a cell-bound inactivator of unintegrated DNA (2).

Similar studies utilizing soluble DNA with eukaryotic cells and animals have established that these systems are capable of taking up DNA. However, very few of the experiments demonstrating transforming activity have been reproduced in independent laboratories (3). This may be due in part to the complex nature of eukaryotic systems in vivo and in vitro, and in part to the relatively recent history of most attempts to transform higher organisms. An example of the complexities which might be encountered is illustrated by the observations of Fox et al on genetic mosaicism in fruit flies (Drosophila) following treatment with Drosophila DNA carrying specific gene markers (4). These workers also noted instances of no apparent effect until two to three generations had passed following treatment of early Drosophila embryos with DNA. These results may be explained by a model in which the DNA fragment associates with but does not integrate into its homologous chromosome segment. The DNA fragment is then believed to replicate in step with the chromosome; however, transcription may alternate between the DNA fragment or the homologous chromosome segment (4).

Many transformation experiments have been performed on eukaryotic cells grown in tissue culture so that extracellular conditions might be better defined than in an intact animal. However, the number of variables in vitro is still extremely large and small variations in technique can cause profound differences in the results. As an example, carefully prepared native DNA samples are only marginally effective in transforming 8-azaguanine-sensitive murine lymphoma cells to resistant ones, while sheared DNA (by sonication) is more active (5).

Even in simple bacterial systems, the outcome of a transformation experiment may be unpredictable. Bacteria must be in a competent state to be transformed. In pneumococci this process is mediated by a specific protein or competence factor (6). This competence factor is highly species specific with no effect on other bacterial strains (7). There have been no reports of similar factors in eukaryotic systems; however, it seems not unreasonable that transformation experiments in higher organisms might require analogous factors.

Cell-to-Cell Transfer

Although transformation utilizing soluble DNA appears to require homology between the donor DNA and the host, at least in the prokaryotes, other mechanisms of gene transfer do not appear to be as stringently controlled. Genes which are transferred by cell-to-cell contact, such as the drug-resistant episomes (R factors), are capable of functioning in a wide range of bacterial hosts. These R factors can be transferred by conjugation to all members of the enterobacteriaceae (8). One possible explanation for the wide host range of these factors is that their genes can multiply autonomously in the host without integration (9). These prokaryotic episomes which are passed by cell-to-cell contact in bacteria have not yet been tested in mammalian systems, although the possibility of such an application has been raised (10). Crown gall tumor, a vegetal cancer, appears to be due to contact between a wounded plant and the bacterium, Agrobacterium tumefaciens. This disease requires plant-bacterial contact. Purified A. tumefaciens DNA alone will not produce a tumor (11). However, induction of the tumor is dependent on the synthesis of bacterial RNA in the plant cells.

Passage of genetic information between eukaryotic cells has been reported following co-cultivation of Chinese hamster cells and mouse Ehrlich ascites tumor cells (12). Similar results have been observed following cell fusion, such as the persistence of human antigen in mouse human hybrids even after complete loss of all human chromosomes (13). Another similar observation was made following fusion of chick erythrocytes and mouse fibroblasts. The chick chromosomes are lost by "premature chromosome condensation," however, some mouse cells retained the functional chicken gene for inosinic pyrophosphorylase (14).

Viral Transmission of Genes

Of all the mechanisms of transfer of genetic information, viral transmission is probably the most common and least species specific. Plant viruses, such as sowthistle yellow vein virus, potato leafroll virus, wound tumor virus, and lettuce necrotic yellow virus are also multiplicative in their insect vectors (15,16). Arbor viruses, of which 200 different varieties have been isolated, multiply in both vertebrates and arthropods (17). Bacteriophages can multiply in taxonomically distant bacterial hosts, and one E.coli bacteriophage, fd, has been reported to be capable of replication in tobacco leaves (18,19). Interestingly, passage of fd through tobacco leaves alters the E.coli host specificity of the phage (19). DNA from the bacteriophage PS8 has been reported to induce tumor in plants. Our laboratory has obtained evidence of transcription and translation of certain genes carried by the E.coli bacteriophage lambda and its transducing derivative, lambda-p-gal, in human skin fibroblasts grown in tissue culture (20,21). Our attempts to search for bacteriophage multiplication in mammalian cells have been delayed by the finding of bacteriophage in the fetal calf serum which is used in cell growth media (22). Eleven lots of serum were obtained from four major commercial suppliers. Phage titers as high as 3.3 times 10^3 pfu/ml were observed on E.coli C (a bacterial strain lacking DNA modification and restriction systems) (23). These phages were heterogenous, as indicated by plaque morphology. It is not yet clear whether the presence of phage is indicative of bacterial contamination during serum processing or if they arise by growth within the animal. Whatever their source, such viruses may have effects on tissue culture experiments, either by altering the cell metabolism, as does phage lambda-p-gal in human fibroblasts; replicating, as does fd virus in tobacco plants; or they might help or interfere with other viral infections.

One of the difficulties in the early work on eukaryotic viruses was the requirement that they cause a gross effect, such as death or tumor formation. With the relatively recent application of biochemical monitoring of the infected host more subtle effects are beginning to appear. Shope papalloma virus which was long associated with wart production in rabbits now appears to be capable of inducing arginase activity in human laboratory workers (24). Herpes simplex, which is usually monitored by its killing effect, is capable of transferring thymidine kinase activity in mouse L cells (25).

Gene transfer by virus transmission can be as stable in the host as genes transferred by any of the other mechanisms, as has been so well documented with viruses which transform mammalian cells, or the studies on lysogenic phages in bacteria, or, most recently, by the thymidine kinase induced in mammalian cells by UV irradiated herpes simplex virus (25).

Barriers to Gene Transfer

In many cases of attempted gene transfer no evidence for a successful transfer is obtained. What processes must genetic material (whether it is packaged as a virus, as a naked nucleic acid molecule, or forced through a cellular appendage) go through to be expressed and, perhaps, multiply in a recipient cell? The potential barriers to such transfer can be divided into several categories: entry into the cell, stability (at least long enough for transcription and/or translation), and, finally, transcription and translation. A larger effect will be observed if multiplication of the genetic material occurs. Multiplication which is syncronous with the cell's division of integration into the cell's genome may assure such stability.

Entrance Barriers

It has been fairly well established that soluble DNA and RNA can be taken up by eukaryotic cells in vitro, although the mechanism of this uptake is unclear (3). The transferred nucleic acids can be detected in nuclei, lysosomes, mitochondria, and cell "sap," as illustrated by the studies of Herrera et al (26) on the uptake of E.coli tRNA by fresh human lymphocytes, murine leukemia cells (L1210), and human lymphoblasts (NC-37). Approximately 20% of this tRNA remains functional and apparently intact as measured by acylation with E.coli aminoacyl-tRNA synthetases, methylation with leukemic cell methylases, and demonstration of ^{14}C-labeled 4S RNA in the cell cytoplasm following addition of ^{14}C-E.coli tRNA (26). The mechanism of distribution of nucleic acids is also complex, as illustrated by the finding that the amount of exogenous DNA which reaches the nucleus is higher at 0°C than at higher temperatures (27). It has been suggested that this effect is due in part to reduced nuclease activity at the lower temperature (27).

Stability

Many phage and plasmid genomes form circles which may enhance their resistance to nuclease activity (28,29). In some bacterial systems foreign genetic material is recognized by its pattern of secondary methylation, and it is then destroyed by an endonuclease which is highly site specific (30). The phage T-3 is apparently able to overcome this restriction system by producing an enzyme early in infection which cleaves the methyl group donor (S-adenosylmethionine) of the host controlled modification and restriction system (30). Other phages, such as F116, are also capable of rendering bacterial DNA genes, which they are transducing, resistant to restriction activity (32). The first indication that eukaryotic systems may also possess similar restriction and modification mechanisms was observed by infecting tobacco leaves with fd bacteriophage DNA. Relative plating efficiencies of the progeny phage, resulting from the tobacco infection, indicated that the DNA was modified in a manner similar to that found in E.coli B (19).

Transcriptional and Translational Barriers

The barriers to successful transcription and translation are difficult to assess at our present level of understanding, especially in eukaryotic cells. Comparisons between R factors and other episomes have shown that in bacteria the host can markedly influence the episomes properties, in particular, their regulatory mechanisms (8,33). However, the apparent universal nature of the genetic code should insure a fair degree of reproducibility in the resulting amino acid sequences.

The production of active viruses in taxonomically distant hosts, such as phage in tobacco plants (18,19), plant virus in insects (15,16), insect virus in mammals (17), and bacteriophage transcription and translation in human cells (20,21), all tend to indicate that whatever barriers there are to transcription and translation they are not insurmountable. This is perhaps further supported by a recent report on the ability of bacterial DNA to "rescue" plants with specific lethal mutations (34). However, these plant experiments are preliminary and do not unequivocally correlate the observed biologic effect with the information content of the DNA used.

Integration

If the transferrred genes can integrate into the host genome, presumably their stability will be greater. It is generally believed that extensive base pairing is required for integration of nucleic acid into a host's genome. However, recent detailed molecular studies on the integration of lambda phage into E.coli indicate that homology is not an absolute requirement. Recombination between the virus attachment site and the bacterial site involves a region of less than 12 base pairs (35). It is interesting to speculate on a statistical basis that there should be "attachment sites" in mammalian cells, since the bacterial attachment site for lambda phage is shorter than 12 bases and mammalian cells contain about 800 times as much DNA as E.coli.

In some cases, incoming genes might require host recombinational systems for integration; however, in the case of lambda virus this is not necessary. Lambda phage carries a gene Int which is utilized for integration of the phage genome into the host genome. The Int gene produce appears to be involved in the recognition of the short sequences in the bacterial and viral attachment sites.

CONCLUSION

As the various mechanisms of genetic exchange are studied in prokaryotic and eukaryotic systems, one begins to wonder if gene transfer events are truly rare. Of 500 strains of E.coli analyzed, 32 carried temperate phages (36). Transmissible drug-resistant episomes are wide-spread in certain strains of bacteria (8). Related to this problem is the question of why bacteria would have competence factors for the uptake and utilization of genetic material, or why eukaryotic cells would take in RNA and DNA without depolymerization (3,26). Recent studies in vertebrate cells indicate that they normally contain genetic information for producing type-C RNA tumor virus (37).

Taxonomic proximity appears to be necessary in prokaryotes for transformation with soluble nucleic acids. This requirement is not as stringent for cell-to-cell passage or viral transmission of genes. These observations seem to indicate that the host range of "genes in transfer" becomes wider as the autonomy of the gene cluster increases (autonomy may be enhanced by the presence of recombination, repair, and replication genes).

The interspecies transfer abilities of viruses make these agents ideal probes into cell processes. The attempts to study the effects of bacteriophage in eukaryotic systems may have given us some insight into the presence of restriction systems in higher plants (18,19), or the possible presence of amber suppression in human cells (21). The growth and quantitation of gene products of viruses in various distant hosts should give information on translational and transcriptional control systems.

If we can learn how to manipulate these gene transfer events with any degree of precision in eukaryotes similar to that achieved in prokaryotes, the field of genetic engineering will move out of the age of selective breeding, which has been one of the pillars of our civilization, into an age of direct genetic manipulation. Hopefully, these techniques will be used primarily in a manner analogous to those employed in selective breeding, to improve mankind's food supply and health. The recent successful transfer of the structural and regulatory genes for nitrogen fixation from Klebsiella pneumonia to E.coli C may be one of the first steps in the development of plants which are capable of fixing their own nitrogen (38).

SUMMARY

Current biological models portray prokaryotic and eukaryotic cells as isolated genetic entities which rarely engage in the uptake of foreign genes. Furthermore, these models assume that taxonomically distant organisms are less likely to have a successful gene transfer. Although the uptake and genetic utilization of soluble DNA require homology in the prokaryotes, other mechanisms of gene transfer are not as stringently controlled. Drug-resistance ractors (R factors) and other episomes may be transferred by cell-to-cell contact, while other genes may be transferred by viruses. The viruses have the largest host range. There are reports of plant viruses which can multiply in insects, insect viruses which multiply in vertebrates, and bacteriophage which cause metabolic alterations in mammalian cells. The current data suggests that the host range of "genes in transfer" becomes wider as the autonomy of the gene cluster increases. This autonomy may be enhanced by recombination, repair, and replication genes.

ACKNOWLEDGMENT

The author would like to thank Drs. T. Friedman and K. Krell for their helpful comments in the preparation of this manuscript.

REFERENCES

1. Jones, D., and Sneath, P. H. A.: Genetic transfer and bacterial taxonomy. Bacteriol. Rev. 34:40, 1970.

2. Chen, K., and Ravin, A. W.,: Mechanism of the deoxyribonucleic acid helping effect during transformation. J. Mol. Biol., 33:873, 1968.
3. Bhargava, P. M., and Shanmugan, G.: Uptake of non-viral nucleic acids by mammalian cells. Edited by Davidson, J. N., and Cohn, W. E. In: Progress in Nucleic Acid Research and Molecular Biology. Academic Press, New York, p 104, 1971.
4. Fox, A. S., Yoon, S. B., Duggleby, W. F., and Gelbart, W. M.: Genetic transformation in Drosophila. In: Informative Molecules in Biological Systems. Edited by L. G. H. Ledoux. North Holland and American Elsevier, New York, p 313, 1971.
5. Roosa, R. A.: Induced and spontaneous metabolic alterations in mammalian cells in culture. In: Informative Molecules in Biological Systems. Edited by L. G. H. Ledoux. North Holland and American Elsevier, New York, p 67, 1971.
6. Tomasz, A.: Model for the mechanism controlling the expression of the competent state in pneumococcus cultures. J. Bacteriol., 91:1050, 1966.
7. Tomasz, A., and Mosser, J. L.: On the nature of the pneumococcal activator substance. Proc. Natl. Acad. Sci. USA, 55:58, 1966.
8. Davies, J. E., and Rownd, R.: Transmissible multiple drug resistance in Enterobacteriaceae. Science, 176:758, 1972.
9. Falkow, S.: Nucleic acids, genetic exchange and bacterial speciation. Amer. J. Med., 39:753, 1965.
10. Roozen, K. J., Fenwick, R. G., and Curtiss, R., III.: Isolation of plasmids and specific chromosomal segments from Escherichia coli K-12. In: Informative Molecules in Biological Systems. Edited by L. G. H. Ledoux. North Holland and American Elsevier, New York, p 249, 1971.
11. Leff, J., and Beardsley, R. E.: Action tumorigene de l'acide nucleique dun bacteriophage present dans les cultures de tissue tumoral de tournesol (Helianthus annuum). C. R. Acad. Sci. (Paris), 270:2505, 1970.
12. Bendich, A., Borenfreund, E., and Honda, Y.: DNA-induced heritable alteration of mammalian cells. In: Informative Molecules in Biological Systems. Edited by L. G. H. Ledoux. Holland and American Elsevier, New York, p 80, 1971.
13. Midgeon, B. R., and Miller, C. S.: Human-mouse somatic cell hybrids with single human chromosome (Group E): link with thymidine kinase activity. Science, 162:1005, 1968.
14. Schwartz, A. G., Cook, P. R. and Harris, H.: Correction of a genetic defect in a mammalian cell. Nature, 230:5, 1971.
15. Sylvester, E. S., and Richardson, J.: Additional evidence of multiplication of the sowthistle yellow vein virus in an aphid vector-serial passage. Virology, 37:36, 1969.
16. Chiu, R., Reddy, D. V. R., and Black, L. M.: Inoculation and infection of leafhopper tissue cultures with a plant virus. Virology, 30:562, 1966.
17. Fenner, F.: The biology of animal viruses. Academic Press, New York, p 21, 1968.
18. Sander, E.: Evidence of the synthesis of DNA phage in leaves of tobacco plant. Virology, 24:545, 1964.
19. Sander, E.: Alteration of fd phage in tobacco leaves. Virology, 33:121, 1967.
20. Merril, C. R., Geier, M. R., and Petricciana, J. C.: Bacterial gene expression in human cells. Nature, 233:398, 1971.
21. Geier, M. R., and Merril, C. R.: Lambda phage transcription in human Fibroblasts. Virology, 27:638, 1972.
22. Merril, C. R., Friedman, T. B., Attallah, A., Geier, M. R., Krell, K., and Yarkin, R.: Isolation of bacteriophage from commercial serum. In Vitro, 8:91, 1972.
23. Aber, W.: Host-controlled variation. In: The Bacteriophage Lambda. Edited by A. D. Hershey. Cold Spring Harbor Laboratory, New York, p 83, 1971.
24. Rogers, S.: Shope papilloma virus: a passenger in man and genome. Nature, 212:120, 1960.
25. Munyon, W., Kraiselbrud, E., Davis, D., and Mann, J.: Transfer of thymidine kinase to thymidine kinaseless L cells by infection with ultraviolet-irradiated herpes simplex virus. J. Virol., 7:813, 1971.

26. Herrera, F., Adamson, R. H., and Gallo, R. C.: Transfer RNA as an informative molecule and its uptake by normal and leukemic human and murine leukocytes. In: Informative Molecules in Biological Systems. Edited by L. G. H. Ledoux, North Holland and American Elsevier, New York, p 378, 1971.
27. Wilczok, T.: Autoradiographic studies on incorporation of heterologous DNA into the Novikoff hepatoma cells. Neoplasma, 9:369, 1962.
28. Burton, A., and Sinsheimer, R. L.: The process of infection with bacteriophage ϕX 174. J. Mol. Biol., 14:327, 1965.
29. Dulbecco, R., and Vogt, M.: Evidence for a ring structure of polyoma virus DNA. Proc. Natl. Acad. Sci. USA, 50:236, 1963.
30. Arber, W.: Host-controlled variation. In: The Bacteriophage Lambda. Edited by A. D. Hershey, Cold Spring Harbor Laboratory, New York, p 83, 1971.
31. Kauffmann, M. H., and Sauerbier, W.: Inhibition of modification and restriction for phages and T_1 by co-infecting T_3. Molec. Gen. Genet., 102:89, 1968.
32. Dunn, N. W., and Holloway, B. W.: Transduction and host controlled modification. The role of the phage. In: Informative Molecules in Biological Systems. Edited by L. G. H. Ledoux, North Holland and American Elsevier, New York, p 223, 1971.
33. Falkow, S., Wolhieter, J. A., Citarella, R. V., and Baron, L. S.: Transfer of episomic elements to proteus. J. Bacteriology, 87:209, 1964.
34. Ledoux, L., Huart, R., and Jacobs, M.: Fate of exogenous DNA in arabidopsis thaliana. In: Informative Molecules in Biological Systems. Edited by L. G. H. Ledoux. North Holland and American Elsevier, New York, p 159, 1971.
35. Davis, R. W., and Parkinson, J. S.: Deletion mutants of bacteriophage lambda. J. Mol. Biol., 56: 403, 1971.
36. Hershey, A. D., and Dove, W.: In: The Bacteriophage Lambda. Cold Spring Harbor Laboratory, New York, p 3, 1971.
37. Todaro, G. J., and Huebner, R. J.: The viral oncogene hypothesis: new evidence. Proc. Natl. Acad. Sci. USA, 69:1009, 1972.
38. Dixon, R. A., and Postgate, J. R.: Genetic transfer of nitrogen fixation from Klebsiella pneumoniae to Escherichia coli. Nature, 237:102, 1972.

7. "GENETIC' TRANSFORMATION OF BACTERIA BY RNA AND LOSS OF ONCOGENIC POWER PROPERTIES OF AGROBACTERIUM TUMEFACIENS. TRANSFORMING RNA AS TEMPLATE FOR DNA SYNTHESIS

M. Beljanski and P. Manigault

Institut Pasteur, Paris, France

INTRODUCTION

Transformation of bacteria by deoxyribonucleic acid (DNA) and transfer of hereditary information (1) has been extensively studied since Avery et al (2) showed that the transforming principle was DNA. Ribonucleic acid (RNA) by itself had not been considered capable of provoking a similar response.

In 1971 we presented data showing that hereditary information could be transferred to different bacterial species by a specific RNA from E. coli K-12 Hfr mutant, showdomycin resistant. Transforming RNA was isolated, purified, and characterized (3,4). It induces a massive transformation of recipient cells. RNase completely destroys the transforming RNA potential, while DNase is without effect (3). Transformants exhibit new and stable biochemical and physiological changes (5). Thus, Agrobacterium tumefaciens (oncogenic for plants), once transformed by E. coli RNA, has lost definitely its oncogenic power and acquired new properties (6).

Appearance of stable transformants in the presence of transforming RNA leads us to a particular approach to this problem. We wanted to find out if some appropriate enzymes could use transforming RNA as a template for RNA and for DNA synthesis. In other words, could the transfer of information from RNA be mediated by two distinct enzymes giving two distinct products? We have demonstrated that polynucleotide phosphorylase (PNPase) of E. coli wild type is capable of recognizing the transforming RNA as a template, synthesizing in vitro a poly AGUC which possesses a particular base ratio characteristic of that RNA (7). Second, we have recently found that bacteria (E. coli) contain an enzyme that, like viral reverse transcriptase, uses transforming RNA as a template for in vitro synthesis of complementary DNA.

SOURCE OF TRANFORMING RNA AND PROCEDURE USED FOR TRANSFORMATION

Transforming RNA has been isolated from E. coli: 1. from preparations of DNA after purification under appropriate conditions (episomal RNA), either from wild-type cells or from showdomycin-resistant mutants (4); 2. from the supernatant of showdomycin-resistant E. coli mutant cultures (mutant M 500 sho-R and mutant ML 30 sho-R); 3. after high speed centrifugation.

Both RNA preparations have a particular base ratio (G+A/C+U = 1.72 - 2.0) and are active in transformation of wild-type E. coli bacteria, regardless of their sex (4).

In order to get the transformants, wild-type E. coli is incubated in fresh synthetic medium (2×10^7 cells/ml for 1 hour at 37°C), supplemented with 0.1 to 2μg of active RNA/ml. The presence of transformants can be easily shown by determining the ratio:ribose/UV absorption at 260 nm of endogeneously synthesized RNAs (see *Note* to Table I). Thus transformation is expressed by the ratio:ribose/UV, that is, more of AMP and GPM nucleotides in the RNAs (Table I), since the ribose of these two nucleotides reacts in orcinol reaction.

TABLE I

Ribose/UV Ratio of Endogenous RNA of E. coli Transformants and Mutant M 500

Recipient bacteria	Transforming RNA	Ribose/UV ratio	Difference (%)
E. coli K-12 Hfr	no	0.64	–
	+ 0.1 μg	0.90	40
	+ 2.0 μg	0.93	45
	+ 2.0 μg + RNase (20 μg)	0.65	1.5
	+ 2.0 μg + DNase (20 μg)	0.91	42
E. coli-RV (8)	no	0.67	–
	+ 0.1 μg	0.89	32
	+ 0.2 μg	0.95	40
	+ 2.0 μg + RNase	0.69	2.8
	+ 2.0 μg + DNase	0.93	38
	+ 2.0 μg + Pronase[a]	0.96	40
	+ 0.1 μg + r-RNA (10 μg)	0.93	45
E. coli mutant M 500 sho-R	no	0.94	–

Note: Exponentially growing bacteria are collected by centrifugation and reincubated at 37°C (2×10^7 cells/ml) in 5 ml of synthetic medium containing glucose and 0.1 to 3 μg of transforming RNA. After 30 min or 2 hr, bacteria are collected and washed three times with 10 ml of 5% TCA solution, and exogeneous RNA extracted by heating the suspension (in 2 ml of 10% TCA) at 100°C for 20 min. After centrifugation supernatant is used for determination of UV absorption at 260 nm and for orcinol reaction (8). Ratio ribose/UV (arbitrary ratio) = divisions read at 670 nm (ribose)/divisions read at 260 nm.

[a]ARN preincubated with 200 μg of pronase for 1 hr.

Figure I shows the quantitative effect of active RNA in transformation of E. coli K-12 Hfr wild type. Transformation does not take place if RNA is pretreated with pancreatic RNase, while DNase has no effect. The modified ratio of endogenous RNA is characteristic of the showdomycin-resistant mutant from which transforming RNA was excreted (Table II). If ribose/UV ratio is close or identical to that of mutant sho-R, all necessary controls are performed with transformants. Ribosomal RNA does not compete with transforming RNA (Table I), and pronase is without effect.

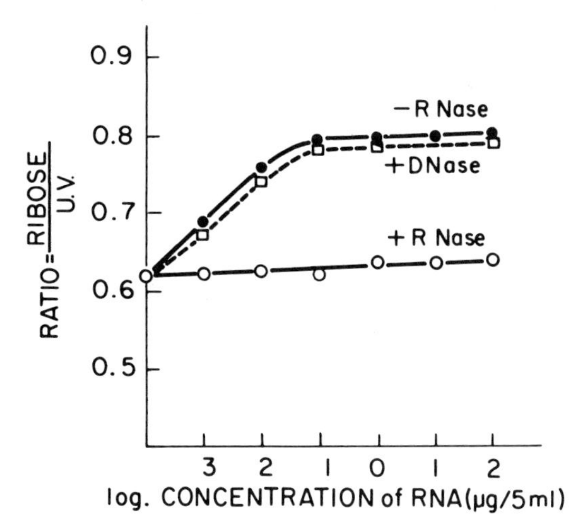

Fig 1. Effect of transforming RNA on change of ribose/UV absorption ratio in E. coli K-12 wild type. Conditions, see *Note* to Table I.

Transformation of wild-type E. coli can also be achieved by simple incubation of a mixture of growing cultures of 98% wild bacteria and 2% mutant sho-R; both strains have the same growth rate at 37°C. After 10–16 hours, the whole population is transformed, that is, it has the characteristics of mutant sho-R.

TABLE II

Nucleotides Composition of r-RNAs of Agrobacterium tumefaciens, Wild-type B_6 and Transformant B_6-Tr-1

Nucleotide	Transforming ARN E. coli ML 30 sho-R	(mol per 100 mol of nucleotides) wild type 23 S	wild type 16 S + 17 S	transformant 23 S	transformant 16 S + 17 S
A	30.3	26.0	25.2	30.6	29.3
G	33.5	30.4	29.8	33.3	31.4
C	18.3	24.7	23.5	19.6	20.6
U	17.8	18.9	21.5	16.5	18.7
G+A/C+U	1.76	1.27	1.22	1.77	1.56
G+C/A+U	1.05	1.20	1.16	1.03	1.08

Note: Ribosomal RNA (23 S, 16 S + 17 S) was isolated by the phenol method from 50 S and 30 S ribosomal subunits after separation on sucrose gradient (5). Purified RNA (1 mg) was hydrolyzed (KOH 0.5 N, 16 hr at 37°C) (60,000 cpm) and the nucleotides were analyzed, using a Dowex 1 X 2 column, 200-400 mesh (5). Transforming RNA was isolated as described from labeled ^{32}P E. coli ML 30 sho-R, separated by gel electrophoresis, hydrolyzed with 0.5 N KOH. Nucleotides were separated by Dowex column and the radioactivity determined.

TABLE III

Repartition of ^{14}C-uracil (Transforming RNA) Incorporated by Wild-type E. coli

Analyzed material	^{14}C-uracil, CPM in TCA precipitable material	
Washed bacteria	18.460	
Washed debris bacteria	3.621	
Membrane fractions	4.990	18.151
Ribosomes (70 S)	1.180	
105.000 x g supernatant	8.360	
DNA isolated directly from washed bacteria (18.370 CPM) and purified as described	1.552	

Note: 200 ml of exponentially growing culture (10^8 cells/ml) were incubated at 37°C with shaking in synthetic medium containing glucose (9) in the presence of 40 μg of ^{14}C-uracil labeled transforming RNA (36,640 cpm). After 1 or 2 hrs incubation, cells were collected by centrifugation at 15,000 x g for 30 min and washed with fresh culture medium. The presence of labeled RNA (TCA precipitable) was determined in the supernatant and in different fractions after labeled bacteria were degraded by grinding with alumina, and different fractions separated.

UPTAKE OF C^{14}-URACIL LABELED TRANSFORMING RNA BY RECIPIENT BACTERIA

^{14}C-uracil labeled RNA, excreted by E. coli mutant M 500 sho-R, was isolated (3) and incubated for 1 hour with recipient strain E. coli K-12 Hfr. Table II shows that half of the labeled RNA was incorporated by recipient bacteria. Remaining RNA, which did not penetrate the cells, is totally precipitable by trichloroacetic acid. This shows that labeled RNA was not degraded before entering the cells. Among different constituents obtained after disruption of labeled cells (Table III), ribosomes contain little radioactivity. "Membrane" fraction and 105,000 X g supernatant contain most of the ^{14}C-uracil. Five to 10% of incorporated ^{14}C-uracil is associated with DNA purified under described conditions (4). Although we cannot assert that all of ^{14}C-uracil RNA was active in transformation, we can conclude that radioactive RNA did penetrate the recipient cells. If this RNA was degraded inside the bacteria, most of the radioactivity should be found in ribosomes which contain roughly 70% of bacterial RNA. This is not the case.

MAIN CHARACTERISTICS OF E. COLI TRANSFORMANTS

1. As shown in Table I, the transformants have the same ribose/UV ratio as that of mutant M 500 sho-R, whose population is homogenous (repeated plating and analysis). Clones of transformants are stable and revertants were not obtained.

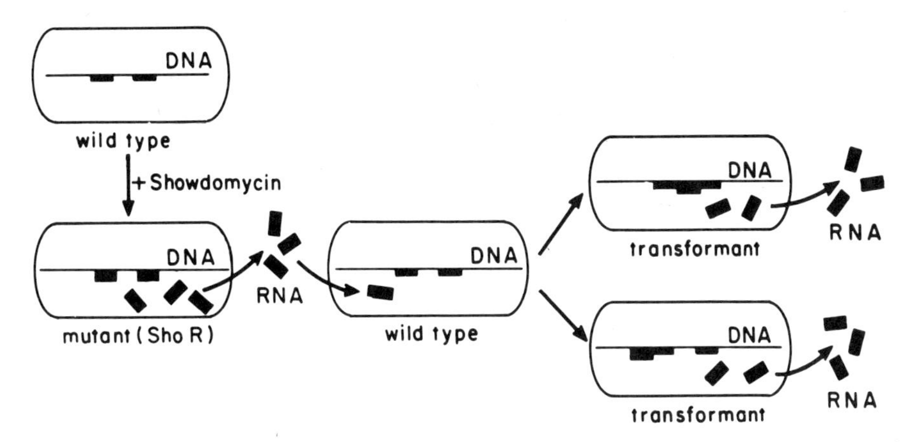

Fig 2. Scheme for excretion of transforming RNA from E. coli mutant sho-R and from transformants.

2. Transformants excrete the transforming RNA into culture medium as does the mutant M 500 sho-R (Fig 2).

3. Both ribosomal RNAs (23 S and 16 S) of transformants contain purine nucleotides in excess over pyrimidines (Table II).

4. The polynucleotide phosphorylase from transformants synthetizes in vitro (from equimolar amount of each of the four XDP) a poly AGUC in which purine nucleotides are in excess over pyrimidine when compared with poly AGUC synthesized by the wild-type enzyme. This is a characteristic of the PNPase of the mutant sho-R (7).

5. Showdomycin resistance is only poorly transferred (4). However, transformants acquire showdomycin resistance much more rapidly than wild-type E. coli.

INTERSPECIFIC TRANSFORMATION

Procedure for Transformation of Agrobacterium tumefaciens

The techniques and criteria used for the measure of the quantitative effect of transforming RNA (RNA excreted by E. coli, ML 30 mutant sho-R), the control of the inhibitory action of RNase and other necessary tests have been done. They are the same as those used for transformation of E. coli, wild type. It should be emphasized that in the case of the Agrobacterium tumefaciens B_6 strain, the incubation time in the presence of transforming RNA takes much longer (6 to 24 hours). The proportion of transformants increases with time, and in certain experiments the whole population was transformed in 16 hours. This looks like "progressive and cumulative transformation" as formulated for tumor induction in plants (10,11). The complete transformants (complete and definitive loss of oncogenic power), plated on solid medium, give colonies twice as large as the wild type. Partial transformants have also been obtained; they are intermediate between those of wild-type and completely transformed cells. Replacement of transforming RNA by a purine-rich nucleotide mixture or synthetic

TABLE IV

Characteristics of Agrobacterium tumefaciens Wild-type B_6 and Transformants

	Conditions of growth		Synthetic growth			
	aerobic	anaerobic	medium 63[a]	medium Stoll[b]	3 keto-lactose formation	Serologic test
Wild type B_6	+	no	no	+	+++	+++
B_6-Tr-4 transformant	+	no	no	+	+++	++
B_6-Tr-1 transformant	+	no	no	+	+ (delayed)	–
E. coli ML 30 sho-R	+	+	+	no	no	no

Note: Test for 3 keto-lactose considered as specific one for Agrobacterium tumefacients (12) was performed on bacteria grown for 36 hr at 30°C on one large spot on solid Stoll medium containing 2% of lactose. The yellow color (3 keto-lactose) appears around grown colonies. Serological test was done with the anti-B_6 serum.

[a]= synthetic medium 63 routinely used for growth of E. coli (9).

[b]Medium Stoll (13) rather specific for A. tumefaciens.

polyribonucleotides does not lead to the appearance of transformants. Spontaneous transformants, such as those we have obtained in the presence of E. coli RNA, never have been described.

Characteristics of Agrobacterium tumefaciens Transformants

The essential and rather specific tests for Agrobacterium tumefaciens are summarized in Table IV. Transformants have kept certain specific characteristics of Agrobacterium tumefaciens and have acquired some new ones described here and elsewhere (6,14). Some striking biochemical changes of transformants are illustrated by Figure 3. The densitometer tracings of ribosomal RNAs and ribosomal proteins of Agrobacterium tumefaciens, wild type, and those of complete transformants show that the importance of the transformation was profound, indeed. In addition, the nucleotide composition of ribosomal RNA of the transformants strongly differs from that of the wild type (Table II). Consequently, we expected a modification in the synthesis or activities of certain enzymes in these transformants.

L-Asparaginase in Transformants of Agrobacterium tumefaciens

L-Asparaginase, which at pH 5.0 degrades l-asparagine into aspartic acid and ammonia, has been studied particularly in Agrobacterium tumefaciens transformed by E. coli RNA (8). The reason for choosing this enzyme was that l-asparaginase of bacterial origin causes certain mammalian tumors to regress and has an antilymphoma and antileukemic effect (16 – 18). Some of the results described (14) are illustrated by Figure 4. It is clear that partial transformants B_6-Tr-4 contain substantially more l-asparaginase (pH 5.0) than wild-type B_6, while in complete transformants B_6-Tr-1 the amount of l-asparaginase is several times greater. Thus, a correlation appears between

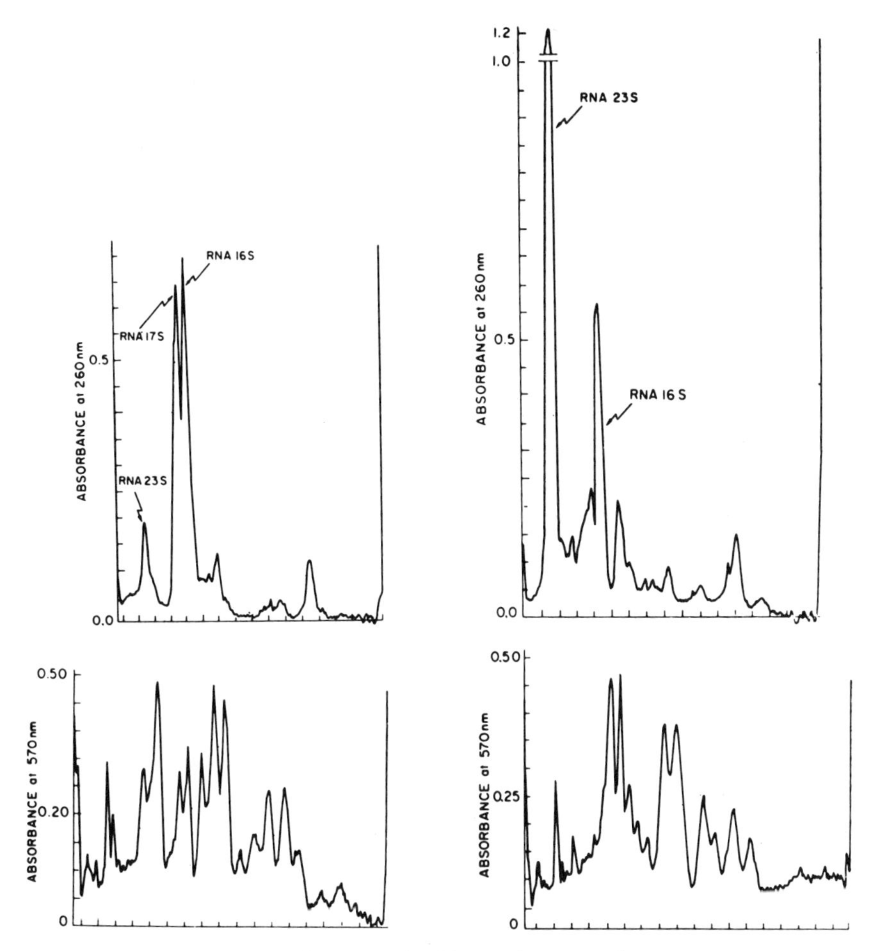

Fig 3. Densitometer tracings of ribosomal RNAs and ribosomal proteins of wild-type B_6 and transformants (A. tumefaciens).

Ribosomal RNA and ribosomal proteins after polyacrylamide gel electrophoresis. Conditions described elsewhere (5). *Above left:* r-RNA of wild-type B_6; *right:* r-RNA of B_6-Tr-1. *Below left:* ribosomal proteins of wild-type B_6; *right:* that of B_6-Tr-1.

loss of oncogenic capacity and great increase of the amount of l-asparaginase in transformants of Agrobacterium t. Commercially purified l-asparaginase, when inoculated with oncogenic strain B_6, causes significant regression (50 - 74%) of tumors in plants. However, one cannot exclude a simple coincidence between loss of oncogenic power and important changes in the activity of l-asparaginase in transformants.

Evidence for Loss of Oncogenic Properties in Agrobacterium tumefaciens Transformed by E. coli RNA

Agrobacterium tumefaciens wild-type B_6, carries a tumor-inducing principle that causes heritable changes in the plant host. Tumors appeared in plants that were inoculated with strain B_6 wild type. The tumor-inducing capacity of partially transformed strain B_6Tr-4A is substantially lower than that of the wild type, as judged by

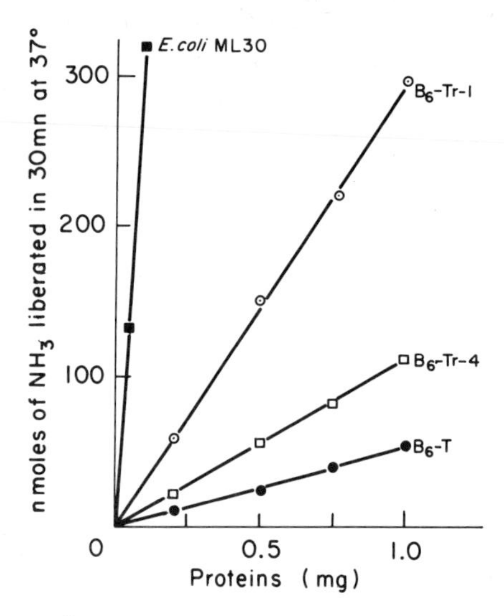

Fig 4. Activity of 1-asparaginase at pH 5.0 in extracts of Agrobacterium tumefaciens wild-type B_6, transformants B_6-Tr-4, B_6Tr-1, and E. coli ML 30 sho-R.

L-asparaginase activity was tested as described (14). After aerobic growth, bacterial culture was left without shaking for 4 hrs at 30°. Bacteria were collected, washed, and sonicated. After centrifugation at 105.000 g for 10 min. supernatant was incubated in the presence of 1-asparagine and the activity determined.

weight of tumors. Tumors did not appear with complete transformants (B_6-Tr-1) (Fig 5). Clones of that strain confirm the homogeneity of the population.

It was suggested (19) that an RNA fraction may be an essential part of the tumor-inducing principle of Agrobacterium tumefaciens. This conclusion came from the observation that tumors do not appear in the presence of RNase, while they do appear normally in the presence of DNase. Our results suggest that a transforming RNA (episomal RNA) could be excreted by oncogenic A. tumefaciens when a plant is inoculated with that bacteria. This RNA could penetrate the plant cells (if it is not destroyed by RNase) and provoke as an autonomous episome the appearance of modified biochemical and biological properties that become inherited in these cells. Eventually, it might be transcribed into DNA which, in turn, would be integrated into the plant genome.

COULD THE TRANSFORMATION OF BACTERIA BY RNA BE DUE TO DNA SYNTHESIZED ON AN RNA TEMPLATE (REVERSE TRANSCRIPTASE)?

a) Transformation of bacteria by transforming RNA from E. coli leading to heritable changes in transformants raised the following question: Could the transfer of information from RNA be mediated by an RNA-dependent DNA polymerase? In the case of oncogenic viruses (20,21), viral RNA is transcribed into DNA by reverse transcriptase contained in viruses (12,14,25,28). On the basis of our results, the search for an enzyme-like "reverse transcriptase" was justified by the fact that E. coli K-12 Hfr wild type do synthesize the transforming RNA and that the showdomycin-resistant mutant of these bacteria excretes a transforming RNA.

Fig 5. Oncogenic power of A. tumefaciens B_6 wild type. Comparison with transformants B_6-Tr-4 and B_6-Tr-1.

Technique for inoculation of bacteria into wounded plants has been described elsewhere (15). Datura stramonium was wounded and bacterial suspension (equal optical density) introduced into the wound. In routine experiments pea seedlings were used to test the oncogenic capacity of Agrobacterium tumefaciens (15). St = sterile treatment of Datura stramonium (no bacteria).

The polymerization of deoxyribonucleoside-5′-triphosphates (d-XTP) can be easily demonstrated when transforming RNA is incubated with an endogenous nucleic acid-free enzyme preparation. Thus, the passage of a 105.000 X g supernatant through a DEAE column gives an enzyme preparation in the presence of which the polymerization of d-XTP is completely dependent on active RNA (Table V). All four d-XTP are required to get a maximal amount of a H3-d-XTP incorporated into a trichloraacetic acid (TCA) precipitable product. Omission of one or two of the d-XTP leads to a considerable decrease in the product formation. One d-XTP does not seem to be polymerized.

DNA synthesis by E. coli transcriptase was examined in the presence of different RNA preparations and E. coli DNA. It should be emphasized that the RNA preparations used did not contain DNA in detectable amounts, while DNA preparations did contain RNA (3,4). The best template for E. coli transcriptase was (Table VI)

TABLE V

^{3}H-d-ATP, Incorporated (20 mn) into Acid-Precipitable Product under Various Conditions

Reaction mixture	ρ moles	Reaction mixture	ρ moles	Inhibition (%)
Complete	411	complete	418	–
Minus transforming RNA	<1	+ DNase 5 μg	<1	99
Minus $MgCl_2$	<2	+ RNase 20 μg	230	40
Minus d-GTP	136	+ RNase 20 μg preinc.	43	90
Minus d-CTP	125	+ showdomycin 50 μg	105	75
Minus d-GTP, d-CTP, d-TTP	<1	+ showdomycin, 50 μg (mutant enz.)	420	0

TABLE VI

Activity of E. coli Reverse Transcriptase in the Presence of Different Templates

Reaction mixture	ρ moles of ^{3}H d-ATP (incorporated in 20 min)	ρ moles of ^{3}H d-TTP (incorporated in 20 min)
Complete with transforming RNA, E. coli M 500	402	656
excreted RNA wild-type E. coli	392	400
ribosomal RNA (23 S + 16 S)	<1	<1
bulk t-RNA	<1	<1
t-RNA^{met}	0	–
A. faecalis 5.5 S RNA	145	136
Poly AG + poly UC	<1	<1
DNA	98	–
DNA treated with RNase	2	–

Note: Reaction mixture contains per 0.2 ml: $MgCl_2$, 2 μM; Tris-HCl buffer pH 7.65, 25 μM; each deoxyribonucleoside-5′-triphosphate 5 n moles + ^{3}H d-ATP or ^{3}H d-TTP (100,000 CPM); transforming RNA, 4 μg; enzyme fraction (DEAE) 60 μg. Incubation 20 min at 37°C. After addition of trichloroacetic acid (TCA) to incubation mixture, the precipitate was filtered on Whatman GC/F glass filter, washed, dried, and radioactivity measured in a Packard spectrophotometer. Enzyme fraction DEAE containing reverse transcriptase activity was obtained by the method described (29). Conditions as described in the note to Table I. Templates 4 μg, ^{3}H d-ATP, and ^{3}H d-TTP (100,000 CPM); DNA was pretreated with 20 μg of pancreatic RNase. Alcaligenes faecalis 5.5 S RNA was isolated as described (30).

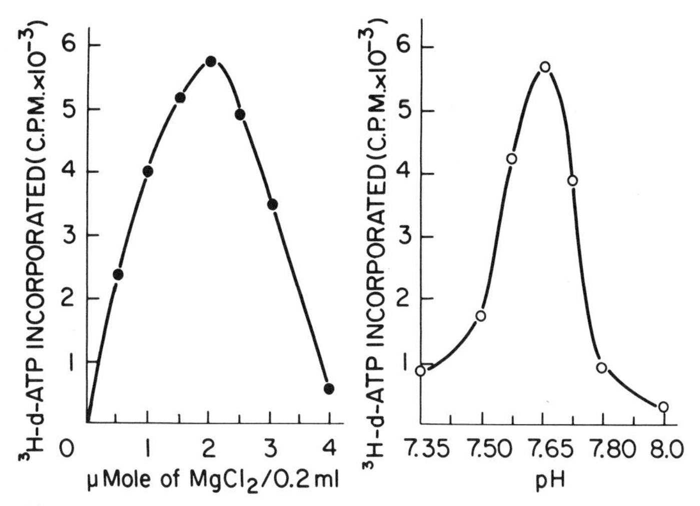

Fig 6. Mg^{++} ions requirement and pH of buffer solution for activity of E. coli reverse transcriptase. Incubation conditions: See *Note* to Table VI.

extracellularly excreted RNA, especially transforming RNA. Ribosomal RNA and transfer RNA are inactive; Alcaligenes faecalis 5.5 S RNA containing rapidly labeled RNA is utilized by E. coli transcriptase for DNA synthesis. Synthetic polyribonucleotides have no effect on enzyme activity. Low enzyme activity observed in the presence of E. coli DNA disappears when the DNA preparation is pretreated with RNase (pancreatic and T_1). Template activity of RNA is destroyed by preincubation with RNase and the synthesis of the deoxypolymer does not take place in the presence of DNase (Table V). Showdomycin inhibits strongly the deoxypolymer formation (Table V).

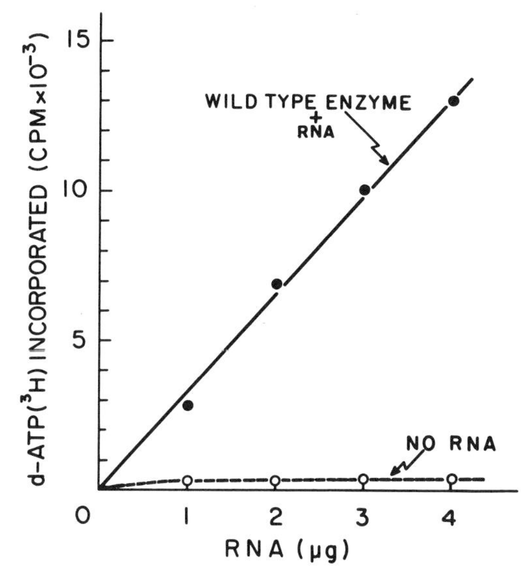

Fig 7. Deoxyribopolymer formation in the presence of different concentrations of transforming RNA.

Conditions: see *Note* to Table VI. Transforming RNA excreted by E. coli mutant M 500 sho-R was used.

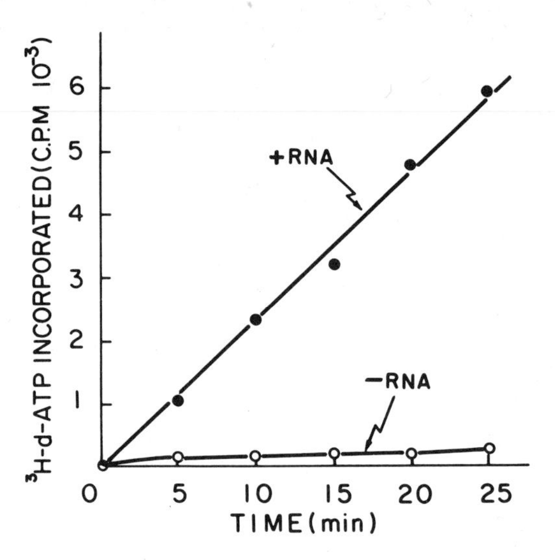

Fig 8. RNA-dependent deoxypolymer formation by E. coli reverse transcriptase in function of time. Conditions as described in the *Note* to Table VI. Transforming RNA from E. coli M 500 sho-R was used.

b) Metal requirements and pH; The maximal incorporation of ^{3}H-product depends upon Mg^{++} ion concentration and on the pH of the buffer (Fig 6). Mn^{++} partially replaces (50%) Mg^{++}.

The formation of ^{3}H deoxypolymer in the absence and presence of different concentrations of transforming RNA and as a function of time are illustrated by Figures 7 and 8. Polymer formation is strictly dependent on RNA acting as template.

c) Properties of synthesized ^{3}H-deoxypolymer: The ^{3}H product synthesized under optimal conditions (20 min.) is not sensitive to RNase, pronase, and KOH. In contrast, it is degraded by pancreatic deoxyribonuclease (35).

TABLE VII

A/U Ratio of Template RNA and dT/dA Ratio of ^{3}H Product Synthesized by E. coli Transcriptase

	mol per 100 mol of nucleotides	A/U ratio	dT/dA ratio of ^{3}H product
E. coli M 500 RNA	A = 29.0 G = 34.0 C = 19.0 U = 18.0	1.57	1.62 (Exp I); 1.54 (Exp II); 1.58 (Exp 3)
A. faecalis 5.5 S RNA	A = 16.2 G = 34.5 C = 32.0 U = 17.3	0.95	0.96 (Exp I); 0.89 (Exp II); 0.99 (Exp 3)

Note: Conditions, see *note* to Table VI.

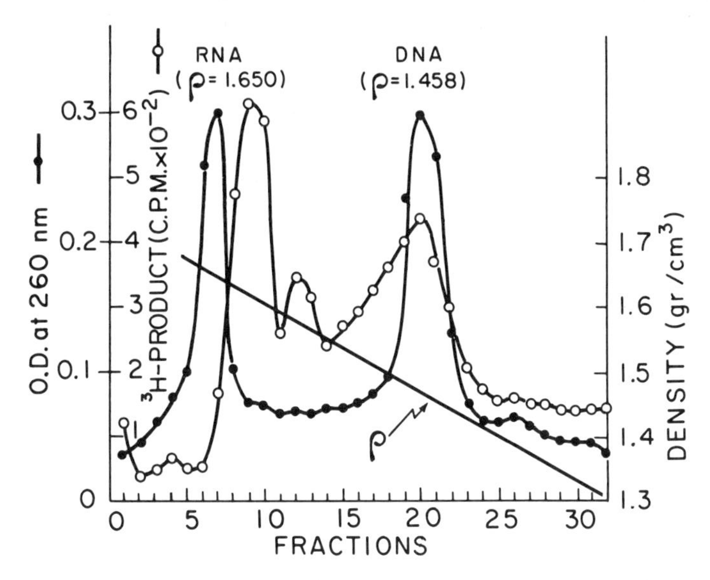

Fig 9. Density of ^{3}H-product synthesized by E. coli reverse transcriptase in the presence of transforming RNA.

^{3}H-product was synthesized as described in the *Note* to Table VI. After 30 min of incubation, the mixture was treated with chloroform (twice), centrifuged, and the aqueous solution dialyzed for 24 hrs at 4°C against 1 1. of Tris buffer 10^{-2} M containing KC1 0.1 M (two changes of buffer solution). Radioactive product is completely precipitable by trichloroacetic acid. Dialyzed ^{3}H product was mixed with saturated C_2SO_4 to a density of 1.550 gr/Cm^3. Internal markers: transforming RNA and E. coli DNA were included and centrifuged at 30,000 rpm at 20°C for 64 hrs in a Spinco SW 39 rotor. Fractions were collected. Absorption at 260 nm and radioactivity were determined.

The data presented in Table VII show the ratio A/U of two template RNAs and that of d-T/d-A of the ^{3}H-deoxypolymer synthesized in vitro by E. coli transcriptase. The ratio d-T/d-A (on average) is identical to that of the A/U of the RNA. On that basis, one can conclude that complementary deoxypolymer was synthesized on the RNA template. To provide evidence that the ^{3}H heterodeoxypolymer is of a DNA nature, we separated it from the enzyme by treatment with chloroform, then dialyzed it and examined it by equilibrium density centrifugation in C_2SO_4. Internal markers (transforming RNA and E. coli DNA) were included with ^{3}H product. Figure 9 shows that the ^{3}H product is overspread in the density regions between RNA and DNA (gradient at pH 7.65). This heterogeneity is expected for RNA-DNA hybrids (23) (27). However, a certain amount of radioactive material is present in the DNA density region, which suggests that some free DNA is present in the gradient. These data and those presented in Table VI show that some specific RNAs, transforming RNA in particular, are the template for the transcriptase by which transfer of information is mediated.

Experiments are in progress to determine if a particular DNA portion can be detected in DNA from wild and mutant bacteria and if this portion is able to hybridize with DNA synthesized by E. coli reverse transcriptase.

CONCLUSIONS

The data presented here summarize our recent work, showing that a transforming RNA excreted by showdomycin-resistant mutants of E. coli penetrates the recipient cells (E. coli wild type) producing a massive transformation. The quantitative effect of transforming RNA on recipient bacteria proves that transformants must arise from the interaction of transforming RNA and bacteria that undergo transformation. Transformants which behave as stable mutants possess new and inherited biochemical characteristics identical to those found in E. coli showdomycin-resistant mutants from which transforming RNA was excreted. However, transfer of the resistance character to recipient cells is very low. It is worth emphasizing that Agrobacterium tumefaciens, transformed by E. coli transforming RNA, has lost completely the oncogenic power for plants and acquired several new properties. In both types of transformants (E. coli and Agrobacterium tumefaciens), we have found a very satisfactory correlation between the chemical properties of transforming RNA (A+G/C+U = 1.75 to 2.0) and those of intracellular RNAs synthesized by transformants (A+G/C+U = 1.75 to 2.0). In addition, as in the E. coli sho-R mutant, the consequence of modified RNAs in transformants is reflected by important changes in ribosomal proteins as well as in certain enzymes. The most intriguing correlation exists between the loss of oncogenic power of transformants and the increase of l-asparaginase pH 5.0 of Agrobacterium t.

The central question raised by our studies is: Is it possible to construct a general picture of the process of transformation of bacteria by RNA?

We propose two possible explanations. First, transforming RNA having particular chemical properties, once inside the recipient cells, could in some way be associated with bacterial DNA in order to be protected as an original "RNA episome," itself (4). This RNA episome can be replicated by a process in which complementarity of bases is not necessary, since the base composition of several RNA species (t-RNA is an exception) in transformants corresponds exactly to that characteristic for transforming RNA. PNPase seems to fulfill this requirement (6), since it has the capacity to replicate in vitro the transforming RNA into a product which has a G+A/C+U ratio close to or identical with that of transforming RNA. Drastic alteration of several RNA species and proteins (ribosomal proteins and enzymes) in transformants obtained by RNA implies that RNA episome must have a strong autonomy, since it escapes the control of the bacterial genome. If this is so, then the RNA episome could be considered as an extrachromosomal determinant. In that regard, it satisfies a number of criteria: it can easily migrate from one cell to another; it induces profound biochemical and physiological changes; and transformants excrete transforming RNA. This last property could easily explain why transformants do not give rise to revertants.

The presence of the RNA episome in small amounts in wild-type bacteria and in large amounts in the showdomycin-resistant mutant (3) strongly suggests that this kind of episome could be considered as a primitive cell determinant which can be expressed only under certain conditions (28).

Another explanation could be that transforming RNA once inside the recipient bacteria can direct DNA synthesis mediated by a reverse transcriptase. Complementary DNA synthesized on the template of transforming RNA could be integrated into a bacterial genome, as suggested by Temin for RNA of oncogenic viruses (2), thus leading to a genetic mutation. This kind of mutation will be drastic if the bacterial genome contains enough sequences, as specified by transforming RNA (G+A/C+U = 1.70 - 2.0). This view is partly supported by the data we have presented in this paper, showing that E. coli K-12 possesses an enzyme, reverse transcriptase, which in the presence of 4 d-XTP and a template (transforming RNA) synthesizes a deoxyribopolymer characterized as being DNA (35). This does not necessarily mean that such DNA has to be incorporated into the cell genome. DNA synthesized on an RNA template may well be helpful in protecting that RNA (transforming RNA for example) which can be freed when needed. Something of that sort could happen even to messenger RNA when it is transcribed into complementary DNA, as has been shown for hemoglobin-purified messenger RNA (25,32,33). Along these lines it should be pointed out that an RNA-DNA "hybrid" preparation (34) can mediate the transfer of sulfamide resistance to pneumococcus, and that both RNase and DNase destroy the transforming principle. It was suggested that during transformation by the RNA-DNA preparation "RNA may act to preserve activity of DNA or vice versa." In addition, the importance of the RNA-DNA hybrid as a template for polydeoxyribonucleotide synthesis by purified E. coli DNA polymerase I has been emphasized (36,37).

The data presented concerning transformation of bacteria by transforming RNA and RNA-directed DNA synthesis constitute a new approach to the study of extrachromosomal inheritance in bacteria and its implication in evolution. As suggested recently (38), a multienzyme complex may be necessary to direct and regulate the biosynthesis of biologically active DNA and RNA in living cells.

SUMMARY

Transforming RNA released from showdomycin-resistant mutants of E. coli is capable of promoting "genetic transformation" of wild-type E. coli and Agrobacterium tumefaciens. Transformants possess profoundly modified biochemical and physiological properties. Thus certain types of transformants of Agrobacterium tumefaciens have definitively lost their capacity to be oncogenic in plants. Evidence for the existence in bacteria of an RNA-directed DNA polymerase is presented. Its possible role in the transformation of bacteria by RNA is discussed.

REFERENCES

1. Hayes, W.: The Genetics of Bacteria and Their Viruses. Blackwell Scientific Publications, London, 1970.
2. Avery, O. T., MacLeod, C. M., and McCarty, M.: Studies on the chemical nature of the substance inducing transformation of pneumococcal types. I. Induction of transformation by a deoxyribonucleic acid fraction isolated from pneumococcus type III. J. Exptl. Med., 79:137, 1944.
3. Beljanski, M., Beljanski, M., and Bourgarel, P.: ARN transformants porteurs de caracteres heréditaires chez E.coli showdomycino-resistant. C. R. Acad. Sc. (Paris), Série D, 272:2167, 1971.
4. Beljanski, M., Beljanski, M., and Bourgarel, P.: "Episome à ARN" porté par l' ADN d'Escherichia coli sauvage et showdomycino-resistant. C. R. Acad. Sc. (Paris), Série D, 272:2736, 1971.

5. Beljanski, M., Bourgarel, P., and Beljanski, M.: Drastic alteration of ribosomal RNA and ribosomal proteins in showdomycin-resistant Escherichia coli. Proc. Nat. Acad. Sci. USA, p 491, 1971.
6. Beljanski, M., Beljanski, M., Manigault, P., and Bourgarel, P.: Transformation of Agrobacterium tumefaciens into a nononcogenic species by an Escherichia coli RNA. Proc. Nat. Acad. Sci. USA, 69:191, 1972.
7. Plawecki, M., and Beljanski, M.: Transcription par la polynucleotide phosphorylase de l' ARN associé à l' ADN d'Escherichia coli. C. R. Acad. Sc. (Paris), Serie D, 273:827, 1971
8. Beljanski, M.: A propos du microdosage du ribose dans les acides nucleiques et leurs dérivés. Ann. Inst. Pasteur, 76:451, 1949.
9. Pardee, A. B., Jacob, F., and Monod, J.: The genetic control and cytoplasmic expression of "inductibility" in the synthesis of β-galactosidase. J. Mol. Biol., 1:165, 1959.
10. Braun, A. C.: Cellular autonomy in crown gall. Phytopath, 14:963, 1951.
11. Braun, A. C., and Mandle, R. J.: Studies on the inactivation of the tumor inducing principle in crown gall. Growth, 12:255, 1948.
12. Bernaerts, M. J., and De Ley, J.: A biochemical test for crown gall bacteria. Nature, 197:406, 1963.
13. Manigault, P., and Stoll, Ch.: Induction aseptique et croissance de tumeurs végétales extemptes de bactéries. Phytopathol.Z., 38:1, 1960.
14. Beljanski, M., Kurkdjian, A., and Manigault P.: Activité d'une l-asparaginase et relation avec le pouvoir oncogene de differents "mutants" d. Agrobacterium tumefaciens. C. R. Acad. Sci. (Paris), Serie D, 274:3560, 1972.
15. Manigault, P.: Intervention dans la plaie d'inoculation de bactéries appartenant à differentes souches d'Agrobacterium tumefaciens. Ann. Inst. Pasteur (Paris), 119:347, 1970.
16. Maral, R., and Werner, G. H.: Antiviral activity of l-asparaginase. Nature New Biology, 232:187, 1971.
17. Crowther, D.: L-asparaginase and human malignant disease. Nature, 229: 1971.
18. Kidd, J. G.: Regression of transplanted lymphomas induced in vivo by mean of normal guinea pig serum. J. Exp. Med., 98:565, 1953.
19. Braun, A. C., and Wood, H. N.: On the inhibition of tumor infection in the crown gall disease with the use of ribonuclease A. Proc. Nat. Acad. Sc. USA, 56:1417, 1966.
20. Temin, H. K., and Mizutani, S.: RNA dependent DNA polymerase in virions of Rous sarcoma virus. Nature, 226:1211, 1970.
21. Temin, H. M.: Homology between RNA from Rous sarcoma virus and DNA from Rous sarcoma virus infected cells. Proc. Nat. Acad. Sc. USA, 52:323, 1964.
22. Baltimore, D.: RNA-dependent DNA polymerase in virions of RNA tumors viruses. Nature, 226:1209, 1970.
23. Spiegelman, S., Burny, A., Das, M. R., Keydar, J., Schlom, J., Travnicek, M., and Watson, K.: DNA characterization of the products of RNA-directed DNA polymerases in oncogenic RNA viruses. Nature, 227:563, 1970.
24. Rokutanda, M., Rokutanda, H., Green, M., Fujinaga, K., Roy, R. K., and Gurgo, C.: Formation of viral RNA-DNA hybrid molecules by the DNA polymerase of sarcoma-leukemia viruses. Nature, 227:1026, 1970.
25. Ross, J., Aviv, H., Scolnick, E., and Leder, P.: In vitro synthesis of DNA complementary to purified rabbit globin m-RNA. Proc. Nat. Acad. Sc. USA, 69:264, 1972.
26. Varmus, H. E., Levinson, W. E., and Bishop, M.: Extent of transcription by the RNA-dependent polymerase of Rous sarcoma virus, Nature, 233:19, 1971.
27. Hurwitz, J., and Leis, J. P.: RNA-dependent DNA polymerase activity of RNA tumor viruses. J. Virology, 9:116, 1972.
28. Gallo, R.: Reverse transcriptase, the DNA polymerase of oncogenic RNA viruses. Nature, 234:194, 1971.
29. Fischer-Ferraro, C., and Beljanski, M.: Nouvelle methode de purification des polypeptide synthetases. European J. Biochem., 4:118, 1968.

30. Beljanski, M., and Bourgarel, P.: Isolement d'un ARN matriciel d' Alcaligenes faecalis. C. R. Acad. Sc. (Paris) (Série D.), 266:845, 1968.
31. Beljanski, M., Beljanski, M., Bourgarel, P., and Chassagne, J.: Synthèse chez les bactéries d'ARN nouveaux n'étant pas la copie de l'ADN. C. R. Acad. Sc. (Paris) Série D, 269:240, 1969.
32. Verma, I. M., Temple, G. F., Fan, H., and Baltimore, D.: In vitro synthesis of DNA complementary to rabbit reticulocyte 10 S RNA. Nature, 235:163, 1972.
33. Kacian, D. L., Spiegelman, S., Bank A., Terada, M., Metafora, S., Dow, L., and Marks, P.: In vitro synthesis of DNA components of human genes for globins. Nature, 235:167, 1972.
34. Evans, A. H.: Introduction of specific drug resistance properties by purified RNA-containing fractions from pneumococcus. Proc. Natl. Acad. Sci. USA, 52:1442, 1964.
35. Beljanski, M.: Synthèse in vitro de l'ADN sur une matrice d'ARN par une transcriptase a'E. coli. C. R. Acad. Sc., (Paris), 274:2801, 1972.
36. Cavalieri, L. F., and Caroll, E.: RNA-dependent DNA polymerase from E. coli: an effect produced by traces of added DNA. Nature, 232:254, 1971.
37. Karkas, J. D., Stavrianopoulos, J. G., and Chargaff, E.: Action of DNA polymerase I of E. coli with DNA-RNA hybrids as templates. Proc. Nat. Acad. Sci. USA, 69:398, 1972.
38. Stavrionopoulos, J. G., Karkas, J. D., and Chargaff, E.: Nucleic acid polymerases of the developing chicken embryo: a DNA polymerase preferring a hybrid templates. Proc. Nat. Acad. Sci. USA, 68:2207, 1971.

8. HOST CELL CHANGES INDUCED BY R FACTORS AND OTHER SEX FACTORS[1]

Tsutomu Watanabe, Yasuko Ogata, Katomi Sugawara, Kazuyo Oda, Tatsuo Saito and Yoshitsuru Yokoyama

Department of Microbiology, Keio University School of Medicine
Tokyo, Japan

INTRODUCTION

Bacterial plasmids are genetic elements which replicate in the cytoplasm of bacterial cells and do not form extracellular independent particles unlike ordinary viruses. Thus they can be regarded as obligate intracellular parasites of bacteria. Some of the bacterial plasmids are transmissible through cell conjugation and therefore are called transmissible plasmids, autotransferable elements, or sex factors. If the plasmids become integrated into host chromosomes and replicate as parts thereof, they are called episomes. Some sex factors are known to behave as episomes. The presence and the transfer of plasmids can be detected through the specific phenotypic changes of host cells which are induced by the plasmids. The detection and identification of plasmids are difficult, if we do not know the specific phenotypic changes which are induced by them. The detection of such plasmids, however, can be made possible by showing the presence of covalently linked closed circular deoxyribonucleic acid (DNA) molecules separable from the DNA of host chromosomes (1).

The phenotypic changes of host cells caused by various sex factors can be divided into specific and common phenotypes. The examples of the specific phenotypes governed by plasmids are colicinogeny by colicin agents and drug resistance by R factors. The common phenotypes are found among various different plasmids. In the present paper, we shall be concerned mainly with the common phenotypic changes of host cells induced by various sex factors.

The most important property of sex factors is that they can transfer themselves and their host chromosomes through causing conjugation. It has been also established that bacterial cells carrying sex factors have specific pili (or fimbriae) on their surface (2). These specific pili are different from the nonspecific pili which are found on plasmid-free bacteria. All the available circumstantial evidence support the view that these specific pili act as organs for the conjugation and transfer of the genetic material (2,3). They are thus called sex pili. Three types of sex pili have so far been recognized through their morphology, adsorbability of sex-specific bacteriophages, and immunological specificity. They are F-like pili, I-like pili, and the pili formed by group N plasmids. F-like pili are formed by F, ColB4, ColV2, ColV3, ColVBtrp, and most fi$^+$ R factors (see below). I-like pili are formed by ColIa, ColIb, and most fi$^-$ R factors. Each of the above three types of sex pili can be further subdivided into several subtypes by their immunological specificity and adsorbability of sex-specific phages (4).

[1] This work was supported in part by United States Public Health Service research grant AI-08078 from National Institute of Allergy and Infectious Diseases.

Bacterial cells carrying sex factors exhibit a particular property called superinfection immunity. The frequencies of conjugational transfer of plasmids to the cells already carrying homologous plasmids are strongly reduced as compared with those to plasmid-free recipient cells. This phenomenon is called superinfection immunity and involves two steps; surface exclusion (or entry exclusion) and incompatibility (or mutual exclusion). Surface exclusion is due to the reduced frequency of conjugation as is evidenced by the morphological observation under a phase contrast microscope. Either donor or recipient cells are vitally stained with triphenyltetrazolium chloride (TTC) and mixed with unstained partner cells; and the mixtures are observed under a phase contrast microscope. We have shown that the frequencies of formation of cell pairs or cell clusters are strongly reduced by the presence of homologous plasmids in the recipients. Superinfection with homologous sex factors by conjugation can occur, although at much reduced frequencies, and thus the clones carrying two homologous plasmids can be produced. The two homologous plasmids in the same cells are genetically very unstable and segregate either plasmid at high frequencies in the course of cell divisions. Otherwise they are recombined to form single recombinant plasmids. The mechanism operating in incompatibility is considered to be the same as that controlling the replication of plasmids. In other words, the control mechanism of the replication of plasmid DNA in pace with host chromosome DNA is thought to be responsible for incompatibility.

It had been assumed that surface exclusion may be due to mechanical interference of conjugation by sex pili of the recipient on the basis of the finding of F^- phenocopy (5), that is the temporary abolishment of fertility by the prolonged growth under strong aeration at stationary phase. In other words, the mechanical removal of sex pili and their inability to regenerate at stationary phase had been assumed to be the mechanism of F^- phenocopy. Later work, however, has shown that sex pili are not involved in the surface exclusion, at least in the superinfection with R factors (6).

Naturally occurring R factors are generally in self-repressed states in the formation of sex pili regardless of their fi types (4). The fact that surface exclusion occurs between homologous R factors indicates that sex pili of the recipient cells are not responsible for the surface exclusion, because only small fractions (usually less than 1%) of the recipient cells have sex pili, and a majority of the recipient cells lack sex pili due to self-repression.

In contrast to the superinfection immunity observed in conjugational transfer of homologous R factors, no immunity was found in the superinfection of homologous R factors by transduction with phage P1 in Escherichia coli, as well as with phage P22 in Salmonella typhimurium (7). This finding suggests that some surface change other than the formation of sex pili of the recipient cells caused by the R factors is responsible for the surface exclusion.

We have been interested in the nature of this surface change, but it is still obscure. We have so far shown that R factors, fi^+ and fi^-, do not alter the sensitivity of the host bacteria to available phages, excluding a possibility of the drastic change of the cell wall. It is known that some fi^- R factors affect the response of host bacteria to some phages, but this phenomenon is not due to the change of phage adsorbability but to host-controlled restriction (8).

We have recently found that R factors and other sex factors confer increased sensitivity to antibiotics flavomycin (also called moenomycin and flavophospholipol) and rifampicin. These results will be reported and their possible significance will be discussed in the present paper.

MATERIALS AND METHODS

Culture Media

Liquid cultures were prepared in Penassay broth (Difco). Plating media were nutrient agar (Difco) and bromothymolblue-lactose-nutrient agar (nutrient agar containing 0.0045% bromothymolblue and 1% lactose).

Drugs

Flavomycin (92% pure sample) was supplied by Hoechst AG, Germany, and rifampicin was a gift of Daiichi Pharmaceutical Co., Ltd., Japan. Other chemotherapeutics used were sulfathiazole (Su) (Takeda), dihydrostreptomycin sulfate (Sm) (Meiji), Chloramphenicol (Cm) (Parke, Davis and Co.), tetracycline chloride (Tc) (Lederle). Rifampicin was dissolved in ethylenglycol to a concentration of 10 mg/ml and then diluted in distilled water. Other drugs were dissolved and diluted in distilled water.

Bacterial Strains

Various substrains of Escherichia coli K-12 were used. They were CSH-2 (met^- pro^-) F^- and F^+, W3110/I (gal $^-\lambda^-$ colI-r) F^-, W3110/I (colIb) (carrying colicin agent Ib), W3110/I (colIb drd) (carrying a derepressed mutant of colIb), W677/PTS (pro^- thr^- leu^- thi^- man^- xyl^- mal^- gal^- lac^- str-r tsx-r) F^-, W2252 ($met^-\lambda^-$ Hfr), W2252 defF (strain with a defective, integrated F mutant derived from W2252 and lacking fertility), AB312 (lac^- thr^- leu^- thi^- λ^- str-r Hfr), AB313 (lac^- thr^- leu^- thi^- str-r Hfr), H3000 (thi^- Hfr), JE2217/S (pil^- str-r) F^-, and B380 (his^- gal^- pil^-) F^-. The strains with a drd mutant of colIb, described above, were isolated and supplied to us by Ohki and Ozeki. Similar mutants were isolated also by Edwards and Meynell. Strains JE2217, JE2571, and B380 do not form nonspecific type 1 pili and were supplied to us by Y. Nishimura and C. C. Brinton, Jr., respectively. Besides these K-12 substrains, Salmonella typhimurium LT-2 wild type was also used.

Strains of R Factors Used

Strains of R factors used in this study are listed in Table I. R factors 222-R_3 and 222-Tc are spontaneous deletion mutants derived from 222. Rl-19 and R64-11 are derepressed (drd) mutants derived from R1 and R64, respectively (9). These drd mutants do not produce cytoplasmic repressors which inhibit the formation of sex pili. Accordingly, they fully develop sex pili and exhibit high conjugal transferability. R1-19 behaves phenotypically as fi^-, because it does not produce a repressor for the forma-

TABLE I

Strains of R Factors Used in This Study

Strain number	Drug resistance marker[a]	*fi* type[b]
222	Su, Sm, Cm, Tc	+
222-R_3	Su, Sm, Cm	+
222-Tc	Tc	+
K	Su, Sm, Cm, Tc	+
S-b	Su, Sm, Cm, Tc	+
N-6	Su, Sm, Tc	+
N-9	Su, Sm, Tc	+
R_6	Su, Sm, Cm, Tc, Km	+
R1	Su, Sm, Cm, Ap	+
R1-19	Su, Sm, Cm, Ap	-
N-1	Su, Sm, Tc	-
N-3	Su, Sm, Tc	-
R-15	Su, Sm	-
S-a	Su, Sm, Cm, Km	-
R64	Sm, Tc	-
R64-11	Sm, Tc	-

[a]Su: sulfonamide, Sm: streptomycin, Cm: chloramphenicol, Km: kanamycin, Tc: tetracycline, Ap: aminobenzyl penicillin.

[b]*fi* fertility inhibition. *fi*$^+$ R factors inhibit the formation of F pili when they are in male strains, while *fi*$^-$ R factors do not have this inhibitory function[13,14].

tion of F pili by the sex factor F in the same cells, although some other *drd* mutants are known which do produce repressors but are insensitive to the repressors (operator-constitutive mutants).

Procedures for Studying Bacterial Sensitivity to Antibiotics

A procedure for studying bacterial sensitivity to antibiotics was to determine the minimal inhibitory concentrations (MIC) of the antibiotics. Each bacterial strain was grown in Penassay broth to an optical density of 0.4 (containing approximately 5 x 10^8 cells/ml) and diluted to 10^{-4} with physiological saline. A standard loopful of this dilution was spread on nutrient agar containing varying concentrations (in twofold dilutions) of a drug. The inoculated plates were incubated at 37°C overnight and the colonies developed were roughly scored. The concentration of the drug two times higher than that on which the number of colonies was about equal to that on a drug-free control plate was taken as the MIC of the drug for this strain. The drug sensitivity of each strain was also studied in broth. Bacteria were grown in Penassay broth containing varying concentrations of the drug and their growth curves were followed turbidimetrically using a Biophotometer (Jouan-Quetin). At the end of 10 hours observation, surviving bacteria were plated on drug-free nutrient agar and their sensitivity to various drugs was studied by replica plating. In order to isolate mutants of R^+ bacteria of higher drug resistance, R^+ bacteria were subcultured on nutrient agar containing increasingly higher concentrations of drug.

For studying the frequency of R transfer, CSH-2 F^- R^+ was used as a donor and W677/PTS as a recipient. Each strain was grown to O.D. 0.4 in broth, and 1 ml of donor culture was mixed with 9 ml of a recipient culture in a 200 ml Erlenmeyer flask, which was then incubated in a 37°C water-bath for 60 min without aeration. The mixed culture was then diluted properly and plated on BTB-lactose-nutrient agar containing 1,000 mcg/ml of Sm plus 25 mcg/ml of Cm or Tc, or 500 mcg/ml of Su. The recipient clones which received the R factor formed lactose-nonfermenting colonies on these selective media. When the effect of a drug was to be tested on the frequency of R transfer, 0.5 ml of varying concentrations of drug solutions was added to 8.5 ml of recipient culture immediately before 1 ml of a donor culture was added to the recipient. Subsequent procedure for determining the frequency of R transfer was identical to that described above.

Experiments with a Chemostat

A chemostat (made by Matsuki Seisakushio Co., Tokyo), which is a modification of the chemostat of Novick and Szilard, was used for continuous flow culture. Either E. coli CSH-2 (222-Tc) alone or a mixture of E. coli CSH-2 (222-Tc) and CSH-2 (R^-) cells was inoculated and grown in Penassay broth containing varying concentrations of flavomycin. The dilution rate was adjusted to approximately 30%/hour in all the experiments. Aliquots of the overflowing culture were taken every 24 hours and after proper dilution were plated on drug-free nutrient agar. The resultant colonies were replica plated onto nutrient agar containing 25 mcg/ml of Tc to determine the ratios of R factor-carrying and R^- cells in the samples.

RESULTS

MIC of Flavomycin for E. coli and S. typhimurium Strains Carrying R Factors or Other Sex Factors

As the results of MIC determination of flavomycin are shown in Table II, many of the E. coli K-12 substrains carrying R factors or other sex factors showed increased sensitivity to flavomycin as expressed by the MIC values. It seems of particular interest that the strains with nonrepressed-type sex factors are, as a rule, more sensitive to flavomycin. Comparable results were obtained also with S. typhimurium LT-2 as a host of the sex factors. Other important findings noted in Table II are that W2252 defF, JE2217 F^-, and JE2571 F^- showed equal flavomycin sensitivity to CSH-2 F^-. The results with JE2217 F^- and JE2571 F^- suggest that type 1 pili do not affect the flavomycin sensitivity of bacteria. However, B380 F^- showed increased sensitivity to flavomycin, unlike the other pil^- mutants.

TABLE II

Minimal Inhibitory Concentrations of Flavomycin for Various Substrains of *Escherichia coli* K-12

Strain	Minimal inhibitory concentrations of flavomycin (μg/ml.)
CSH-2 F^-	20
CSH-2 (222)	5
CSH-2 (222-R_3)	2.5
CSH-2 (222-Tc)	1.25
CSH-2 (K)	5
CSH-2 (S-b)	10
CSH-2 (N-6)	1.25
CSH-2 (N-9)	2.5
CSH-2 (R_6)	10
CSH-2 (R1)	10
CSH-2 (R1-19)	5
CSH-2 (N-1)	5
CSH-2 (N-3)	10
CSH-2 (R-15)	5
CSH-2 (S-a)	2.5
CSH-2 (R64)	5
CSH-2 (R64-11)	1.25
CSH-2 F^+	1.25
W2252 (Hfr)	1.25
W2252 *def*F	20
AB312 (Hfr)	1.25
AB313 (Hfr)	1.25
H3000 (Hfr)	2.5
W3110I F^-	20
W3110/I (Col Ib)	5
W3110/I (Col Ib *drd*)	1.25
JE2217 F^-	20
JE2571 F^-	20
B380 F^-	5

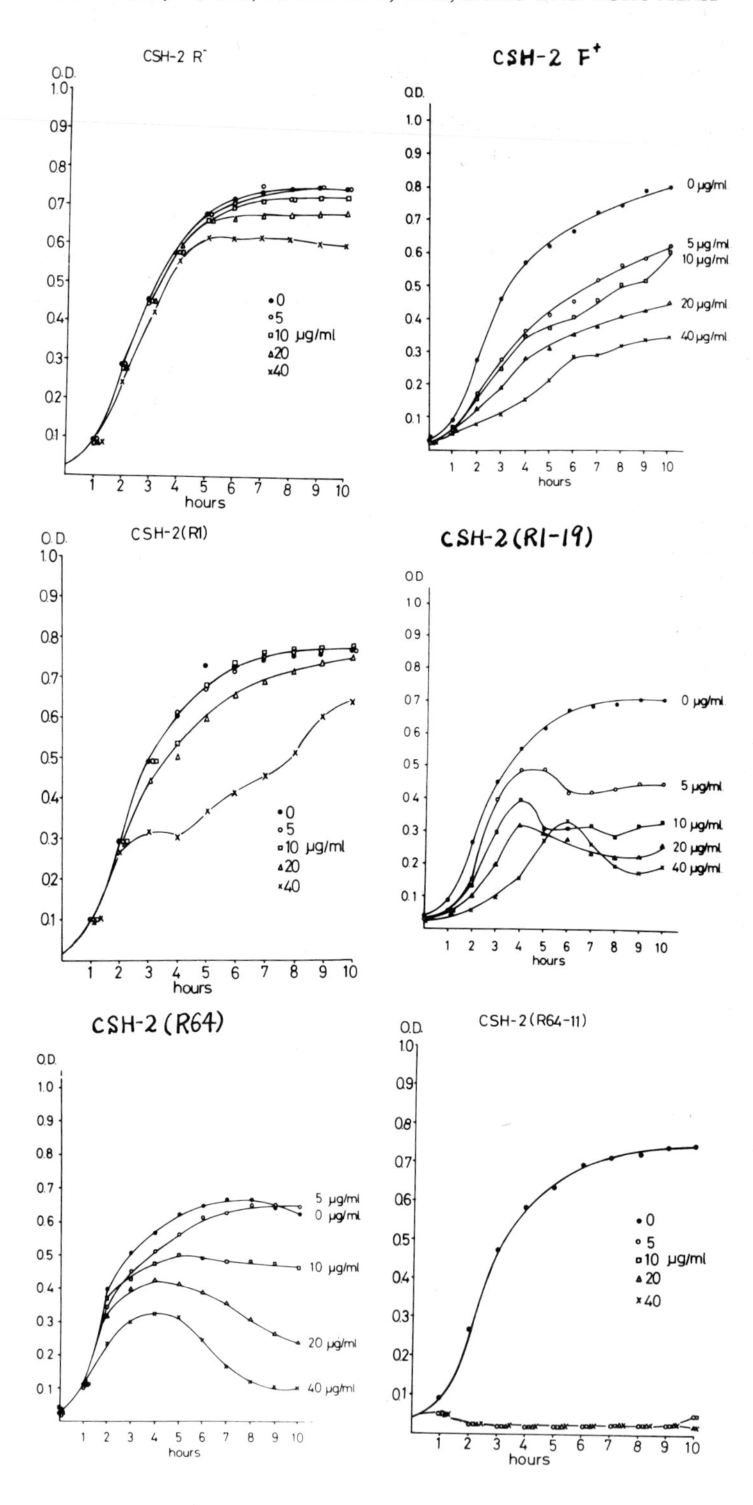

Fig 1. Growth curves of E. coli K-12, CSH-2 strains carrying R factors or other sex factors in broth containing varying concentrations of flavomycin.

Growth Curves of E. coli and S. typhimurium Strains Carrying R Factors or Other Sex Factors in Broth Containing Varying Concentrations of Flavomycin

Typical growth curves of E. coli CSH-2 strains with and without R factors or other sex factors are shown in Figure 1. These figures and similar data with CSH-2 strains again carrying other R factors, not shown here, again indicate clearly that R factors and other sex factors increase the flavomycin sensitivity of their host bacteria with an exception of W2252 defF regardless of their fi types. It is characteristic that reduction in OD of bacterial cultures occurs during their growth in the inhibitory concentrations of flavomycin, suggesting bacterial lysis by this antibiotic. In fact, reduction of viable cell counts was found in parallel with the reduction in OD. The reduced growth rates of bacteria with drd mutant R factors, especially R1-19, observed in these experiments, were already reported by Meynell and Datta (9) and a similar fact was observed with a drd mutant of colIb. Experiments with S. typhimurium LT-2 as a host of R factors have also indicated increased sensitivity of the host bacteria to flavomycin as a part of the data shown in Figure 2.

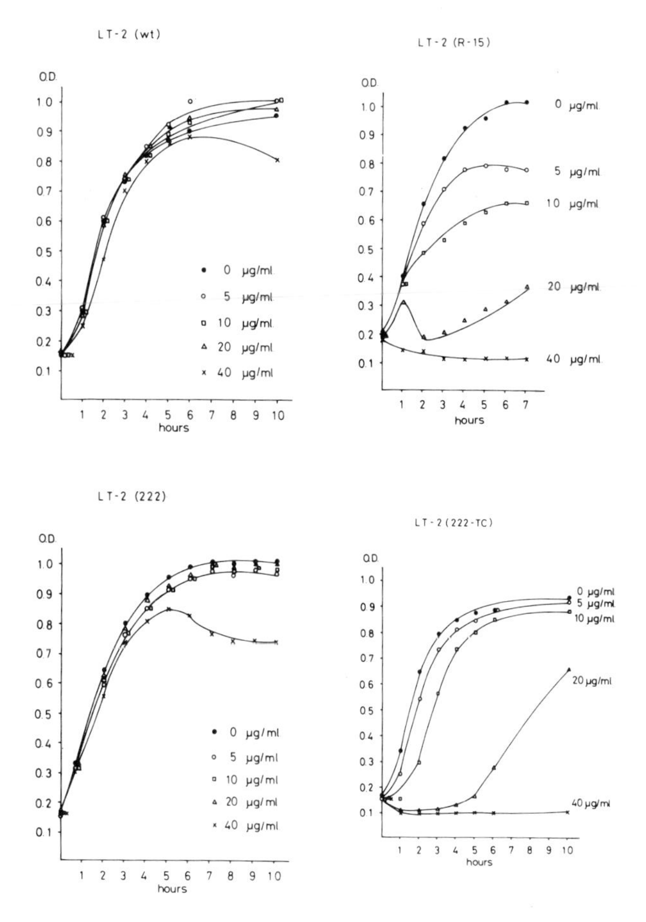

Fig 2. Growth curves of S. typhimurium LT-2 strains carrying R factors or other sex factors in broth containing varying concentrations of flavomycin.

TABLE III

Frequencies of Transfers of R1-19 and R64-11 between *Escherichia coli* Strains in the Presence of Flavomycin

Flavomycin concentration (μg/ml.)	Frequency of transfer (per introduced donor cell)[a]	
	R1-19	R64-11
0	2.0×10^{-1}	9.3×10^{-1}
5	1.9×10^{-1}	6.9×10^{-1}
10	1.6×10^{-1}	1.3×10^{-1}
20	1.3×10^{-1}	4.3×10^{-2}
40	5.0×10^{-2}	1.5×10^{-2}

[a]0.5 ml of varying concentrations of flavomycin solution was added to 8.5 ml. of a broth culture of *E. coli* W677/PTS (O.D. 0.4) and, immediately thereafter, 1 ml. of a broth culture of *E. coli* CSH-2 (R1-19) or CSH-2 (R64-11) (O.D. 0.4) was added. The mixture was incubated at 37°C for 60 min and, subsequently, varying dilutions of the mixed culture were plated on BTB-lactose-nutrient agar containing 1,000 μg/ml. of Sm plus 25 μg/ml. of Cm or Tc.

TABLE IV

Killing Effect of Flavomycin on *Escherichia coli* Strains with and without R Factors

Flavomycin concentration (μg/ml.)	Surviving bacteria/ml.[a]		
	CSH-2 (R1-19)	CSH-2 (R64-11)	W677/PTS
0	5.3×10^{8}	7.9×10^{8}	2.3×10^{8}
5	4.9×10^{7}	3.7×10^{6}	2.5×10^{8}
10	3.2×10^{7}	1.9×10^{6}	2.3×10^{8}
20	9.8×10^{6}	9.0×10^{5}	1.5×10^{8}
40	9.2×10^{6}	8.0×10^{5}	4.0×10^{7}

[a]8.5 ml. of a broth culture of *E. coli* W677/PTS (O.D. 0.4), 1 ml. of broth and 0.5 ml. of varying concentrations of flavomycin solution were mixed and incubated at 37°C for 60 min. In the series with CSH-2 (R1-19) and CSH-2 (R64-11), 1 ml. of a broth culture (O.D. 0.4), 8.5 ml. of broth and 0.5 ml. of varying concentrations of flavomycin solution were mixed and incubated at 37°C for 60 min.

Effect of Flavomycin on the Transfer Frequency of R Factor

Relatively high concentrations of flavomycin seemed to reduce to some extent the net transfer frequency of R factor (Table III), but this antibiotic is bactericidal to both donor and recipient, and the donor cells are more easily killed by this antibiotic than the recipient, as shown by the reconstruction experiments (Table IV), suggesting that flavomycin probably does not strongly inhibit the transfer of R factors, if at all.

All of the surviving bacteria after treatment of R^+ bacteria with varying concentrations of flavomycin in broth at 37°C for 10 hours still retained R factors, indicating that flavomycin probably does not effectively eliminate R factors, if at all (the frequencies of R^- bacteria, if any, among survivors were less than 1% with all the R factors studied).

Transferability of the Acquired Flavomycin Resistance of E. coli CSH-2 Carrying R Factor 222

Several mutants resistant to 150 mcg/ml of flavomycin were isolated from E. coli CSH-2 (222) in multiple-step selections on flavomycin-containing nutrient agar. The R factors of these mutants as well as the original R factors were transferred to E. coli W677/PTS, but no difference in flavomycin sensitivity was found among the recipient strains which received these R factors, indicating that the mutations of R^+ bacteria to flavomycin resistance were not mediated by R factors.

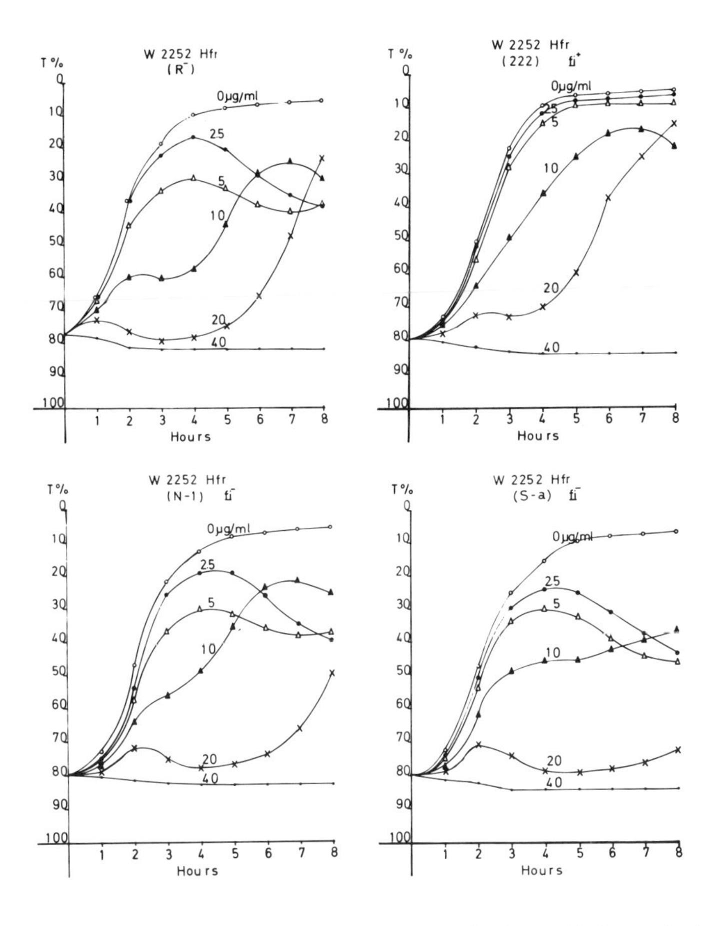

Fig 3. Growth curves of E. coli W2252 Hfr with and without R factors in broth containing varying concentrations of flavomycin.

Growth Curves of E. coli W2252 Hfr with and without R Factors in Broth Containing Varying Concentrations of Flavomycin

E. coli W2252 strains with *fi*$^+$ or *fi*$^-$ R factors and without R factors were grown in Penassay broth containing varying concentrations of flavomycin. As the results shown in Figure 3 indicate, fi$^-$ R factors N-1 and S-a did not appreciably alter the flavomycin sensitivity of their host W2252 Hfr, whereas fi$^+$ R factor 222 reduced the flavomycin sensitivity of W2252 Hfr to the level of E. coli CSH-2 F$^-$ (222).

Growth of E. coli CSH-2 (222-Tc) Alone and a Mixture of CSH-2 (222-Tc) and CSH-2 R$^-$ in Broth Containing Varying Concentrations of Flavomycin in a Chemostat

Typical data are shown in Figure 4. As seen in these figures, R$^-$ cells increased more rapidly than R$^+$ cells in the presence of flavomycin and finally became predominant.

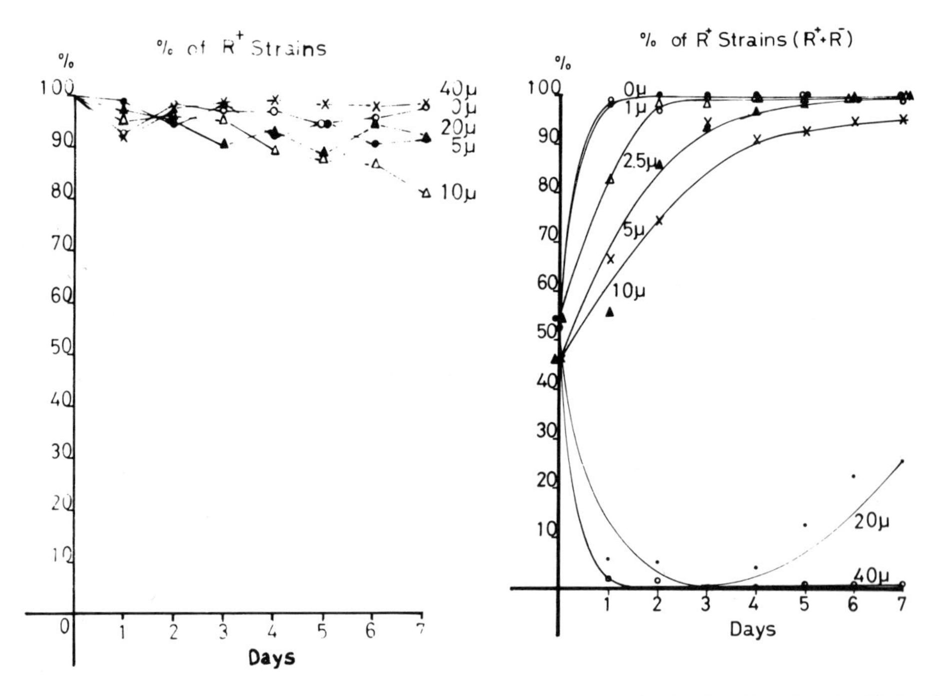

Fig 4. Growth of E. coli CSH-2 (222-Tc) alone and a mixture of CSH-2 (222-Tc) and CSH-2 R$^-$ in broth containing varying concentrations of flavomycin in a chemostat.

Growth Curves of E. coli and S. typhimurium Strains Carrying R Factors or Other Sex Factors in Broth Containing Varying Concentrations of Rifampicin

E. coli K-12 substrains and S. typhimurium LT-2 strains with and without R factors and other sex factors did not show clearly different MIC values of rifampicin. Therefore, their growth curves were followed in Panassay broth containing varying concentrations of rifampicin. Typical growth curves of E. coli CSH-2 strains with and without R factors or other sex factors are shown in Figure 5. This figure and

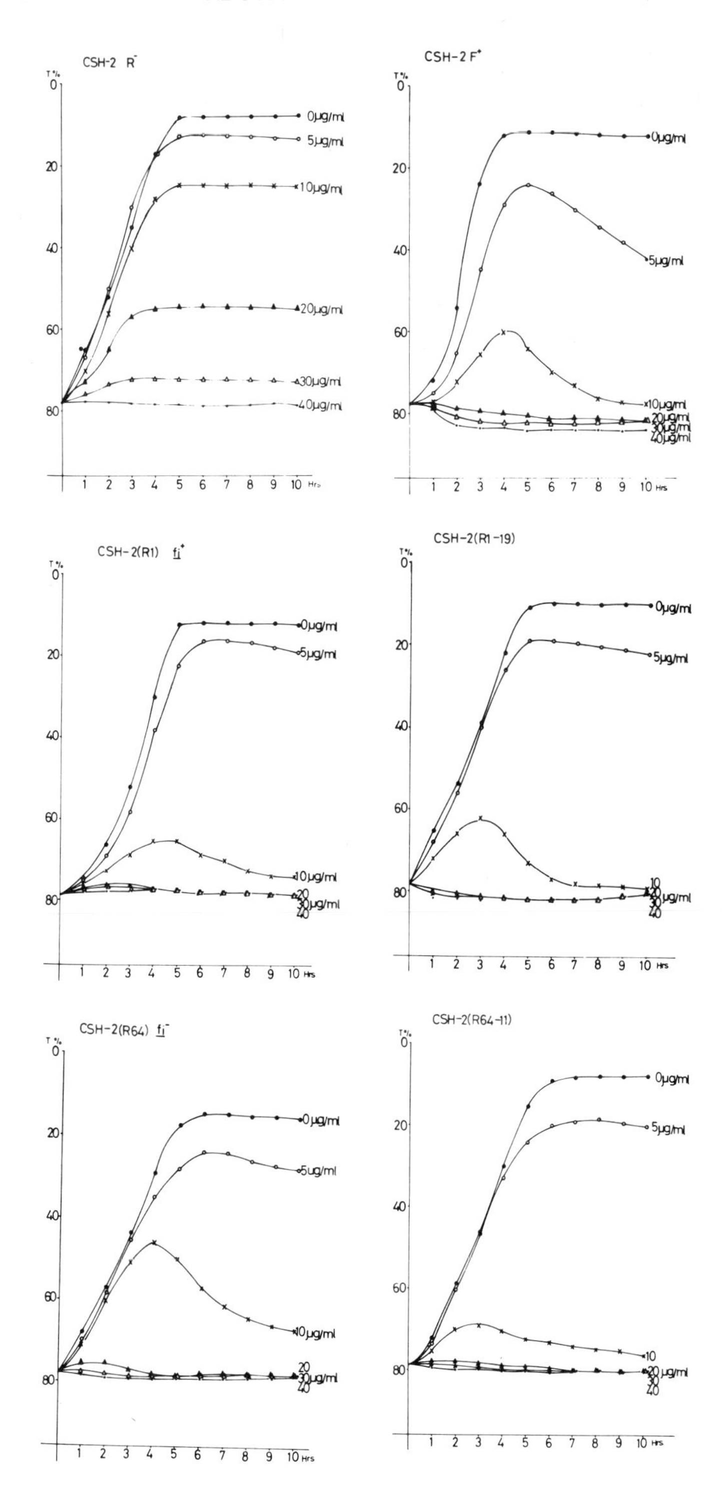

Fig 5. Growth curves of E. coli K-12, CSH-2 strains carrying R factors or other sex factors in broth containing varying concentrations of rifampicin.

similar data with CSH-2 strains carrying other R factors, not shown here, indicate that R factors and F and colicin agents increase the rifampicin sensitivity of their host CSH-2, although the degrees of rifampicin sensitivity conferred by the sex factors are less pronounced than those of flavomycin sensitivity conferred by the same sex factors. It is noted that derepressed mutants derived from self-repressed R factors, R1 and R64, respectively R1-19 and R64-11, confer slightly higher rifampicin sensitivity upon their host CSH-2 than the self-repressed wild-type plasmids. The results of similar studies with S. typhimurium LT-2 strains with R factors are shown in Figure 6. As seen in these figures, the rifampicin sensitivity of LT-2 is not appreciably increased by the presence of these R factors.

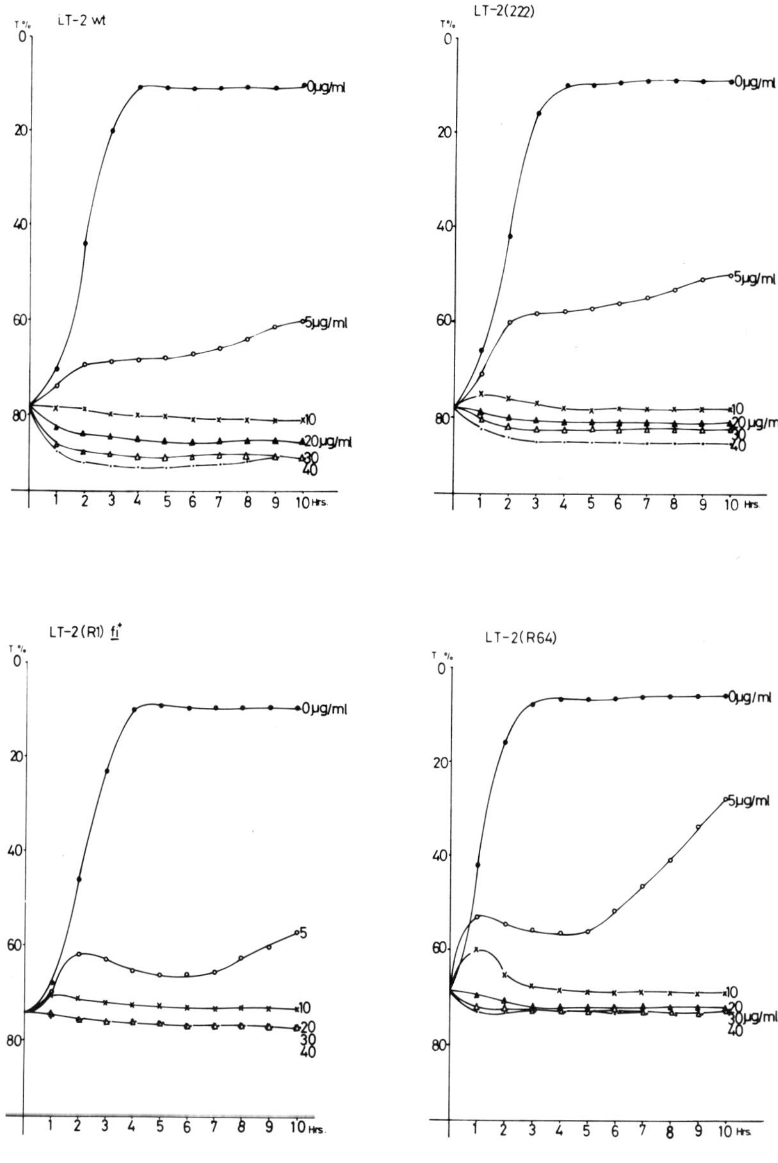

Fig 6. Growth curves of S. typhimurium LT-2 strains carrying R factors or other sex factors in broth containing varying concentrations of rifampicin.

DISCUSSION

It has been reported above that R factors, F, and colicin agent Ib increase the flavomycin and rifampicin sensitivity in E. coli. The increase in flavomycin sensitivity by R factors was found also in S. typhimurium, whereas the rifampicin sensitivity of S. typhimurium was not clearly increased by R factors. The fact that nonrepressed sex factors give more marked flavomycin sensitivity to their host bacteria than self-repressed sex factors suggests that the presence of sex pili may be responsible for the increased flavomycin sensitivity of bacteria. In fact, fi^+ R factor in an Hfr strain of E. coli reduced the flavomycin sensitivity of the host to the level of F^- strains carrying the same fi^+ R factor, apparently by inhibiting the formation of F pili. The fact that defF without fertility does not change the flavomycin sensitivity of the host bacteria also seems to support this view. Type 1 pili apparently have no connection with the flavomycin sensitivity of bacteria. The increased flavomycin sensitivity of E. coli B380 F^- may be due to an unknown, additional mutation of bacteria. Prophage λ also seems to have no influence on the flavomycin sensitivity of host bacteria, in view of the data shown in Table II. It is interesting to note that Mitsuhashi et al. (10), independently of our work, found that macarbomycin, an antibiotic which is chemically related to but definitely distinct from flavomycin, has a similar action on sex-factor-carrying E. coli and that they also assumed that the increased macarbomycin sensitivty of E. coli may be due to the presence of sex pili.

The fact that self-repressed sex factors confer increased flavomycin sensitivity upon their host bacteria suggests that the presence of the sex factors in self-repressed states, in other words, in the absence of sex pili, causes increased flavomycin sensitivity. It is assumed that bacteria with self-repressed sex factors have sex pilin (precursor protein of sex pili) in their cytoplasmic membranes in amounts comparable to the bacteria with nonrepressed sex factors. If this were the case, the presence of sex pilin in cytoplasmic membranes should cause increased flavomycin sensitivity, although its level is lower than conferred by the presence of sex pili. Alternatively, it is also possible that only the cells with sex pili in the population of cells with self-repressed sex factors acquires increased sensitivity to flavomycin, where we have to postulate that such "derepressed" cells arise rather frequently in order to account for our data.

The mechanism of action of flavomycin on Staphylococcus aureus was shown to be due to the inhibition of cell wall synthesis. Lysis of bacteria in a hypotonic medium by a similar mechanism may probably be occurring also in E. coli with and without sex factors, given the reduction in turbidity of bacterial cultures during their growth in flavomycin-containing broth. The presence of sex pili or pilin of the cell surface may probably somehow facilitate the permeation of flavomycin into the cells. Alternatively, the presence of sex factors (particularly those in nonrepressed states) may induce some other mechanism of increased sensitivity of the cells to flavomycin simultaneously with the formation of sex pili.

The increase of rifampicin sensitivity of bacteria caused by sex factors showed different features from flavomycin. First, the levels of the increase in rifampicin sensitivity in E. coli were not as marked as those in flavomycin sensitivity caused by the same R factors. Second, S. typhimurium with R factors did not show clear increase in rifampicin sensitivity. Third, nonrepressed sex factors increased the rifampicin sensitivity to levels only slightly higher than those caused by self-repressed sex factors.

The failure of R factors to induce the clear increase in rifampicin sensitivity of S. typhimurium may suggest that the increased rifampicin sensitivity observed in E. coli-carrying sex factors is probably a product of cooperation of sex factors with the host genome. The decrease in levels of rifampicin resistance of the rifampicin-resistant mutants of E. coli has already been reported with some R factors (11) and has been assumed to be due to the increase of permeability to refampicin by R factors. A similar mechanism is conceivable also in our work with rifampicin-sensitive E. coli. Here again, the presence of sex pili, and probably also sex pilin, may somehow facilitate the permeation of rifampicin. It is known that rifampicin interferes with DNA-dependent RNA polymerase (12,13). The apparent bacterial lysis by rifampicin observed in the present investigation is difficult to interpret on this basis.

As has been described in the Introduction, it has been established that sex factors, also in a self-repressed state, in conjugational recipient cells cause some change of the host cell surface which leads to surface exclusion, namely, the inhibition of effective cell contact formation between donor and recipient cells. This change cannot be the formation of sex pili as we have already pointed out. It is reasonable to assume that this change is due to the accumulation of sex pilin in the cytoplasmic membrane, which, as we have discussed above, is probably responsible for the increased sensitivity to flavomycin and rifampicin. Dowman and Meynell (14) reported that E. coli and S. typhimurium with sex factors show increased sensitivity to some surface actants such as tris + EDTA without the intervention by sex pili. Their results may be explainable by the accumulation of sex pilin. The finding by Tomoeda, et al (15-17) that sex factors can be "eliminated" by sodium dodecyl sulfate is now accounted for by the selective growth of sex factor-minus spontaneous segregants or self-repressed mutants without sex pili in the presence of this chemical. Thus, the mechanisms of entry exclusion and increased sensitivity to certain antibiotics and surface actants of bacteria with sex factors, nonrepressed and derepressed, now seem to be explainable unitarily by the formation of sex pili and the accumulation of sex pilin.

The fact that bacteria with R factors are much more sensitive to flavomycin than R^- bacteria is not only interesting from the view point of molecular biology but also seems to offer a great practical advantage to this antibiotic as a feed additive, together with its low toxicity and its property of not being absorbed through the intestines hardly at all. We note in particular that R^+ bacteria with sex pili (and therefore with conjugal donor ability) have more marked sensitivity to flavomycin. Bacteria with R factors can mutate to flavomycin resistance, but this resistance is not mediated by R factors and is likely due to chromosomal gene mutations. Our kinetic studies with a chemostat have shown that flavomycin reduces the relative numbers of R^+ bacteria when they are grown together with R^- bacteria in the presence of flavomycin. Spontaneous R^- segregants were selected for when R^+ bacteria were grown in the presence of flavomycin. We may expect that this antibiotic reduces R^+ bacteria also in animal intestines, although in vivo studies should be performed before we can draw a definitive conclusion on this point.

SUMMARY

The phenotypic changes of host bacteria commonly caused by transmissible plasmids, particularly those which seem to be due to the changes in the host cell surface, were presented. Among them, the increased flavomycin sensitivity seems to be directly related to the presence of sex pili or pilin. The increased rifampicin sensitivity also seems to be related to sex pili, although the data with rifampicin are not as clear cut as with flavomycin. We may assume that sex pili somehow help the permeation of these antibiotics into the cells.

On the other hand, it is clear that surface exclusion is not directly related to sex pili. The mechanism of surface exclusion might be due to some unknown cell surface change which interferes with the effective cell contact formation by sex pili. Recent unpublished data suggest that the cells with self-repressed plasmids contain pilin, which is a component protein of sex pili, in the vicinity of the cytoplasmic membrane in almost comparable amounts to the cells with derepressed plasmids. In other words, it is suggested that the step of morphogenesis of sex pili is repressed in self-repressed plasmids. If this is true, the pilin molecules accumulated in the vicinity of the cytoplasmic membrane may be responsible for entry exclusion.

From the public health point of view, flavomycin seems to have an advantage as a feed additive for animals, since it is shown that R^+ bacteria, especially those with conjugational transferability, are more sensitive to flavomycin than R^- bacteria. The use of antibiotics for animal feed additives has resulted in the increase of R^+ bacteria in our environment and has caused a public health problem.

REFERENCES

1. Clowes, R. C.: Molecular structure of bacterial plasmids. Bact. Rev., 36:361, 1972.
2. Brinton, C. C., Jr.: The structive function synthesis and genetic control of bacterial pili and a molecular model for DNA and RNA transport in gram negative bacteria. Trans. N.Y. Acad. Sci., Ser. II, 27:1003, 1965.
3. Brinton, C. C., Jr.: The properties of sex pili, the viral nature of "conjugal" genetic transfer systems, and some possible approaches to the control of bacterial drug resistance. Crit. Rev. Microbiol., 1:105, 1971.
4. Datta, N., Lawn, A. M., and Meynell, E.: The relationship of F type piliation and F phage sensitivity to drug resistance transfer in R^+ F^- Escherichia coli K-12. J. Gen. Microbiol., 45:365, 1966.
5. Lederberg, J., Cavalli, L. L., and Lederberg, E. M.: Sex compatibility in Escherichia coli. Genetics, 37:720, 1952.
6. Watanabe, T., Nishida, H., Ogata, C., Arai, T., and Sato, S.: Episome-mediated transfer of drug resistance in Enterobacteriaceae. VII. Two types of naturally occurring R factors. J. Bact., 88:716, 1964.
7. Watanabe, T., Sakaizumi, S., and Furuse, C.: Superinfection with R factors by transduction in Escherichia coli and Salmonella typhimurium. J. Bact., 96:1796, 1968.
8. Watanabe, T., Arai, T., Nishida, H., and Sato, S.: Episome-mediated transfer of drug resistance in Enterobacteriaceae. X. Restriction and modification of phages by fi^- R factors. J. Bact., 92:477, 1966.
9. Meynell, E., Meynell, G. G., and Datta, N.: Phylogenetic relationships of drug resistance factors and other transmissible bacterial plasmids. Bact. Rev., 32:55, 1968.
10. Mitsuhashi, S., Iyobe, S., Hashimoto, H., and Umezawa, H.: Preferential inhibition of the growth of Escherichia coli strains carrying episomes. J. Antibiot., 23:319, 1970.
11. Riva, S., Fietta, A. M., and Silvestri, L. G.: Effect of R factors on rifampicin resistance in Escherichia coli. Nature New Biology, 234:56, 1971.

12. Wehrly, W., Nüesh, J., Knüsel, F., and Staehelin, M.: Action of rifamycins on RNA polymerase. Biochem. Biophys. Acta, 157:215, 1968.
13. Sippel, A., and Hartmann, G.: Mode of action of rifamycin on the RNA polymerase reaction. Biochem. Biophys. Acta, 157:218, 1968.
14. Dowman, J. E., and Meynell, G. G.: Pleiotropic effects of de-repressed bacterial sex factors on colicinogeny and cell wall structure. Molec. Gen. Genet., 109:57, 1970.
15. Tomoeda, M., Inuzuka, M., Kubo, N., and Nakamura, S.: Effective elimination of drug resistance and sex factors in Escherichia coli by sodium dodecyl sulfate. J. Bact., 95:1078, 1968.
16. Inuzuka, N., Nakamura, S., Inuzuka, M., and Tomoeda, M.: Specific action of sodium dodecyl sulfate on the sex factor of Escherichia coli K-12 Hfr strains. J. Bact., 100:827, 1969.
17. Adachi, H., Nakano, M., Inuzuka, M., and Tomoeda, M.: Specific role of sex pili in the effective eliminatory action of sodium dodecyl sulfate on sex and drug resistance factors in Escherichia coli. J. Bact., 109:1114, 1972.

9. DISSOCIATION AND REASSOCIATION OF THE TRANSFER FACTOR AND RESISTANCE DETERMINANTS OF R FACTORS AS A MECHANISM OF GENE AMPLIFICATION IN BACTERIA[1]

R. Rownd, D. Perlman, H. Hashimoto,[2] S. Mickel,[3] E. Applebaum and D. Taylor

Laboratory of Molecular Biology and
Department of Biochemistry
University of Wisconsin
Madison, Wisconsin 53706

INTRODUCTION

Drug-resistance factors (R factors) are a class of extrachromosomal genetic elements in bacteria which confer resistance to a variety of antibiotics. Genetic (1 - 4) and physical (5 - 10) analyses of R factors have shown that these agents are composed of two distinguishable units: a resistance transfer factor (RTF) and a unit referred to as r-determinants. RTF mediates the infectious transfer of multiple-drug resistance among bacteria through the synthesis of sex pili which are involved in the process of bacterial mating; r-determinants carry drug-resistance genes which direct the synthesis of a number of enzymes which specifically inactivate the antibiotics to which the R factors confer resistance. In Escherichia coli and Serratia marcescens RTF and r-determinants appear to be stably associated in the form of a composite structure (an R factor), although segregant R factors which have lost specific combinations of drug-resistance genes are observed at low frequency (6,9,11). In Proteus mirabilis there is considerable evidence that RTF and r-determinants dissociate and reassociate in such a way as to regulate the number of copies of r-determinants (drug-resistance genes) per host cell (5,6,10,12). Most of the studies on this mechanism of gene amplification in P. mirabilis have been carried out using the R factor NR1. This R factor confers resistance to chloramphenicol (CM), streptomycin/spectinomycin (SM/SP), sulphonamide (SA), and tetracycline (TC). At rapid growth rates there are multiple copies of NR1 per cell in P. mirabilis which are selected at random for replication from a multicopy pool during the bacterial division cycle (6,13,14). NR1 replication continues for several hours in stationary phase after host chromosome replication has ceased (5,6,13-15).

[1] This work was supported by U.S. Public Health Service research grant GM 14398 and by Research Career Development Award GM 19206 (to R.R.) from the National Institute of General Medical Sciences. D. Perlman, S. Mickel, E. Applebaum, and D. Taylor were supported by U.S. Public Health Service Training Grants to the Laboratory of Molecular Biology and to the Department of Biochemistry at the University of Wisconsin.

[2] Present Address: Department of Microbiology, Gunma University School of Medicine, Maebashi, Japan.

[3] Present Address: Department of Pathology, University of Colorado Medical Center, Denver, Colorado 80220.

VARIABILITY OF THE DENSITY PROFILE OF NR1 DNA IN P. MIRABILIS

In many of the studies on episomes and plasmids in Enterobacteriaceae these extrachromosomal genetic elements have been transferred to host strains having a chromosomal DNA base composition which is different than that of the extrachromosomal DNA. In this situation the extrachromosomal DNA appears as a satellite DNA band to the chromosomal DNA in a CsCl density gradient, since the buoyant density of DNA is a linear function of its base composition. P. mirabilis has been most commonly employed as the host strain in our studies on R factors. Its base composition (40% G+C) is significantly different from the base composition of all R factors which have been examined (10).

In P. mirabilis the nature of the density profile of NR1 DNA in a CsCl gradient depends on the conditions under which the host cells are cultured (5,6,12,13). When R^+ P. mirabilis is cultured in drug-free medium for a long period, NR1 DNA forms a single satellite band of density 1.712 g/ml. The ratio of R factor to chromosome DNA is about 8% of the P. mirabilis chromosomal DNA (1.700 g/ml) in stationary phase cultures (Fig 1A). After prolonged growth in medium containing any of the drugs to which NR1 confers resistance (with the exception of TC), a much larger satellite band of density 1.718 g/ml is observed in the NR1 DNA density profile, which is usually markedly skewed toward the less dense side (Fig 1C). As illustrated in Figure 1, these two types of density profiles are interconvertible, depending on the

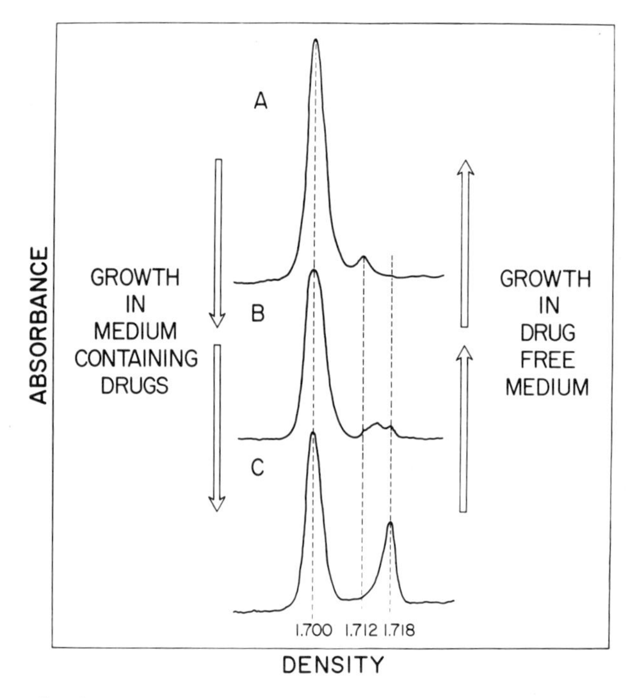

Fig 1. Systematic changes in DNA density profiles of the R factor NR1 in P. mirabilis, depending on whether the host cells are cultured in drug-free medium or medium containing appropriate drugs. The band of density 1.700 g/ml is P. mirabilis chromosomal DNA. More detailed analyses of the changes in the NR1 DNA density profiles during the transition and the back transition have been presented elsewhere (5,6,12).

conditions of cell culture. A broad and diffuse intermediate density band (Fig 1B) has been observed on the transition between these two states. This intermediate density NR1 DNA appears to consist of a collection of molecules having a broad spectrum of density between 1.712 and 1.718 g/ml. In the following discussion, this replacement of the 1.712 g/ml NR1 DNA band by the 1.718 g/ml NR1 DNA band, which results from growth of the bacteria in medium containing appropriate drugs, will be referred to as the "transition." The reversal of this process during growth in a drug-free medium will be referred to as the "back transition."

Many of our experiments on the transition in R^+ P. mirabilis have been carried out at drug concentrations which permit cell growth in the presence of the antibiotic, although at a slower rate than in a drug-free medium. These experiments have shown that with the exception of TC all of the other antibiotics to which NR1 confers resistance can cause the transition when included in the medium at sufficiently high concentrations. In general, the extent of the conversion of the R factor DNA to the 1.718 g/ml form is greater when the cells are cultured in medium containing a higher concentration of antibiotics or after a larger number of cell doublings in medium containing a given concentration of antibiotics. Thus, the greater the degree of "drug challenge," the greater the extent of the transition. As will be discussed below, transitioned cells have considerably higher levels of resistance to drugs which cause the transition (minimal inhibitory concentrations and specific activities of the drug-inactivating enzymes) than nontransitioned cells.

The back transition appears to occur in two phases. There is an initial 5- to 10-fold reduction in the area of the 1.718 g/ml NR1 DNA band within about ten generations of growth in a drug-free medium which is accompanied by the appearance of a broad intermediate density NR1 DNA band. Thereafter, the shift in the distribution of NR1 DNA to the 1.712 g/ml form occurs much more slowly, usually requiring over 200 cell doublings. There is no difference in the growth rate of transitioned and nontransitioned cells in a drug-free medium, so that the changes in the density profile of the NR1 DNA do not appear to be due to the selective outgrowth of nontransitioned cells in the population (16).

MOLECULAR STATE OF R FACTORS IN P. MIRABILIS

These systematic changes in the density profile of NR1 DNA in P. mirabilis appear to reflect the dissociation and reassociation of the RTF and r-determinants of the R factor under different growth conditions, as illustrated schematically in Figure 2 (5,6,12,17). There is now evidence that several different R factors can dissociate to varying degrees into these two elements when harbored by P. mirabilis (5 - 9,12). Both RTF and r-determinants appear to be capable of autonomous replication in the dissociated state in this host. The TC-resistance genes of the R factor NR1 appear to reside on the RTF component (RTF-TC) which has a density of 1.711 g/ml. The other drug-resistance genes (CM, SM/SP, and SA) reside on r-determinants (1.718 g/ml). When an RTF-TC and r-determinants are united, the composite structure appears to replicate under the control of the RTF-TC replication system. There is also considerable evidence that molecules consisting of repeated tandem sequences of r-determinants are formed in P. mirabilis that can be incorporated into individual R factors to form polygenic molecules harboring repeated sequences of r-determinants as illustrated in

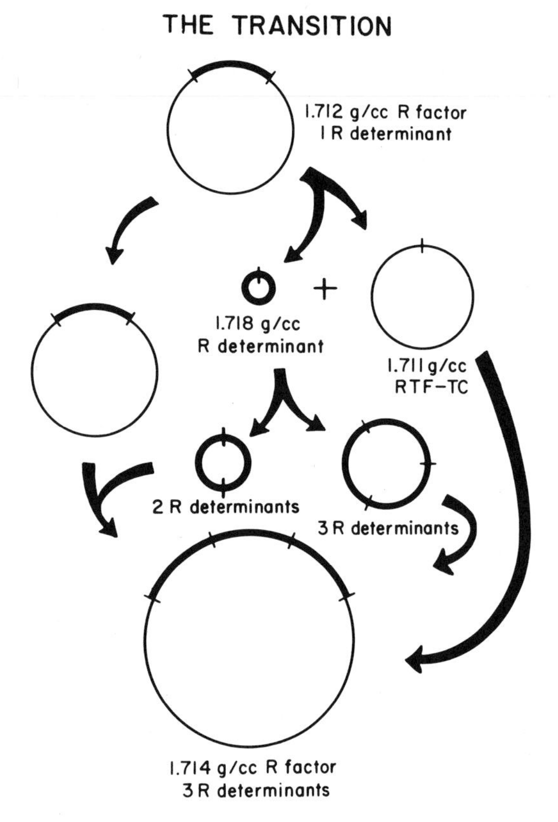

Fig 2. Schematic diagram illustrating the dissociation and the reassociation of RTF-TC and r-determinants in P. mirabilis and the density increase that accompanies the recombination of multiple copies of r-determinants into R factors. After the recombination of a large number of r-determinants, the R factor DNA would have essentially the same density as r-determinant DNA (1.718 g/ml), since most of the mass of the DNA is due to r-determinants.

Figure 2. The extra copies of r-determinants required for this formation of tandem sequences are provided by the autonomous replication of r-determinants in the dissociated state.

These observations explain the different density profiles which have been observed for NR1 DNA in P. mirabilis, as illustrated in Figure 1. During growth in a drug-free medium the majority of R factors in the cell population have only one r-determinant, so that NR1 DNA appears as a single satellite band of density 1.712 g/ml. Since only a minority of the R factors are in the dissociated state (17), the amount of r-determinant DNA (1.718 g/ml) is too low in proportion to register as a detectable satellite band in the NR1 DNA density profile. During growth of R^+ P. mirabilis in medium containing drugs to which r-determinants confer resistance, there is selection for cells harboring the largest number of copies of r-determinants; these cells grow more rapidly, since they would have the highest level of drug resistance. This state of affairs can be achieved by the incorporation of multiple copies of r-determinants into individual R factors. The addition of extra copies of r-determinants (1.718 g/ml) to an RTF-TC (1.711 g/ml) increases the density of R factor DNA. R factors containing only a few copies of r-determinants have an intermediate density and give rise to the broad intermediate density band seen in Figure 1B. The density of R factors containing a large number of copies of r-determinants approaches a limiting value of 1.718 g/ml, since most of the mass of the R factor DNA is due to r-determinants. The proportion of the NR1 DNA in this state is considerably increased relative to the 1.712 g/ml NR1 DNA band present in cells grown in a drug-free medium, since the incorporation of additional copies of r-determinants into R factors would increase their size. This state would be achieved after prolonged growth in media containing appropriate drugs (Fig 1C).

The shift in the density of NR1 DNA from the 1.718 g/ml form to the 1.712 g/ml form appears to be essentially the reverse of the process just described. During growth of a transitioned culture of R^+ P. mirabilis in a drug-free medium, r-determinants are dissociated from R factors harboring repeated tandem sequences of these elements, resulting in the formation of less dense and smaller R factors. This results in a shift in the density distribution of the NR1 DNA toward a limiting value of 1.712 g/ml. Since there is no difference in growth rates between transitioned and nontransitioned cells (16), the decrease in the proportion of the 1.718 g/ml band which results from growth in a drug-free medium indicates that r-determinants are diluted from the cells after dissociation. This finding suggests that autonomous r-determinants undergo little or no replication in P. mirabilis during balanced exponential growth. Any extra copies of r-determinants due to dissociation are therefore diluted passively by cell division.

The inital rapid decrease in the proportion of the 1.718 g/ml NR1 DNA band during the growth of a transitioned culture in drug free medium could reflect an instability of R factor molecules containing a large number of tandem sequences of r-determinants. Such molecules must contain a large degree of internal genetic homology due to the repeated sequences of r-determinants and, therefore, would be expected to dissociate rapidly. The intermediate density NR1 DNA contains only a few copies of r-determinants and the dissociation of r-determinants from these structures would be expected to occur more slowly.

STRUCTURE OF R FACTOR DNA IN P. MIRABILIS

A number of the experiments described above on the molecular state of R factors in P. mirabilis led to definite predictions about the structure of NR1 DNA molecules in the nontransitioned and transitioned states. Our investigations on the structure of NR1 DNA molecules in P. mirabilis, using sedimentation in sucrose gradients and electron microscopy, have yielded results which are consistent with these expectations (17). The major fraction of the 1.712 g/ml form of NR1 DNA (Fig 1A) has been found to be circular molecules of molecular weight 75 X 106. These molecules have a contour length of 37 microns, which corresponds to a composite structure consisting of an RTF-TC (29 – 30 microns) and a single copy of r-determinants (8 microns). In a number of experiments a fraction of the NR1 DNA has been found to be linear molecules having a length of 37 microns. It is not yet known whether these linear structures actually exist in situ or whether they are artifacts produced during DNA isolation. These findings, that the major fraction of the 1.712 g/ml form of NR1 DNA consists of an RTF-TC plus a single copy of r-determinants, have caused us to revise our earlier suggestion that R factor dissociation in P. mirabilis is extensive during growth in a drug-free medium (5,6,12). It now appears that there is a low frequency of dissociation of NR1 into an RTF-TC and r-determinants in P. mirabilis during balanced growth in a drug-free medium.

Our characterization of the 1.718 g/ml form of NR1 DNA (Fig 1C) isolated from transitioned P. mirabilis cells has shown that the major fraction of this DNA is linear in structure, heterogeneous in size, and of molecular weight similar to P. mirabilis chromosomal DNA which has been isolated under the same conditions (ca. 100 to 150 times 10^6). Since the linear chromosomal DNA most likely results from the fragmentation of the bacterial chromosome, these findings suggest that the large, linear 1.718 g/ml NR1 DNA molecules are probably breakage products of larger circular R factor molecules. The observed large size of 1.718 g/ml NR1 DNA is not unexpected, since our enzymological data indicate that as many as 20 copies of r-determinants may be incorporated into the R factor NR1 during growth in media containing appropriate drugs. The resulting polygenic R factors would have molecular weights in the range of 200 to 400 million and would be subject to shear breakage during isolation and handling, just as is the P. mirabilis chromosomal DNA. While this interpretation lacks definite proof, the larger size of NR1 DNA in transitioned cultures is consistent with the increase in size expected from the incorporation of multiple copies of r-determinants into R factors.

Because of the problems associated with the breakage of 1.718 g/ml NR1 DNA, we have focused our attention on the intermediate density NR1 DNA (Fig 1B). These molecules should be smaller in size, since they would contain fewer copies of r-determinants. Using electron microscopy, it has been possible to identify large circular NR1 molecules (that therefore cannot be breakage products) that fit the size distribution expected for molecules consisting of an RTF-TC + *n* r-determinants, where *n* is an integer having values as high as six. Thus these experiments clearly show that multiple copies of r-determinants can be incorporated into R factors in P. mirabilis, which results in an increase in the density of the NR1 DNA. Circular molecules consisting of multimeric sequences of r-determinants themselves have also been observed in the 1.718 g/ml region of the NR1 DNA density profile, using electron microscopy.

LEVEL OF DRUG RESISTANCE OF TRANSITIONED CELLS

As mentioned previously, the proportion of the 1.718 g/ml NR1 DNA band in transition experiments with R^+ P. mirabilis depends on both the drug concentration in the growth medium and the number of generations of growth in media containing drug. According to our previous discussion, this variation should be due to a variation in the average number of copies of r-determinants contained in R factors. If this interpretation is correct, there should be a corresponding variation in the specific activity of r-determinant gene products. There is a linear gene dosage relationship between the percentage of the 1.718 g/ml NR1 DNA band and the specific activity of chloramphenicol acetyltransferase (5,6), the enzyme which inactivates CM (18,19).

The higher level of resistance of transitioned cells to drugs that cause the transition when included in the growth medium is also manifest in the ability of individual cells to form colonies on plates containing appropriate antibiotics (16). As shown in Table I, nontransitioned cells do not form colonies on Penassay-agar plates

TABLE I

Level of Drug Resistance of Nontransitioned and Transitioned Proteus mirabilis ØS-3 NR1

State of transition	Growth of single cells of diluted culture on penassay agar plates containing														
	No drugs	CM (μg/ml)				SM (μg/ml)						TC (μg/ml)			
		50	100	200	400	12.5	25	50	100	200	400	12.5	25	50	100
Not transitioned	77	81	79	-	-	82	-	-	-	-	-	49	22	-	-
Transitioned with CM	144	137	133	127	97	138	140	97	32	-	-	125	121	-	-
Transitioned with SM	66	57	90	80	77	62	73	78	63	37	-	76	9	-	-

Note: Nontransitioned cells and cells which had been transitioned using either CM or SM were appropriately diluted and inoculated on Penassay agar plates containing either no drugs or various concentrations of CM, SM, or TC. The ability of single cells to form clones on plates containing the drugs can be ascertained by comparing the number of colonies formed on the drug-free plates with the number of colonies formed on the plates containing the different concentrations of CM, SM, or TC.

containing more than 100 μg/ml CM or 6.25 μg/ml SM. Transitioned cells, however, have much higher levels of resistance to these antibiotics (Table I). It should be noted that transitioned cells produced by growth in medium containing either CM or SM acquire an increased level of resistance to both antibiotics simultaneously. This indicates that both the CM and the SM drug resistance genes of the R factor NR1 reside on the same physical unit (r-determinants) which is amplified during growth in medium containing appropriate drugs.

The level of resistance of both nontransitioned and transitioned cells to TC is essentially the same (Table I). As mentioned previously, growth of R^+ P. mirabilis in medium containing TC does not result in the transition to the 1.718 g/ml form of NR1 DNA. These findings suggest that the TC resistance genes of NR1 reside on the RTF component of the R factor (16). According to the previous discussion, there is no increase in the number of copies of any of the genes which reside on RTF due to the transition.

MECHANISM OF THE TRANSITION

Our investigations on the mechanism of the transition have revealed two possible ways of accounting for the replacement of the 1.712 g/ml NR1 DNA band by the 1.718 g/ml band during growth in media containing appropriate drugs.

As shown in Table I, the majority of nontransitioned cells are not able to form colonies on plates containing greater than 6.25 μg/ml SM. However, 10^{-2} to 10^{-3} of the cells in a nontransitioned population do form colonies on plates containing

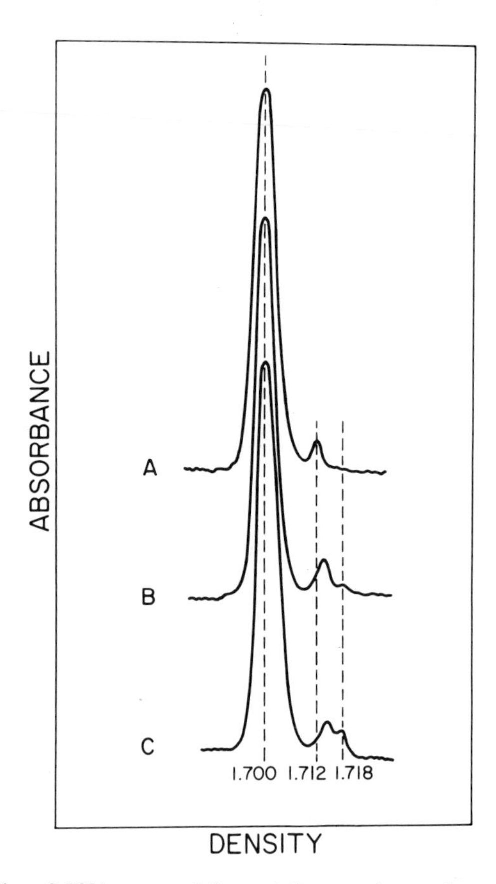

Fig 3. Density profiles of DNA prepared from stationary phase cultures of nontransitioned and partially transitioned cells of P. mirabilis harboring the R factor NR1. A culture of nontransitioned cells in drug-free Penassay broth was plated on drug-free plates. Spontaneously occurring partially transitioned cells in the nontransitioned population were isolated by replica plating onto plates containing 25 μg/ml SM. DNA was prepared from stationary phase cultures of the cells in drug-free Penassay broth.

A: density profile of DNA prepared from the original culture of nontransitioned cells.

B and C: density profiles of DNA prepared from two different partially transitioned clones isolated by replica plating.

25 μg/ml SM. This suggested that a subpopulation of transitioned cells might exist in a nontransitioned population, even when the cells are cultured in a drug-free medium. Using replica plating it has been possible to isolate clones from a nontransitioned population which are resistant to 25 μg/ml SM without previous exposure of the cells to drugs. Whereas the DNA prepared from the nontransitioned population showed only the 1.712 g/ml band characteristic of nontransitioned cells (Fig 3A), the DNA prepared from the clones resistant to 25 μg/ml SM has a higher density, even though the cells had been cultured in a drug-free medium (Figs 3B and 3C). It should be noted that most of the NR1 DNA in the density profiles shown in Figures 3B and 3C is of intermediate density between 1.712 and 1.718 g/ml and its proportion is considerably smaller than the 1.718 g/ml NR1 DNA observed after prolonged growth in media containing appropriate drugs. Thus these cells are most properly described as "partially transitioned" and it would appear that they can be converted to a more

highly transitioned state by growth in a medium containing drugs. By mixing partially transitioned cells isolated by replica plating, as described above, with nontransitioned cells it is readily shown that the former outgrow the latter in a medium containing 100 μg/ml CM. These findings suggest that the changes in the NR1 DNA density profile during growth in a medium containing lower concentration of CM (25 to 100 μg/ml) are due primarily to the more rapid growth of preexisting partially transitioned cells, which are thus enriched in the cell population, and the further amplification of the number of copies of r-determinants contained in individual R factors (16).

At CM concentrations which are high enough to prevent the growth of nontransitioned cells (250 μg/ml), the transition appears to occur by a different mechanism (20). Under these conditions nontransitioned cells undergo a growth lag for 20 to 30 hours and then begin to grow (Fig 4). During this lag period the CM in the medium is inactivated at a rate of approximately 5 μg/ml/hour/10^8 cells. Just about the time growth of the culture resumes, a 1.718 g/ml band appears in the NR1 DNA density profiles which increases in proportion at a rapid rate during the first few hours of outgrowth. In Figure 4 the letters which identify the NR1 DNA density profiles

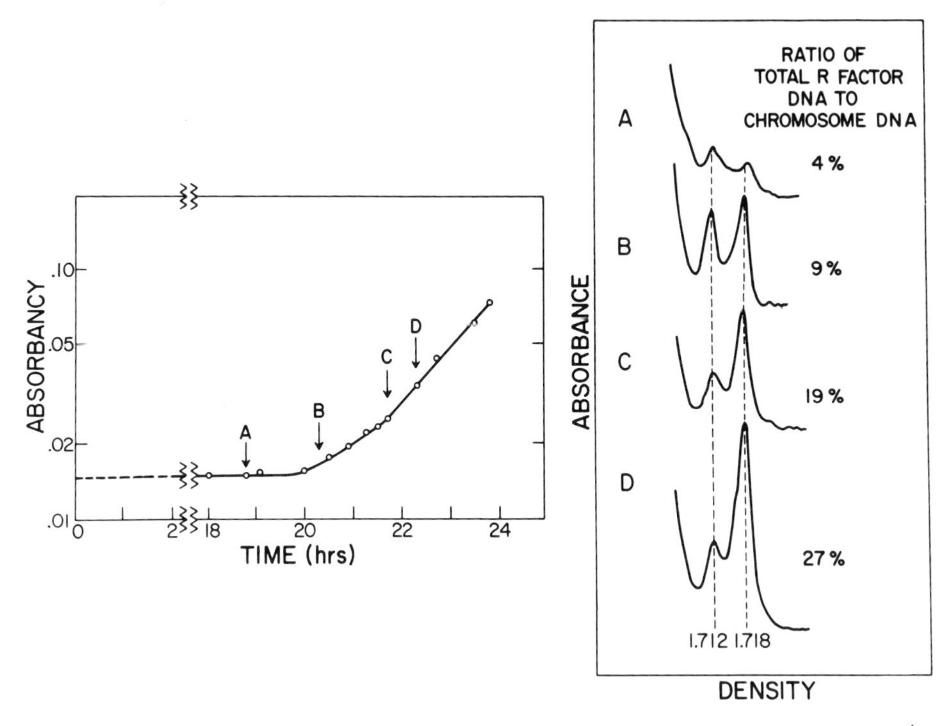

Fig 4. Forced transition of the R factor NR1 in P. mirabilis. A nontransitioned culture of R^+ P. mirabilis in late exponential phase in Penassay broth was diluted one-hundred fold into Penassay broth containing 250 μg/ml CM. Growth of the culture was monitored by absorbancy measurements at 650 mμ. DNA was prepared from samples of the cells harvested at the times indicated on the growth curve at the left. The letters identifying the R factor DNA density profiles shown on the right correspond to the times at which the cells were harvested during the growth curve shown at the left. The P. mirabilis chromosomal DNA density profiles are not included in this figure in order to emphasize the R factor DNA distribution in the density gradient. The ratio of total R factor DNA to the chromosome DNA for each sample is shown on the right.

shown on the right correspond to the time at which the cell samples were taken for DNA isolation during the growth curve shown on the left. Electron microscopy and sucrose gradient sedimentation have shown that the 1.718 g/ml DNA in the profile shown in Figure 4A consists of monomers, dimer, trimers, tetramers, and higher multimetric forms of r-determinant DNA. At longer times after outgrowth, the 1.718 g/ml DNA becomes so large in size that it is impossible to isolate it without breakage for the reasons discussed previously. Eventually, the 1.712 g/ml band disappears from the density profiles. Preparative CsCl density gradient experiments in which the original 1.712 g/ml NR1 DNA has been radioactively labeled indicate that it has recombined with the r-determinant DNA to form R factors harboring repeated tandem sequences of r-determinants. Thus, at higher drug concentrations it appears that there is a rapid amplification of autonomous r-determinants following the 20 to 30 hour lag period. This results in a very rapid "forced transition" during the outgrowth period. Under these conditions almost all of the original cells are converted to the transitioned state. These and other experiments have suggested that subjecting the cells to conditions of stress preconditions them to undergo the forced transition during the outgrowth period. This phenomenon of R factor dissociation and r-determinant amplification is a possible mechanism by which a small proportion of partially transitioned cells arise in a

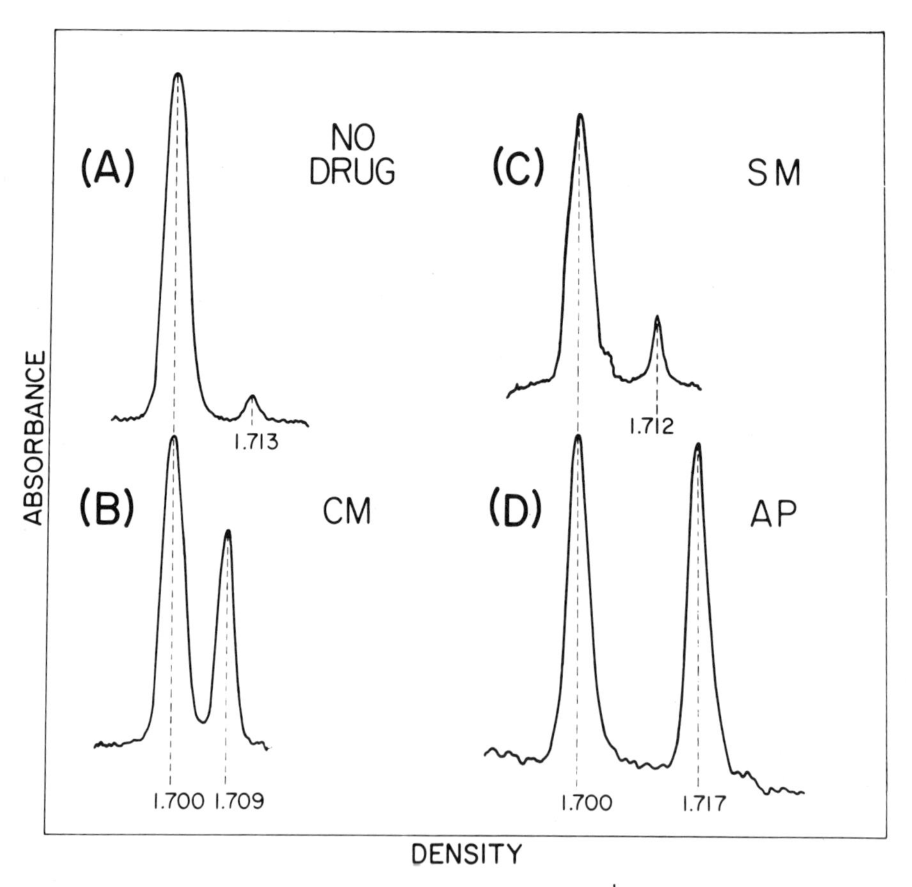

Fig 5. Transition of the R factor NR84 during growth of R^+ P. mirabilis in Penassay broth containing either no drugs (*A*), CM (*B*), SM (*C*), or AP (*D*).

nontransitioned population. Even if a cell population is never exposed to an antibiotic, occasional cells may undergo a "physiological stress" which results in a forced transition response.

OTHER R FACTORS ALSO UNDERGO THE TRANSITION IN P. MIRABILIS

A number of other R factors have been found to undergo the transition in P. mirabilis during growth in a medium containing appropriate drugs. The behavior of the R factor NR84 which confers resistance to ampicillin (AP), CM, SM/SP, SA, and TC is of particular interest, since the nature of the NR84 DNA density change depends on the drug added to the growth medium (Fig 5) (21). When P. mirabilis harboring NR84 is cultured in a drug-free medium, NR84 DNA appears as a satellite band of density 1.712 g/ml, which is about 5% of the chromosomal DNA (Fig 5A). Growth in medium containing CM usually results in a decrease in the density of NR84 DNA to values as low as 1.709 g/ml and an increase in its proportion (up to 65% of the chromosomal DNA) (Fig 5B). Growth in a medium containing SM does not result in a change in the density of NR84 DNA, but in some experiments its proportion has increased to as much as 20% of the chromosomal DNA (Fig 5C). Growth in a medium containing AP results in an increase in the density of NR84 DNA to 1.717 g/ml and a very large increase in its proportion (up to 100% of the chromosomal DNA) (Fig 5D). Growth in TC has no apparent effect on the NR84 DNA density profile. It is interesting to note that growth of P. mirabilis harboring NR84 in medium containing CM or SM increases the resistance of the cells to only the antibiotic which was added to the growth medium. However, growth in AP results in a simultaneous increase in the level of resistance to CM, SM, and AP.

These findings may be explained by a model similar to the mechanism illustrated in Figure 2, except that it is neccssary to postulate that the CM, SM, and AP-resistance genes all reside on separate r-determinants, each of which is capable of autonomous replication. The TC-resistance genes of NR84 may reside on the RTF, as in the case of the R factor NR1. Although we have not as yet carried out experiments on the molecular structure of NR84 DNA in the nontransitioned and the transitioned states, these preliminary experiments suggest that the composite structure of NR84 may consist of as many as four independent replicons.

R FACTORS DO NOT UNDERGO THE TRANSITION IN ESCHERICHIA COLI AND SERRATIA MARCESCENS

Our studies using strains of E. coli and S. marcescens harboring NR1 have shown that there is no change in the density of NR1 DNA during growth in a medium containing CM (6). In experiments similar to those shown in Table I it has been found that there is no increase in the level of resistance of these strains to CM, SM, SP, or TC which results from prolonged growth in medium containing CM (22). Our available evidence suggests that NR1 exists as a composite structure in E. coli and S. marcescens and that RTF-TC and r-determinants do not dissociate and reassociate in these genera as they do in P. mirabilis. This indicates that the host plays an important role in determining whether R factors are able to undergo the transition. In future studies it should be of interest to examine whether mutants of E. coli and S. marcescens can be isolated in which this mechanism of gene amplification does occur.

CONCLUSIONS

The central theme of this symposium has been the ability of exogenous nucleic acids to change the genetic, biochemical, and other functional characteristics of cells. A variety of experimental systems have been described which amply demonstrate how both DNA and RNA can function in this capacity in both prokaryotic and eukaryotic cells. Our own experiments using the bacterium P. mirabilis have shown how the dissociation and the reassociation of the components of one of the classes of extrachromosomal elements in bacteria, drug-resistance factors or R factors, can provide a novel means by which the number of copies of drug-resistance genes (r-determinants) per cell can be regulated. E. coli and S. marcescens, on the other hand, are apparently unable to increase their complement of drug-resistance genes by the dissociation and reassociation of RTF-TC and r-determinants under different growth conditions. In all three bacterial species the number of copies of R factors per cell can also be increased approximately five-fold by the continued replication of R factors in the stationary growth phase, which is accompanied by a corresponding increase in the specific activity of some R factor gene products. In P. mirabilis the specific activity of chloramphenicol acetyltransferase of R factors may be varied over a hundred-fold range as a result of the gene amplification mediated by the dissociation and reassociation of RTF-TC and r-determinants and the replication of R factors in stationary phase (6). Presumably these variations reflect a comparable difference in the level of R factor messenger RNA in nontransitioned and transitioned P. mirabilis cells.

The experiments which we have described for the dissociation and reassociation of the transfer factor and r-determinants of the R factor NR1 have been described in greater detail elsewhere and we will make only several general comments on our experiments at this time.

Over the past dozen years there has been a dramatic increase in the incidence of pathogenic enteric strains which harbor R factors (3,4,23 - 25). This increase almost certainly has its origin in the widespread administration of antibiotics in animal husbandry and in the chemotherapeutic treatment of infection in human and animal subjects and is clearly a problem of considerable medical importance. While our studies are not directly concerned with the clinical aspects of R factors, they do illustrate a potential mechanism by which R^+ strains can acquire much higher levels of resistance to antibiotics. A number of R factors which have been examined are able to amplify the number of copies of r-determinants in P. mirabilis (26). While R factors do not appear to undergo the transition in all species of the Enterobacteriaceae, it is possible that either a bacterial host or an R factor may mutate to a state in which they do. Should this occur, the effectiveness of chemotherapy in the treatment of multiple drug-resistant pathogens would become even more limited.

In recent years it has become evident that a substantial fraction of the DNA of most higher eukaryotes contains highly repetitive DNA sequences (27). The biological function of these repeated DNA sequences is presently not known, although their widespread occurrence is suggestive of a functional role which is of general significance to cellular growth and development. In the case of the genes coding for ribosomal RNA in higher eukaryotes, there is definitive evidence that these genes are present in a multicopy state situated in tandem sequences on one or more chromosomes (28 - 30). Moreover, there is evidence that ribosomal RNA genes can undergo a process of amplification during oocyte maturation in many eukaryotes (31,32). While these systems of gene amplification and the formation of repeated gene sequences are entirely different phenomena than the amplification of r-determinants in our experiments, the dissociation and reassociation of the transfer factor and r-determinants of R factors should provide an interesting and comparatively simple model system for studying mechanisms of gene amplification and the formation of tandemly repeated gene sequences in cells, as well as the reversibility of these processes.

R factors also represent an interesting model system for the study of DNA replication and its control. Many R factors are composed of more than one replicon or units capable of autonomous replication. When these independent replicons are associated to form a composite structure, it seems most likely that replication is controlled by only one of them. For example, our present evidence indicates that R factors replicate under the control of the RTF-TC replication system when in the composite form, even when the R factors contain multiple tandem sequences of r-determinants. In many ways the association of an r-determinant with an RTF to form an R factor is analogous to the integration of a temperate bacteriophage into a bacterial chromosome. In both situations one replicon becomes subordinate to another, and it is an interesting question whether related processes are involved in these different systems. R factors such as NR84 may consist of up to four different replicons which may replicate under the control of only one of them when in the composite form. It will be of considerable interest to elucidate the mechanisms involved in this control as well as the factors which result in the dissociation and subsequent amplification of particular r-determinants under the different growth conditions. Clearly, these events can lead to dramatic changes in the genetic, biochemical, and physiological characteristics of cells.

REFERENCES

1. Watanabe, T.: Infective heredity of multiple drug resistance in bacteria. Bacteriol. Rev., 27:87, 1963.
2. Watanabe, T.: The concept of R factors. *In*: Symposium on Infectious Multiple Drug Resistance. Edited by S. Falkow. U.S. Govt. Printing Office, Washington, D.C., p 5, 1967.
3. Mitsuhashi, S.: Transferable Drug Resistance Factor R. Univ. of Tokyo Press. Tokyo, 1971.
4. Anderson, E. S.: The ecology of transferable drug resistance in the enterobacteriaceae. Ann. Rev. Microbiol., 22:131, 1968.
5. Rownd, R., Watanabe, H., Mickel, S., Nakaya, R., and Gargan, B.: The molecular nature and the control of the replication of R-factors. *In*: Progress in Antimicrobial and Anticancer Chermotherapy, vol. 2. University of Tokyo Press, p 535, 1970.
6. Rownd, R., Kasamatsu, H., and Mickel, S.: The molecular nature and replication of drug resistance factors of the Enterobacteriaceae. Ann. N.Y. Acad. Sci., 182:188, 1971.
7. Falkow, S., Haapala, D. K., and Silver, R. P.: Relationships between extrachromosomal elements. *In*: Ciba Foundation Symposium on Bacterial Episomes and Plasmids. Edited by G. E. Wolstenholme and M. O'Connor. Little, Brown, Boston, p 136, 1969.
8. Cohen, S., and Miller, C.: Non-chromosomal antibiotic resistance in bacteria. II. Molecular nature of R-factors isolated from Proteus mirabilis and Escherichia coli. J. Mol. Biol., 50:671, 1970.

9. Cohen, S. N., Silver, R. P., Sharp, P. A., and McCoubrey, A. E.: Studies on the molecular nature of R factors. Ann. N.Y. Acad. Sci., 182:172, 1971.
10. Davies, J. E., and Rownd, R.: Transmissible multiple drug resistance in Enterobacteriaceae. Science, 176:758, 1972.
11. Kontomichalou, P., Mitani, M., and Clowes, R.: Circular R factor molecules controlling penicillinase synthesis, replicating in *Escherichia coli* under either relaxed or stringent control. J. Bacteriol., 104:34, 1970.
12. Rownd, R., and Mickel, S.: Dissociation and reassociation of RTF and r-determinants of the R factor NR1 in Proteus mirabilis. Nature New Biology, 234:40, 1971.
13. Rownd, R.: The molecular nature and the control of the replication of R factors. *In:* Symposium on Infectious Multiple Drug Resistance. Edited by S. Falkow. U.S. Govt. Printing Office, Washington, D.C., p 17, 1967.
14. Rownd, R.: Replication of a bacterial episome under relaxed control. J. Mol. Biol., 44:387, 1969.
15. Kasamatsu, H., and Rownd, R.: Replication of R factors in Proteus mirabilis: replication under relaxed control. J. Mol. Biol., 51:473, 1970.
16. Hashimoto, H., and Rownd, R.: Increased level of drug resistance due to the incorporation of multiple copies of r-determinants into the R factor NR1 in Proteus mirabilis. Submitted to J. Bact.
17. Perlman, D., Mickel, S., Kasamatsu, H., and Rownd, R.: Structure of R factor NR1 DNA in Proteus mirabilis. Manuscript in preparation.
18. Suzuki, Y., and Okamoto, S.: The enzymatic acetylation of chloramphenicol by the multiple drug-resistant Escherichia coli carrying R factor. J. Biol. Chem., 242:4722, 1967.
19. Shaw, W. V.: The enzymatic acetylation of chloramphenicol by extracts of R factor resistant Escherichia coli. J. Biol. Chem., 242:687, 1967.
20. Perlman, D., and Rownd, R. Forced transition of the R factor NR1 in Proteus mirabilis. Manuscript in preparation.
21. Applebaum, E. A., Nakaya, R., and Rownd, R.: Unpublished results.
22. Taylor, D. P., and Rownd, R.: Unpublished results.
23. Datta, N.: Infectious drug resistance. Brit. Med. Bull., 21:254, 1965.
24. Lebek, G.: Ueber die Entstehung mehrfachresistenter Salmonellen. Ein Experimenteller Bietrag. Zentr. Bakteriol. Parasitenk., Abt. 1, 188:494, 1963.
25. Smith, D. H., and Armour, S. E.: Transferable R factors in enteric bacteria causing infection of the genitourinary tract. Lancet, 7453:15, 1966.
26. Nakaya, R., and Rownd, R.: Unpublished results.
27. Britten, R. J., and Kohne, D. E.: Repeated sequences in DNA. Science, 161:529, 1968.
28. Wallace, H., and Birnstiel, M. L.: Ribosomal cistrons and the nucleolar organiser. Biochim. Biophys. Acta, 114:296, 1966.
29. Brown, D. D., and Weber, C. S.: Gene linkage by RNA-DNA hybridization. I. Unique DNA sequences homologous to 4 S RNA, 5 S RNA, and ribosomal RNA. J. Mol. Biol., 34:661, 1968.
30. Wensink, P. C., and Brown, D. D.: Denaturation map of the ribosomal DNA of Xenopus laevis. J. Mol. Biol., 60:235, 1971.
31. Brown, D. D., and Dawid, I. B.: Specific gene amplification in oocytes. Science, 160:272, 1968.
32. Brown, D. D., and Dawid, I. B.: Developmental genetics. Ann. Rev. Genetics, 3:127, 1969.

10. DISCUSSION

Dr. Merril:
(Bethesda, Md.)

The contaminating phage that is present in fetal calf serum is a serious problem and something that everyone who is concerned with tissue culture will have to deal with in his own way.

I don't think the experimental results we obtained with the lambda P gau can be explained by contamination. We would grow a flask of cells which would then be split into three parts. All the subsequent flasks were fed with the same media. One flask was infected with lambda P gau, one was infected with lambda P gau with the point mutation, the transferase genes, and one flask was not infected at all. This is a typical type of experiment.

So the only effect that I can see that contaminating viruses might have had is their interference with the experiments. They might cut down the amount of material produced as many viruses do in E. coli, or they might act as helpers in some occasions and amply the results. But they in and of themselves cannot explain the results.

Furthermore, every culture that was grown was tested for bacterial and micoplasma contamination and, of course, all the experiments that were reported were negative for these contaminants.

Dr. Beers:

Is it possible that these phages are the classical virus looking for a disease?

Dr. Merril:

That is a possibility, I think. I think it is something that will eventually have to be looked into.

At this point, we don't really know what the origin of these phages is. You know, it would seem to be rather simple to just go out and get fetal calf serum on one's own, but it turns out that it is not so easy. There is another system that is present in serum from cows and other animals. That is complement plus antibody which will lyse bacteria.

Fetal calves supposedly have about 1/20th the amount of complement present in an adult cow. Most of these cows are being grown for slaughter and the serum of the fetal calf is just a by-product.

Many of the animals that are slaughtered today are put in feeding pens and given steroids. It has been shown, not on cows but on mice, that steroids very often will knock out the amount of C-6 factor of complement required for bacteria lysis.

Dr. Gillespie: (Baltimore, Md.)
Dr. Beljanski, have you tried essentially to determine whether you have a product primer complex or a product template complex with in vitro action?

Dr. Beljanski: (Paris, France)
This is just the first part of our work. We don't have more data and we are just starting to work with this problem so I don't know.

Dr. Gillespie:
Have you looked for an RNA ligase in the mutants?

Dr. Beljanski:
That is a very interesting question and I would like very much to do this kind of experiment, but I am in an institute in which when I need something this is refused. It seems to me that this ligase question is an extremely important one and I promise you if we have some help next time we will answer you.

Dr. Watanabe: (Tokyo, Japan)
R factors increase the carbomycin sensitivity of E. coli; and a group in Italy has also done some work with rifampicin. They have used rifampicin-resistant mutants of E. coli. They introduced R factors into these mutants and have shown that they become more sensitive to rifampicin.

The third point which I forgot to mention is that flavomycin inhibits the synthesis of the cell wall with Staphylococcus aureus.

I found that some people misunderstood what I mentioned in my talk. I meant that flavomycin induces the formation of spheroplasts both in R factor-carrying bacteria and R-minus bacteria; but R factor-carrying bacteria form spheroplasts in much lower concentrations of flavomycin.

Dr. Helinski:
Is it your view that the disappearance of these giant and conceivably circular molecules very rich in 17-18S material occurs through recombinational events involving many, many monomer units, or conceivably through a replication scheme? The monomer rather than being duplicated to yield two monomer circle units now generates very large circles via congrowing circle model.

Dr. Rownd: (Madison, Wis.)
We have no data which would distinguish between the two obvious possibilities of recombination and replication. As I pointed out, the phenomenon at the lower drug concentration appears to be more of a selective outgrowth of partially transitioned cells which exist in the population. This requires so much growth that it is not possible to monitor in any molecular terms the size of the units which are being added to the R factors. The fact that we do see intermediate density materials suggests under these conditions there are going to be a few that are added in order to form these transitioned molecules.

On the other hand, at very high drug concentrations, under the conditions of this forced transition we have been able to monitor the molecular events because they occur comparatively rapidly. There is also a relatively large amount of material to work with. We have observed the dimers, trimers, and tetramers which are present.

The fact that we are getting units which consist of an odd number of r-determinants tends to suggest that we simply are not getting the replication or the duplication of the structure. The failure of the two replicas to separate to produce a molecular structure which is twice as big would lead one to expect to find even numbers of r-determinants in these structures. We have identified very definitely that there can be an odd number which is present in these species. That would suggest that there must be some contribution due to recombination.

On the other hand, Dr. Helinski has evidence with Proteus mirabilis that one can form polygenic colicine factors in the host as well. In effect, these are due to errors in replication.

I think that both are real possibilities, but we really don't have any definite information.

Dr. Helinski: Isn't there available a good recombination-deficient Proteus strain to test this?

Dr. Rownd: No, but Dr. Hashimoto has now returned to Japan and I think he is going to tackle this sort of problem.

Dr. Helinski: Consider the appearance of these cells very rich in the 17-18S satellite material after many generations of growth and in the presence of high concentrations of a drug. Is it your view that these take over the population because of selective pressure against those 499 out of 500 cells that have a normal satellite picture? Growth of the 1/500th with 17-18S material is facilitated. Is it possibly an effect on the normal cell, inducing it to generate these 17-18S DNAs?

Dr. Rownd: It may be some of both types. It's easy to show, for example, that transitioned cells will rapidly outgrow nontransitioned cells. You can take, for example, galactose-positive transitioned cells and mix them with galactose-negative nontransitioned cells, or the reverse, and dilute them into drug-free medium or medium containing chloramphenicol or streptomycin. What you see is that transitioned cells will outgrow the others in a population where there is no difference in growth rate in a drug-free medium. So that is certainly a possible mechanism.

Now, at the lower drug levels, if stress is causing this type of release of the r-determinant and its autonomous replication, you must remember these cells are resistant to chloramphenicol, even though it does affect their growth rate. So it may be at the lower nonstressing drug levels that these partially-transitioned cells in the population may then be able to outgrow the other cells because they are already on their way to a transition stage. In contrast, the cells which are in the nontransition stage would not dissociate the r-determinants.

On the other hand, we cannot exclude the possibility that simply growth in a drug accumulates transitioned cells. Certainly both mechanisms would be possibilities for converting the whole population.

Dr. Bokrath: (Indianapolis, Ind.) Dr. Beljanski, you say that the E. coli you described were resistant to showdomycin. Have you studied several independently isolated resistant mutants and found them all to have this characteristic, or would you say that it was a coincidence that you got one mutant that did it? Are all mutants resistant to showdomycin characterized as you have described?

Dr. Beljanski: All mutants E. coli cultured in the presence of showdomycin, on a solid medium at 46°C reveal completely different mutant colonies. So you can really distinguish one, at least, which is not resistant to showdomycin.

Dr. Bockrath: Do you select colonies that can grow in the presence of showdomycin?

Dr. Beljanski: Well, we incubate for one or two hours Y type E. coli with showdomycin.

Dr. Bockrath: You incubate at 46°C?

Dr. Beljanski: No, the incubation temperature is 37°. To distinguish very easily colonies of mutant cells we have found that you can grow them up to 46°C, where you can see very nice differences between the normal and mutant forms.

What is interesting is that at 37 degrees this showdomycin-resistant mutant can grow in the presence of 3% potassium chloride or sodium chloride. The wild type does not grow.

Dr. Bockrath: What concentration of showdomycin do you use?

Dr. Beljanski: You can use from 5 to 100 micrograms per ml. Only one thing is very bad with showdomycin: It is extremely expensive. Just to give you an idea, one gram costs about $1,000.

Dr. Kanter: (Durham, N.C.) Dr. Watanabe, I may have misunderstood you, but did you suggest that the presence of pili might decrease the resistance of cells to flavomycin and rifampicin?

Dr. Watanabe: Yes, that is exactly what I meant.

Dr. Kanter: Couldn't you test this possibility by looking at the effects of flavomycin and rifampicin on males converted to female phenocopies? I think a phenocopy is a cell carrying an R factor but no pili.

Dr. Watanabe: The trouble with flavomycin is its bacteriocidal effect.

Dr. Kanter: There is no other way to get a female phenocopy?

Dr. Watanabe: I have never tried it, but I am afraid you cannot do this kind of experiment.

Dr. Cordaro: (Baltimore, Md.) Dr. Merril, I find it amazing that you have found phage in a serum. If you assumed that the people who prepared the serum were fairly fastidious, you would have to say that the phage were generated by spontaneous lysis of lysogenic bacteria in the veins of the fetal calf.

If this were true, lysogenic salmonella strains that liberate about 10^3 phage particles per 10^9 bacteria per ml would mean that the calf was suffering from septicemia. One wonders how it existed that long. Unless there is really something that I have missed, I can't see where the phage are coming from.

Dr. Merril: Neither can we, to be quite frank. But we have talked to the commercial companies that are preparing this and we have bought lots of serum. At this point I think we have bought 15 lots of serum from four different commercial suppliers and every one of them was contaminated.

Most of the suppliers are fairly secretive about the way they get this serum. It is usually obtained in the slaughter house, a subsupplier. They claim that in some cases they take the serum by cardiac puncture of the fetal calf. If that is the case, then it is a real puzzle.

As I said in the talk, because virus is able to kill bacteria, it is merely an operational definition of one of the things that virus can do. That may or may not tell its full host range. So that is one possibility.

The other possibility is that they are being a lot more sloppy than we think they are. We have to realize that all of the fetal calf serum we can buy commercially is apparently filter-sterilized after it is collected. That would get rid of any bacteria that might be present.

If you could imagine that they collected a big pool of it before they filter it and it is stored fairly warm, quite a bit would indicate a lot of bacterial growth. It would also indicate that there would be other bacterial products like endotoxins in the serum which may or may not do your cells any good. But I think it is a real problem and we are going to have to find a solution to it.

Audience: Professor Rownd, have you ever looked for naladixic acid containing RTFs resistant to naladixic acid?

In other words, if you were to take strains like this and grow them, do you see this kind of distribution on your gradients?

Dr. Rownd: We have not done an experiment of that type. To my knowledge, R factors do not carry naladixic acid resistance.

Audience: How about resistance to the rifampicins?

Dr. Rownd: I don't believe so, to my knowledge.

Dr. Kriehl: (Salt Lake City, Utah) Dr. Beljanski, do your bacteria, which have been transformed by RNA, retain the resistance to showdomycin that the original ones had?

Dr. Beljanski: This is a good question. I am sorry I did not explain that point.

When we isolate this transforming RNA from the culture medium and do our experiments for transformation, usually we use a mutant which is resistant to, let's say, 200 micrograms of showdomycin per ml. Surprisingly enough, the resistance to showdomycin was just partially transferred to the transformants. Instead of being resistant to 200 micrograms they are resistant to 40 micrograms, and I really don't know why.

Dr. Beers: I would like to address a question to the panel with respect to the specificity of viral infections and the relationship that may have to the general phenomenon of the transmission of R factors.

Is the specific host virus relationship determined primarily by immunological factors or by the machinery of the cell which is responsible for the replicating process, itself, or a combination of the two?

Is there any evidence, for example, in the transmission of R factors from one species to another that immunological factors play any particular role? If not, is this a particularly unique situation we are experiencing?

Dr. Rownd: To my knowledge, nobody has looked for immunological differences between R-minus and R-plus. One would imagine, since some of the R factor gene products which have been looked at constitute a very significant fraction of the total protein of the cell, that there would be some very substantial immunological differences. For example, chloramphenicol-acetyl transferase in E. coli comprises about one-half of one percent of the total soluble protein.

Now, in the case of Proteus in a transition culture you can get the activity increased from twenty- to thirty-fold. Although we haven't really measured the fraction of the total soluble protein, this would mean there is a very substantial percentage of the soluble protein in this one enzyme.

There are also three or four other enzymes on the r-determinant which we are not monitoring and presumably these would also increase. I think there are now, according to the latest count, on the order of ten genes which are involved in mating alone on the RTF components. There should be a very substantial number of new antigens which could be added to an R-plus cell.

Does this answer it?

Dr. Beers: That answers it partly, yes.

Dr. Merril, with respect to the production of the bacteriophage, if it is by other than a bacteria, how can you account for such a broad specificity range? Would you consider or have you made an attempt to look for immunological responses to such a foreign system in a calf plasma?

Dr. Merril: Actually, I will have to answer that question in a number of parts.

First of all, from looking through the literature, I have an impression, that in the prokaryotes transformation seems to be related to different species. As you go from naked DNA into cell-to-cell contact (viruses), it appears that the host range of these genes in transfer, if you want to call them that, becomes wider as the gene cluster begins to carry more factors that give it autonomy. Examples are replication genes and recombinational genes. Lambda will integrate into this small site in E. coli chromosome, which apparently is less than 12 nucleotides, and it doesn't do it by general recombination. It does it by carrying in the information for its own enzymes. One of the enzymes apparently is an end gene product. We are not really sure whether it is an enzyme, but it carries a specific recombination at that site.

One thing I should have pointed out during the talk is that if that site is only 12 nucleotides in length, and the mammalian cell has about 800 times as much DNA as an E. coli, statistically you would expect there to be sites in a mammalian cell similar to the lambda integration site on the E. coli, just on a random basis.

As far as immunological reactions to these viruses go, one could imagine if these viruses were present at the time the fetus was formed, then there may not be any immunological response at all, because it would never be recognized as non-cell.

Dr. Beers: Dr. Rownd, are the dissociations of the transfer factor resistance determinants species specific or can you mix them?

Dr. Rownd: As I pointed out in my talk, we have been able to detect only dissociation in Proteus mirabilis. We have looked in several other generic groupings. One is the E. coli, the other Salmonella and finally the Escherichia marcessant.

However, it should be very simple to select mutants which undergo this process, since there is a huge increase in resistance. One would expect that a phenotype of a mutant of coli which undergoes this process would acquire very large increases in resistance to all of the drugs to which the r-determinant confers resistance; so it should be very easy to select for such a type. We have not carried out such experiments, but it is one of the things we will look at in the future.

Dr. Watanabe: I think that the dissociation is, perhaps, occurring also in E. coli and other bacteria, not only in Proteus. But the reason why you don't observe the two components is, perhaps, because mutual exclusion or incompatibility is very stringent in bacteria other than Proteus. In Proteus mutual exclusion or incompatibility may not be operating.

Dr. Merril: Yes, it does. We have done such experiments. It is just as stringent as in E. coli.

Dr. Rownd: I would like to disagree with that statement because, on the molecular level, a number of groups have looked at R factors in E. coli. If the molecules dissociate, they should give smaller units. Again, one must look only at the closed circular DNA. We have been above to find no evidence that dissociation does actually occur.

I might point out one thing which came to mind when I heard Dr. Helinski's comments this morning about looking at the closed circular DNA fraction. Mr. Fred Morris, a graduate student in my laboratory, has shown in Proteus mirabilis the R factor DNA as a separate satellite band. As a result of that, we could look at all of the R factor DNAs.

In nontransitioned Proteus mirabilis harboring NR-1, with one r-determinant in the composite structure, only 30% of this material is co-valently closed circular when the cells are cultured in Penassay broth. If you culture the same cells in a minimum medium with a very long doubling time, about 70% is closed circle DNA.

If you isolate a segregant which has lost the r-determinant and culture the cells in Penassay broth, the RFT TC segregant is 100% closed circular. Originally with the R factor the DNA is 30% co-valently closed.

So, the presence of the r-determinant in the composite structure results in the fact that 70% of the DNA exists in the nonco-valently closed circular form. This may well be related to the ability of the R factor to dissociate into the two composite structures. Unless there is evidence to the contrary I don't think it necessarily a safe assumption that you are looking at all of the DNA of a cell by focusing on the co-valently closed circular fraction.

Experiments by Mr. Fred Morris in my laboratory have shown a derivative of NR-1, which we call R-12, has changed in a number of properties. It does not undergo transition. There are four times as many copies per cell as there are in the case of NR-1. From this R factor, which is identical in size with NR-1, one can isolate 100% of DNA as co-valently closed circular DNA, pH 10. If you do an experiment at pH 11, using the same isolation procedure, you get zero percent of co-valently closed circular DNA. The product contains only one break which is in a unique strand of duplex and, as alluded to by Dr. Helinski this morning in his system, in the case of R-12 the linear strand appears to be attached to a membrane component. So it is a completely open question, in my mind at least, whether the structure of the R factor DNA in a cell is co-valently closed circular or whether it exists in some intermediate state which can be converted one way or the other, either to the open circular or to the closed circular form, depending upon the isolation conditions that one uses in the experiment. It is very difficult in this type of experiment to rule out the possibility that somehow you are breaking the DNA, although we have carried out all sorts of controls to try to rule this out. Of course, negativity in the way of controls gives you more circumstantial evidence, but we have no reason to believe that we are breaking the DNA. If we are, it would be, in my opinion, quite singular that we should find only one break in a unique strand of a duplex.

So in summary, I think assuming that 100% plasma DNA in a cell is co-valently closed circular is not justified, at least in the type of experiment which we have done in our laboratory.

Dr. Watanabe: I recall the phenomenon which you named, relaxed control and stringent control of R factor DNA replication in Proteus. You mentioned in the stationary phase the R factor DNA becomes maybe 60 to 100 per cell. Do you still believe this will really happen?

Dr. Rownd: I didn't go into this in my talk; it is reported in studies which we have already published on replication.

Many of the DNA profiles which I showed to you displayed very large bands which are characteristics of stationary phage cultures. In the case of NR1, the area of the satellite band which one observes in exponential phase is two-and-a-half-fold lower than what one observes in stationary phase.

Actually, we have shown in independent experiments that the R factor continues to replicate in stationary phage cultures and undergoes about a five-fold increase in the R factor pool size. There are no other changes in the molecular types analogous to the transition phenomenon. This is an aspect which we have referred to as relaxed control.

Our earlier estimates as to the number of copies of R factors present in Proteus evidently were too great, because we were assuming at that time that F factors would have a size in the region of what has been guessed from experiments carried out in E. coli, that is, something on the order of 40,000,000 to 60,000,000 daltons.

Now we believe that some of these structures have molecular weights in the region of 200,000,000 to 300,000,000 daltons because of these tandem sequences of r-determinants. As a result we do not think there is a very much larger number of copies present in Proteus.

Dr. Watanabe: About how many copies of DNA molecules are present in the Proteus chromosome?

Dr. Rownd: Probably on the order of between one and two, as closely as we can measure. It is not a large number.

Part III

11. ENDOGENOUS RNA-DIRECTED DNA POLYMERASE ACTIVITY IN NORMAL CELLS[1]

Howard M. Temin, Chil-Yong Kang and Satoshi Mizutani

McArdle Laboratory
University of Wisconsin
Madison, Wisconsin

INTRODUCTION

The DNA provirus hypothesis states that RNA tumor viruses replicate through a DNA intermediate. This hypothesis is supported by the biology of RNA tumor viruses, the results of experiments with inhibitors of nucleic acid synthesis, the results of experiments with nucleic acid hybridization, and the absence of evidence for double-stranded RNA, RNA complementary to virion RNA, or RNA replicase in RNA tumor virus-infected cells (1). The most striking evidence for the DNA provirus hypothesis is the existence of an RNA-directed DNA polymerase in the virions of all infectious RNA tumor viruses (2).

The protovirus hypothesis states that RNA-directed DNA synthesis is a mechanism for genetic change in normal cells and that neoplasia and RNA tumor viruses arise as a result of derangements of this normal genetic system (3,4). A major prediction of the protovirus hypothesis is that RNA-directed DNA polymerase activity exists in normal cells. This paper will present evidence for the existence in normal cells of RNA-directed DNA polymerase activity unrelated to known viruses.

All infectious virions of RNA tumor viruses contain in their core, which is surrounded by a lipid-containing envelope, endogenous RNA-directed DNA polymerase activity. The envelope must be disrupted for full DNA polymerase activity. The endogenous DNA polymerase activity is sensitive to RNase, resistant to DNase, and partially resistant to actinomycin D (Fig 4). The product is DNA and hybridizes to the large RNA of the virion. The primer is also RNA (2).

ENDOGENOUS DNA POLYMERASE ACTIVITY IN INFECTED CELLS

Chicken cells after exposure to Rous sarcoma virus (RSV) become transformed and produce progeny virus. When these infected cells are disrupted with detergent and a high-speed pellet fraction prepared, endogenous RNA-directed DNA polymerase activity is found (5). This activity in infected cells bands at a higher density than virion cores and is RNase-resistant, whereas the activity of virion cores is

[1] Supported by Public Health Service Research Grant CA07175 from the National Cancer Institute; Training Grant TO1 CA 5002 from the National Cancer Institute; and Grant VC-7 from the American Cancer Society. H. M. Temin holds Research Career Development Award KO3-CA08182 from the National Cancer Institute. C.-Y. Kang is a research fellow of the National Cancer Institute of Canada.

RNase-sensitive. The DNA product of the cell activity hybridizes to the RNA of the RSV virion, and, therefore, the RNA template of this activity is probably virus RNA. These particles in RSV-infected chicken cells appear to be a precursor of virion cores. Their difference in density and RNase-sensitivity from virion cores may be related to maturation processes during the formation of the RSV virion.

Rat cells exposed to some strains of Rous sarcoma virus become transformed into neoplastic cells, but do not release any virus or virus-like particles. These cells still contain the entire genome of the Rous sarcoma virus, which can be rescued by fusion with chicken cells. When these cells are disrupted and a high-speed pellet fraction prepared, endogenous RNA-directed DNA polymerase activity is also found (6). This activity bands at a lower density than virion cores, but is RNase-sensitive. Much to our surprise, we found that the DNA product of this endogenous RNA-directed DNA polymerase activity from rat cells transformed by Rous sarcoma virus did not hybridize to RNA from Rous sarcoma virus or RNA from two murine C-type viruses, and that the endogenous DNA polymerase activity was not neutralized by an antibody which neutralized Rous sarcoma virus virion DNA polymerase activity (Fig 1).

These results establish that neither the RNA template nor the DNA polymerase of this endogenous RNA-directed DNA polymerase activity from rat cells infected with Rous sarcoma virus is related to Rous sarcoma virus. These findings suggested the hypothesis that Rous sarcoma virus infection increases the activity of a preexisting endogenous RNA-directed DNA polymerase system. This hypothesis was tested in two ways: 1. It was shown that the DNA product of the endogenous RNA-directed DNA polymerase activity from rat cells transformed by Rous sarcoma

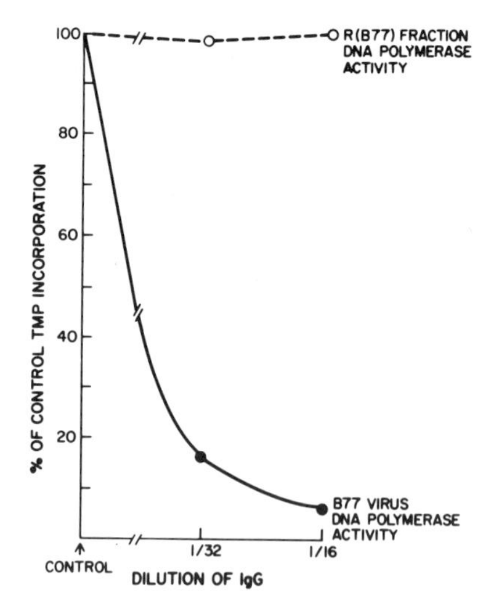

Fig 1. Attempted neutralization of rat cell endogenous DNA polymerase activity by antibody against avain myeloblastosis virus DNA polymerase. 25 μl of diluted IgG (7) and 25 μl of detergent-activated virus (8 μg protein) or the high-speed pellet fraction from rat cells infected with Rous sarcoma virus (100 μg protein) were incubated at 37° for 30 min. The residual endogenous DNA polymerase activity was then determined. The radioactivity of the TCA-precipitated fraction was determined after incubation for 30 min at 40°. 100% endogenous DNA polymerase activity for the fraction from rat cells infected with Rous sarcoma virus was 1,450 counts/min; 100%, for B77 virus was 8,000 counts/min. o----o, rat cells infected with B77 virus. •——•, B77 virus.

TABLE I

Endogenous DNA Polymerase Activity in Nuclei and Cytoplasm of Chicken Cells[a]

	Endogenous activity: ^{3}H-TMP incorporated
Nuclei	14,000 cpm/mg protein/30 min
Cytoplasm	12,900 cpm/mg protein/30 min

[a]Five-day chicken embryos were washed twice with phosphate-buffered saline and once with Tris buffer containing 0.25 M sucrose, 0.025 M KCl, 0.005 M $MgCl_2$, and 0.002 M 2-mercaptoethanol; homogenized by eight strokes in a tightly fitting Dounce glass homogenizer; and centrifuged at 100 x *g* for 5 min. The supernatant fraction was further centrifuged at 650 x *g* for 10 min. The pellet fraction from the second centrifugation, containing nuclei, was washed and treated with Nonidet and dithiothreitol. The disrupted nuclear fraction and the cytoplasmic fraction were centrifuged at 45,000 x *g* for 1 hr, and the pellets were further purified on a discontinuous sucrose gradient and assayed for endogenous DNA polymerase activity by the standard method of Temin and Mizutani (9). Protein was determined by the Lowry method after removal of dithiothreitol. The specific activity of the ^{3}H-TTP used for all DNA polymerase assays was 13.4 C/mmole.

virus hybridized to RNA from normal rat cells; and 2. it was shown that uninfected rat cells contained an endogenous RNase-sensitive DNA polymerase activity (6).

ENDOGENOUS DNA POLYMERASE ACTIVITY IN UNINFECTED CHICKEN CELLS

We decided to look for a similar activity in uninfected chicken cells because we have better knowledge of possible contaminating viruses in chicken cells. When we prepared a similar high-speed pellet fraction from lysed chicken embryos or cells, either positive or negative for avian leukosis virus group-specific antigens, we also found endogenous DNA polymerase activity (8). This endogenous DNA polymerase activity could be separated by successive density and velocity sucrose gradients from much of the exogenous DNA-directed DNA polymerase activity. Roughly, equal amounts of this activity could be isolated from the nuclei and cytoplasm of chicken cells (Table I).

In an attempt to test the hypothesis that this endogenous DNA polymerase activity was the result of complexing one of the soluble DNA polymerases of the cell with nucleic acid during fractionation, a reconstitution experiment was carried out (Table II). Addition of soluble DNA polymerases to disrupted cells and subsequent fractionation did not lead to an increase in endogenous DNA polymerase activity. This result suggests that this activity is not an artifact of fractionation. This hypothesis is further supported by the results shown below (Table VII), which indicate that the DNA polymerase which is responsible for the endogenous activity is different from the large soluble DNA polymerases of chicken cells.

The requirements for this endogenous DNA polymerase activity from normal uninfected chicken cells were compared with the requirements for endogenous DNA polymerase activity of disrupted Rous sarcoma virus virions (Table III). Both activities have similar requirements, but the chicken endogenous activity prefers Mg^{++}, whereas the B77 virus activity slightly prefers Mn^{++}.

TABLE II

Attempt to Reconstitute Endogenous DNA Polymerase Activity[a]

Fraction	Endogenous activity: ^{3}H-TMP incorporated (counts/min)	
	Chicken embryo cells	Chicken embryo cells mixed with soluble DNA polymerase
8,000 rev/min supernatant fluid	500	450
45,000 rev/min supernatant fluid	100	100
45,000 rev/min pellet	3,700	3,200

[a]0.5 ml, containing 2 mg protein, of the soluble fraction from normal 7-day chicken embryos was added to 4.5 ml of the 5-day chicken embryos disrupted as described in the legend to Table I. This amount of soluble fraction incorporated 116,000 cpm/hr in a DNA polymerase reaction templated by 25 μg of calf thymus DNA. Low-speed and high-speed fractions were then prepared and tested for endogenous DNA polymerase activity in a 30 min reaction.

TABLE III

Requirements for DNA Polymerase Activity of a Fraction from Chicken Cells and of Disrupted B77 Virus[a]

Reaction conditions	^{3}H-TMP incorporated (counts/min)	
	Chicken cells	B77 virus
Complete	8,600	79,000
- Mg^{++}	400	160
- Mg^{++} + Mn^{++}[b]	3,600	91,000
- dCTP	1,700	6,400
+ oligo$(dC)_{12\text{-}18}$, 500 p moles/reaction	8,750	74,700
+ oligo$(dT)_{12\text{-}18}$, 500 p moles/reaction	8,200	78,500
+ yeast RNA, 25 μg/reaction	8,800	73,900
+ yeast RNA, 25 μg/reaction + oligo$(dC)_{12\text{-}18}$, 500 p moles/reaction + oligo$(dT)_{12\text{-}18}$, 500 p moles/reaction	8,300	64,000
+ native calf thymus DNA, 25 μg/reaction	328,300	1,417,600

[a]DNA polymerase activity was assayed (9) for 30 min at 41°C with the addition or omission of components as indicated (8). The chicken fraction contained about 600 μg protein and the B77 virus 80 μg in a reaction volume of 125 μl.

[b]1.5 mM $MnCl_2$ was substituted for the 10 mM $MgCl_2$.

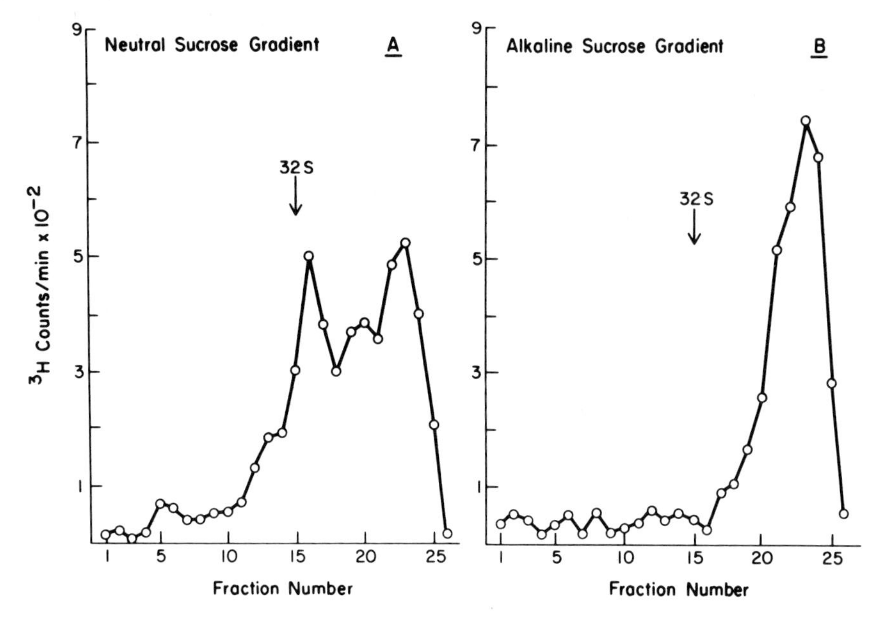

Fig 2. Nature of product. Tritium-labeled product was purified with sodium dodecyl sulfate and Bacovin from a 5-min reaction of the chicken endogenous DNA polymerase activity. The purified product was analyzed in a 5 – 20% sucrose gradient at pH 7 (A) or in a similar gradient made in 0.03 M NaOH (B). The gradient was centrifuged at 40,000 rpm for 2.5 hr. Fractions were collected and acid-insoluble radioactivity determined. The arrow represents the position of DNA from purified T7 bacteriophage.

The size of the product of the chicken endogenous DNA polymerase activity was 5 to 30S in neutral sucrose gradients and around 5S in alkaline gradients (Fig 2). The product was resistant to alkali and RNase, but sensitive to DNase. It banded in cesium chloride and cesium sulfate equilibrium density gradients at the density of DNA. It was labeled with both ^{3}H-dTTP and ^{3}H-dCTP. Therefore, the produce is DNA.

An experiment was also conducted to determine whether there was any RNA polymerase activity in this fraction. Incorporation of labeled UTP was found (Fig 3). But this incorporation was the result of terminal transferase activity, because it was unaffected by the removal of CTP or ATP or by the addition of actinomycin D. No DNA-directed RNA polymerase activity was found. In addition, it was shown (Table IV) that the chicken fraction contained low levels of DNase activity, almost no RNase activity for single-stranded RNA, and significant levels of RNase for hybrid RNA. The last activity may be RNase H (10,11).

RNA IS A TEMPLATE FOR THE ENDOGENOUS DNA POLYMERASE ACTIVITY

The sensitivity of the chicken endogenous DNA polymerase activity to treatment with RNase, DNase, and actinomycin D was compared with that of disrupted RSV virions (Fig 4). It was found that endogenous DNA polymerase activity from normal uninfected chicken embryos and that from disrupted Rous sarcoma virus virions were sensitive to RNase, resistant to DNase, and, initially, partially resistant to actinomycin D. In addition, it was shown that the addition of oligo $(dC)_{10}$ or oligo

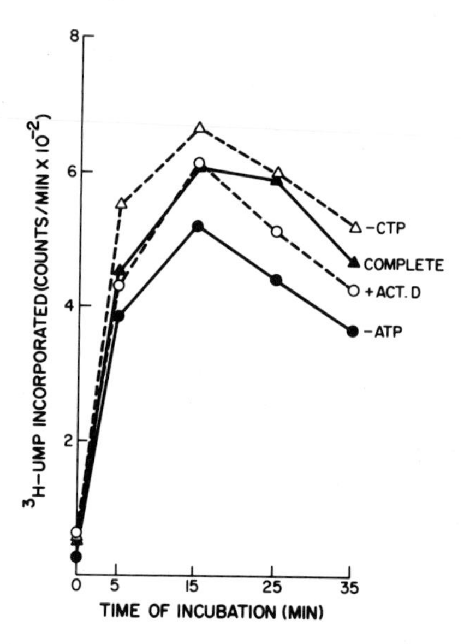

Fig 3. Incorporation of labeled UMP by chicken fraction. A fraction from chicken cells (70 μg protein) containing approximately 1000 counts/min endogenous DNA polymerase activity per 25 μl was incubated in 0.1 ml Tris buffer containing 0.4 mM EDTA; 10 mM dithiothreitol; 10 m moles each of ATP, CTP, and GTP; 10 mM $MgCl_2$; 20 mM KC1; 0.25 μg phosphoenol pyruvate; 10 μg pyruvate kinase; and 2.5 μCi of ^{3}H-UTP (specific activity 15 Ci/m mole). In parallel reactions, 5 μg of actinomycin D was added or CTP or ATP were omitted, as indicated.

$(dT)_{10}$ as primers did not restore the activity after RNase treatment (data not shown). Therefore, it appears that the endogenous DNA polymerase activity from normal uninfected chicken cells is RNA-directed, as is that of the Rous sarcoma virus virion.

Further confirmation of this hypothesis was gained by hybridization of the DNA product of the endogenous DNA polymerase activity from normal uninfected

TABLE IV

Nuclease Activities in Chicken Fraction with Endogenous DNA Polymerase Activity[a]

Substrate	Enzyme preparation		
	Chicken fraction (110 μg/ml)	RNase A (50 μg/ml)	DNase 1 (50 μg/ml)
^{3}H-T7 DNA	56[b]	105[b]	7[b]
^{3}H-chicken RNA	91[b]	3[b]	99[b]
^{32}P-λmRNA-λDNA hybrid	50[b]	100[b]	ND[c]

[a]The indicated components were mixed in a solution containing 10mM $MgCl_2$, 20 mM KCl, and 20 mM Tris-HCl (pH 8.0) for 30 min at 40°C.

[b]Percent of TCA-insoluble counts remaining. 100% was 15,300 cpm for the T7 DNA, 184,700 cpm for the ^{32}P-λmRNA-λDNA hybrid. The λRNA-DNA hybrid was prepared by Dr. H. Lozeron, University of Wisconsin.

[c]Not done.

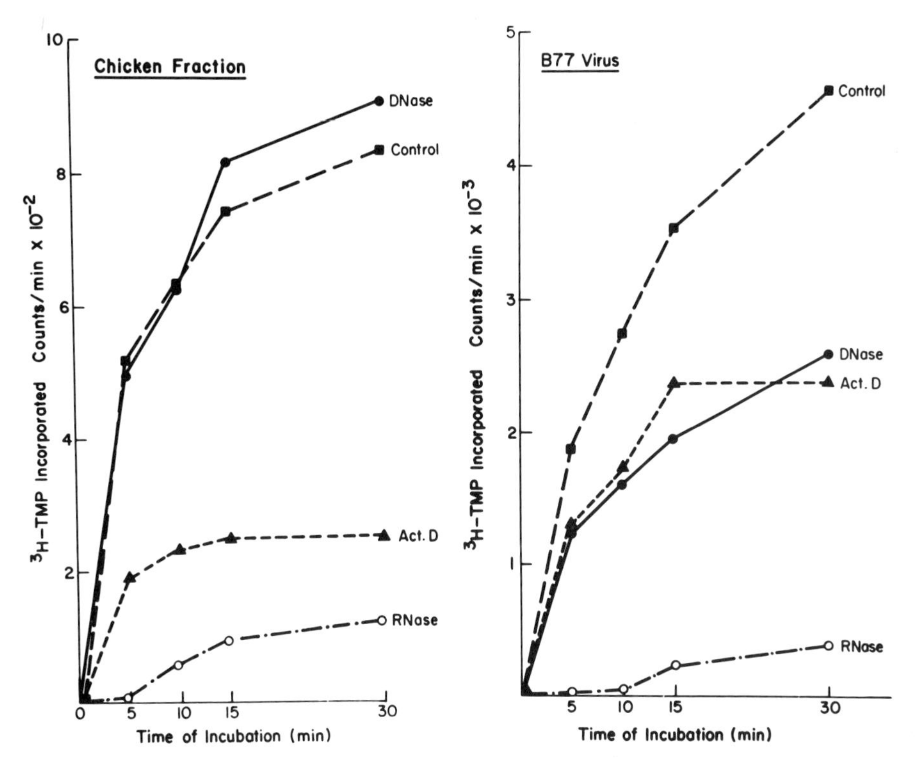

Fig 4. Effects of nucleases and actinomycin D on endogenous DNA polymerase activity. The chicken endogenous DNA polymerase activity (60 μg protein) and disrupted B77 virus (4 μg protein) were treated with water (■), 50 μg per ml of boiled pancreatic ribonuclease (Worthington, New Jersey) (○), or 50 μg per ml of electrophoretically purified pancreatic deoxyribonuclease (Worthington, New Jersey) (●) for 30 min at room temperature. In a parallel incubation, 25 μg of actinomycin D (▲) was added to 1 ml of reaction mixture. Endogenous DNA polymerase activity was then determined in a standard assay (8).

chicken embryos to RNA isolated from the same fraction (Fig 5). Although only a low level of activity was found in the RNA region of the gradient with 200 μg of RNA per 0.4 ml, it was possible (by using more RNA, 800 - 1000 μg/0.4 ml) to hybridize 40% of the DNA counts. However, DNA isolated from the RNA region of the gradient still hybridized inefficiently (Table V). This result suggests that there was a low concentration of RNA templates.

In order to demonstrate that this hybridization was to template RNA, rather than to RNA which was not a template but had the same sequence as another nucleic acid template, attempts were made to isolate RNA-DNA hybrid intermediates. Early attempts failed, probably because of the presence in the chicken fraction of RNase activity for hybrid RNA (Table IV). However, when products of a 2-minute reaction were isolated by sodium dodecyl sulfate and phenol and analyzed in sucrose gradients, a peak sedimenting at approximately 35 - 40S was found (Fig 6). This peak disappeared when the sample was pretreated with either heat or RNase in a low salt concentration. Furthermore, a portion of the same product of the initial reaction banded in equilibrium cesium sulfate density gradient centrifugation at the density of RNA (data not shown). These results demonstrate that the DNA product was synthe-

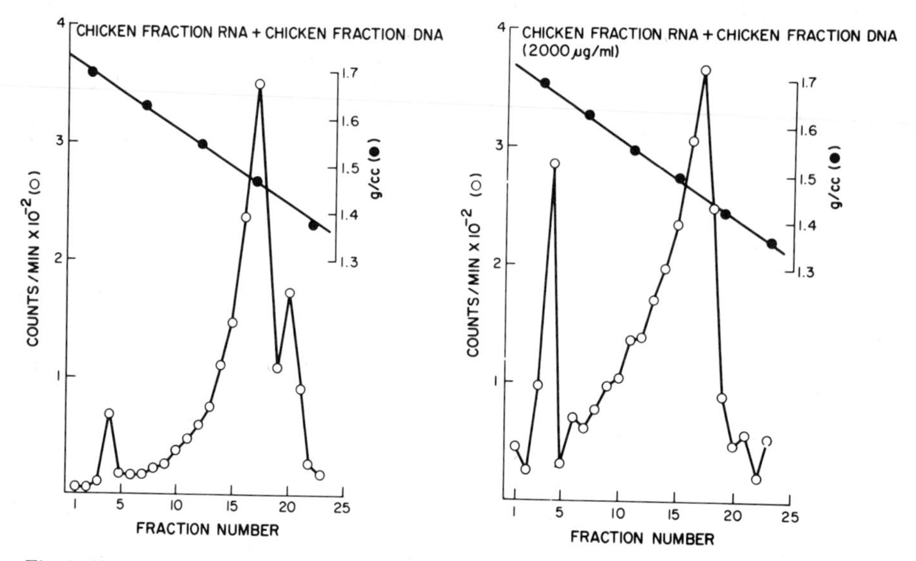

Fig 5. Hybridization of the DNA product of the chicken endogenous DNA polymerase activity with RNA from the same fraction. RNA and DNA were extracted as described in the legend to Figure 2. The DNA was treated with 0.2 N NaOH for 12 hr and then neutralized. 200 or 800 μg of RNA from the chicken fraction were annealed in 0.4 ml to about 4000 counts/min of tritiated product from a 5-min reaction of the chicken endogenous DNA polymerase activity. Annealing was at 68° for 5 hr in 2 times SSC. After being annealed, the samples were analyzed in equilibrium cesium sulfate density gradients. The density of alternate fractions was determined by refractometry. (Part of data from [8].) Open circles are cpm; filled circles, density.

TABLE V

Distribution of ^{3}H-labeled Chicken Endogenous DNA Polymerase Product after Rehybridization to RNA from the Same Chicken Fraction[a]

Position in gradient of starting DNA (gm/cc)	Percent counts/min recovered as hybrid[b]	nonhybrid[c]
1.65 - 1.68	30	70
1.56 - 1.59	40	60
1.44 - 1.47	10	90

[a]DNAs banding at the densities of 1.65-1.68 gm/cc, 1.56-1.59 gm/cc, and 1.44-1.47 gm/cc were pooled separately from gradients like those of Figure 5. The samples were then treated with 0.2 N NaOH, each was reannealed with 200 μg of RNA from the chicken fraction with endogenous DNA polymerase activity and analyzed by cesium sulfate equilibrium density gradient centrifugation.

[b]Defined as the percent of counts/min which banded in a Cs_2SO_4 gradient at densities greater than 1.53 gm/cc.

[c]Defined as the percent of counts/min which banded in a Cs_2SO_4 gradient at densities less than 1.53 gm/cc.

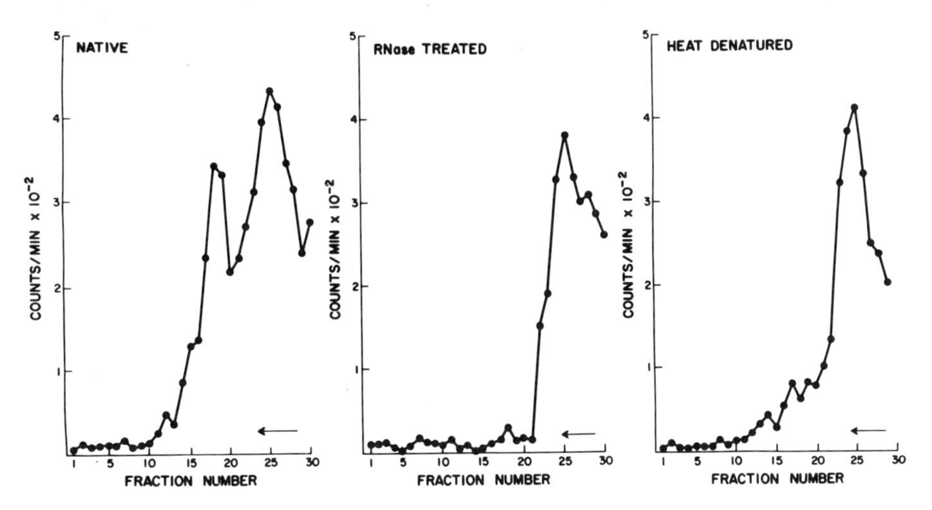

Fig 6. Sucrose gradient centrifugation of initial product. A standard chicken endogenous DNA polymerase reaction was carried out in the presence of ^{3}H-TTP and ^{3}H-dCTP (specific activity, 22.6 Ci/mmole) at 40°C for 2 min. DNA synthesis was stopped by the addition of 100 mM EDTA and 0.1% SDS. The product DNA was then extracted twice with water-saturated distilled phenol, and the aqueous phase was analyzed on a 5 to 20% linear sucrose gradient made in 0.1 M phosphate buffer, pH 7.0, after treatment with buffer (native), 50 μg/ml of RNase A in low salt (RNase treated) at 40°C for 30 min, or incubation in a boiling water bath (heat denatured) for 15 min. The gradients were centrifuged at 40,000 rpm for 2.5 hr in a SW 50.1 rotor, and approximately 0.17-ml fractions were collected from the bottom of the centrifuge tubes. Trichloroacetic acid-insoluble radioactivity was determined and plotted after subtraction of the background. The arrow indicates the direction of the centrifugal force.

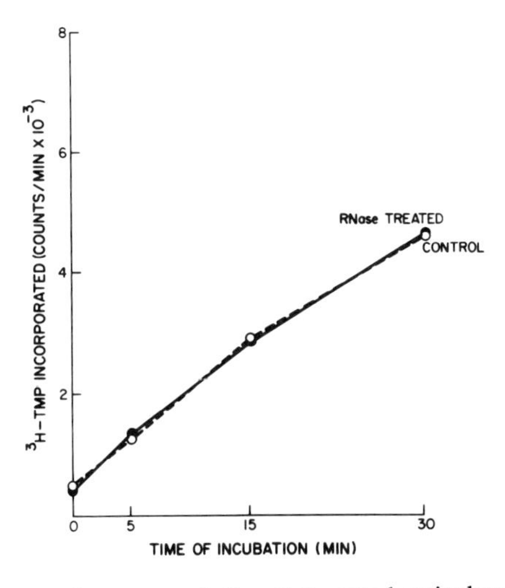

Fig 7. Endogenous DNA polymerase activity of disrupted reticuloendotheliosis virus. Reticuloendotheliosis virus (strain T) was grown in chicken embryo fibroblasts in culture and purified by centrifugation (8). Nonidet-disrupted virus (15 μg protein) was treated with 50 μg per ml of boiled ribonuclease A at 37°C for 30 min or with water, and the endogenous DNA polymerase activity was measured.

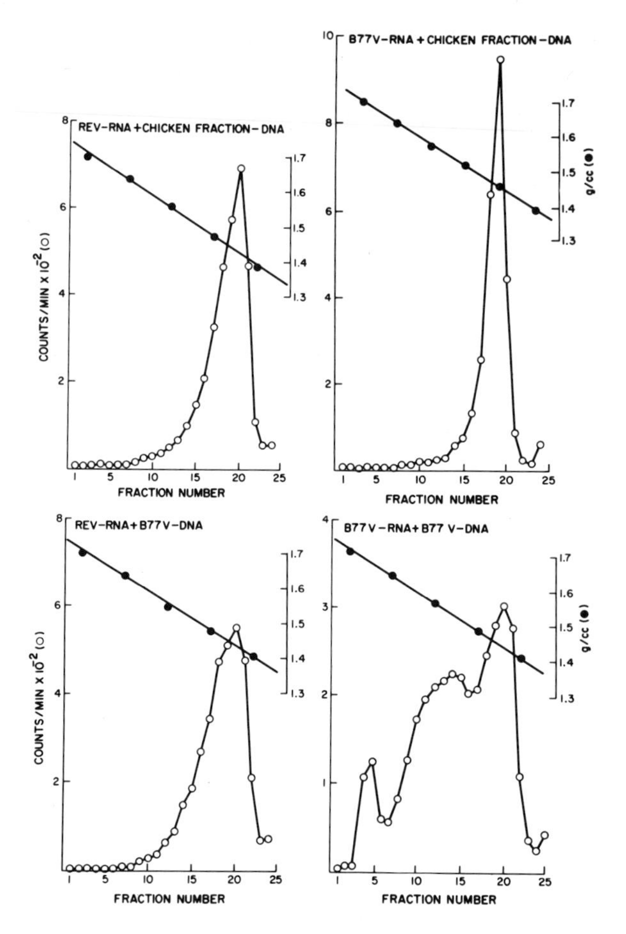

Fig 8. Hybridization of the products of chicken endogenous DNA polymerase activity and B77 virus endogenous DNA polymerase activity to RNA isolated from reticuloendotheliosis virus or B77 virus. 4000 counts/min of the indicated ^{3}H-labeled DNA product was mixed with either 5 μg of RNA from reticuloendotheliosis virus (REV) or 1.5 μg of RNA from B77 virus, hybridized, annealed, and analyzed by equilibrium Cs_2SO_4 density gradient centrifugation.

sized on an RNA template (see [2] references to similar experiments with RNA tumor viruses).

The results of these experiments establish that this chicken endogenous DNA polymerase activity, like that of RNA tumor virus virions, is RNA-directed (templated). The nature of the primer is under investigation.

RELATION TO VIRUSES

To determine whether this endogenous RNA-directed DNA polymerase activity in normal uninfected chicken cells is related to virus polymerase activity, experiments were carried out on the nature of the RNA template and DNA polymerase. Two viruses were used for comparison: the Rous sarcoma virus and the reticuloendotheliosis virus. Reticuloendotheliosis virus has been shown to have DNA polymerase activity (12). We confirmed that there was DNA polymerase activity in purified preparations of

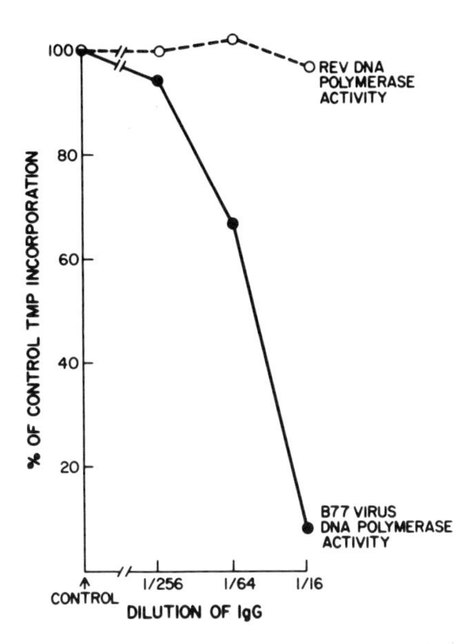

Fig 9. Neutralization of endogenous DNA polymerase activity of reticuloendotheliosis and B77 viruses with antibody to avian myeloblastosis virus DNA polymerase. The experiment was carried out as described in the legend of Figure 1. 100% endogenous DNA polymerase activity for reticuloendotheliosis virus derived from chicken plasma (10 μg protein) was 2700 counts/min; 100% activity for B77 virus (8 μg protein) was 7900 counts/min.

reticuloendotheliosis virus. However, we found that this DNA polymerase activity was not RNase-sensitive (Fig 7). It probably was directed by DNA in the virus preparation.

Product DNAs prepared from reactions of the chicken endogenous DNA polymerase activity and Rous sarcoma virus virion endogenous DNA polymerase activity were annealed with RNA from RSV and reticuloendotheliosis virus (Fig 8). It was found that the DNA product from the chicken endogenous DNA polymerase activity did not anneal with either of the viral RNAs. Figure 5 shows a positive control. The DNA polymerase product of the Rous sarcoma virus virions, in addition, was found not to anneal with RNA from reticuloendotheliosis virus.

This experiment established that the RNA template of the endogenous RNA-directed DNA polymerase activity from normal uninfected chicken embryos was not related to RNA of the only two types of chicken viruses known to have RNA and DNA polymerase.

Next the relationship of the virus DNA polymerases to that of the chicken endogenous DNA polymerase activity was studied. Nowinski, et al (7) have prepared an antibody to the purified avian myeloblastosis DNA polymerase. This antibody neutralized DNA polymerase activity of B77 virus but not that of reticuloendotheliosis virus (Fig 9). However, there is still some relationship between the DNA polymerase of reticuloendotheliosis virus and that of Rous sarcoma virus because, as seen in Figure 10, reticuloendotheliosis virus was able to partially block the antibody activity. Further experiments are being carried out to define this relationship more fully.

When this antibody to purified avian myeloblastosis virus DNA polymerase was used in similar tests with the endogenous RNA-directed DNA polymerase activity

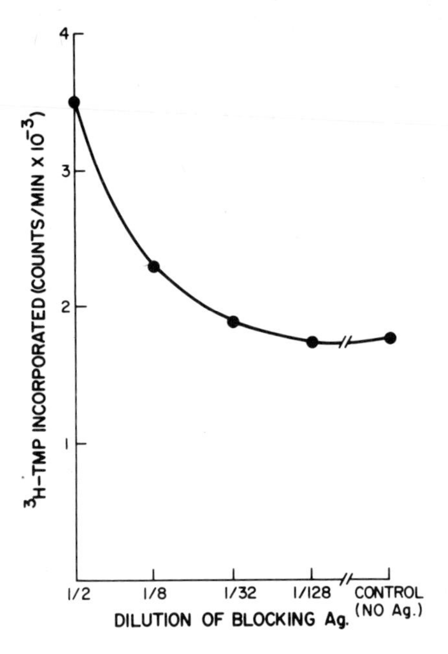

Fig 10. Antibody blocking power of reticuloendotheliosis virus for antibody to avian myeloblastosis virus DNA polymerase.

Serial four-fold dilutions of Nonidet-disrupted reticuloendotheliosis virus (10 μg protein) containing approximately 2700 counts/min of endogenous DNA polymerase activity in 25 μl were made and mixed with 25 μl of a 1:32 dilution of antibody (IgG) to avian myeloblastosis virus. The mixtures were incubated at 37° for 30 min and then at 41°C for 4 hr to destroy the reticuloendotheliosis virus endogenous DNA polymerase activity. The residual neutralizing activity of the antibody was tested by adding 25 μl of disrupted B77 virus (7 μg protein) containing 6700 counts/min endogenous DNA polymerase activity to each reaction mixture. The residual endogenous DNA polymerase activity in each sample was checked after incubation at 37° for 30 min.

from normal uninfected chicken embryos, it was found that the DNA polymerase activity was not neutralized, and that the chicken fraction with endogenous DNA polymerase activity was unable to block the antibody activity (Fig 11). Polymerase solubilized from this fraction was also not neutralized (data not shown). Therefore, the DNA polymerase of the endogenous RNA-directed DNA polymerase activity from normal uninfected chicken cells appeared to be different from the DNA polymerases of the two possible contaminating viruses.

We conclude that the endogenous RNA-directed DNA polymerase activity from normal uninfected chicken cells is not related in template or polymerase to known avian RNA tumor viruses.

RELATION TO SOLUBLE CELLULAR DNA POLYMERASES

Experiments were then performed to determine whether the DNA polymerase of the endogenous RNA-directed DNA polymerase activity from normal uninfected chicken cells was related to the soluble DNA polymerases of chickens. Mizutani (13) has partially purified several soluble DNA polymerases from normal chicken embryos. Antibody prepared against one of these partially purified DNA polymerases was active

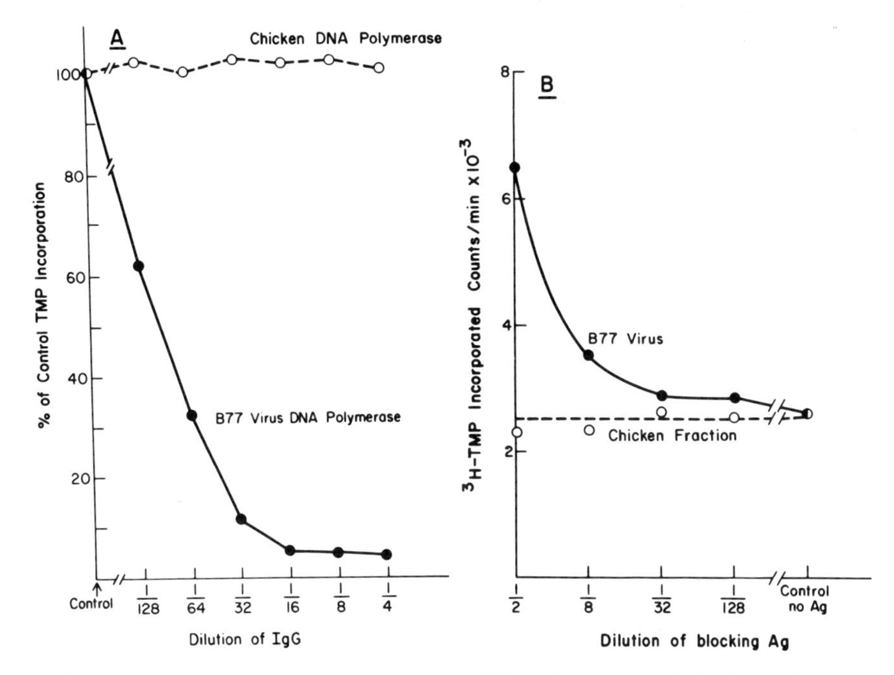

Fig 11. Neutralization of chicken endogenous DNA polymerase activity by antibody against avian myeloblastosis virus DNA polymerase and the antibody-blocking power of this chicken fraction. Tests were carried out as described in the legend to Figures 9 and 10. 100% of B77 virus endogenous DNA polymerase activity (10 μg protein) was 10,500 counts/min, and 100% of chicken endogenous DNA polymerase activity (400 μg protein) was 5600 counts min. For the blocking tests, dilutions of chicken endogenous DNA polymerase activity (300 μg protein) containing approximately 4500 counts/min in 25 μl and dilutions of B77 virus (6 μg protein) containing approximately 6500 counts/min of endogenous DNA polymerase activity in 25 μl were used (8).

in partially neutralizing the activity of all of the DNA polymerases (Table VI). However, when the DNA polymerases were solubilized from the fraction from normal uninfected chicken embryos with endogenous RNA-directed DNA polymerase activity, one polymerase was found to differ in its S value from the other cellular DNA polymerases. This polymerase, pellet-S, was not neutralized by the antibody which partially neutralized all the large soluble chicken DNA polymerases (Table VII).

Therefore, the endogenous RNA-directed DNA polymerase activity from normal uninfected chicken embryos involves an insoluble DNA polymerase different from the large soluble chicken DNA polymerases.

Experiments are now under way to characterize the nature of the DNA polymerase.

CONCLUSIONS

The existence of this endogenous RNA-directed DNA polymerase activity in uninfected cells has numerous implications, especially in terms of the protovirus hypothesis (3,4). The protovirus hypothesis states that RNA-directed DNA synthesis is a mechanism involved in cellular differentiation, that carcinogens act on this process to

TABLE VI

Neutralization of Soluble Chicken DNA Polymerases[a]

DNA polymerase		Percent activity remaining
S-value	Purification	
10[b]	DEAE-B	0.5
7	DEAE-B	5
10	DEAE-A	10
5	DEAE-A	20
10	unabsorbed to PC	15
10	nuclear	20

[a]DNA polymerases were purified 100- to 200-fold from the soluble fraction of 7-day chicken embryos by successive chromatography on phosphocellulose and DEAE-cellulose, and sedimentation in a glycerol gradient. Antibody was prepared in a rat to the 10S polymerase purified on phosphocellulose and eluted from DEAE-cellulose at 0.3 M NaCl. IgG was prepared from immunized and control rats, incubated with polymerases (about 20 μg/ml) for 1 hour at room temperature, and the rate of residual calf thymus DNA-directed polymerase activity measured for 20 min at 40°. 100% activity was from 15,000 to 20,000 counts/min incorporated/hr.

[b]Used for immunization.

TABLE VII

Neutralization of DNA Polymerases from Cells and Virus by Antibodies to DNA Polymerases from Cells and Virus[a]

DNA polymerase	Antibody to DNA polymerase of	
	AMV[b]	Chicken[c]
RSV[d]	15[f]	130
Chicken[c]	110	5
Pellet-L[e]	165	10
Pellet-S[e]	115	115

[a]Neutralization and residual DNA polymerase activity were measured as described in the notes to Table VI.

[b]Avian myeloblastosis virus (7).

[c]10 S DEAE-B (Table VI) (13).

[d]B77 strain of avian sarcoma virus.

[e]DNA polymerases were isolated from a pellet fraction by KCl extraction and purified by chromatography on phosphocellulose and centrifugation in a glycerol gradient. L was about 10 S; S was about 3-4 S.

[f]Percent activity relative to control.

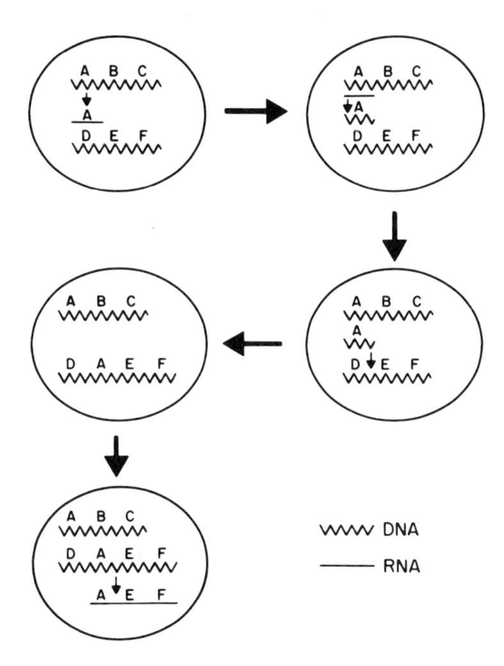

Fig 12. Model of protovirus hypothesis. By use of RNA-directed DNA synthesis a gene can be translocated and a recombinant RNA can be produced.

give rise to the genes for neoplastic transformation, and that RNA tumor viruses originated from this genetic system. The protovirus hypothesis is especially suited to explain gene amplification and recombination or reassortment (Fig 12).

SUMMARY

Virions of RNA tumor viruses and cells tranformed by RNA tumor viruses contain endogenous RNA-directed DNA polymerase activity. In Rous sarcoma virus-transformed chicken cells, this endogenous RNA-directed DNA polymerase activity appears to be in virion precursors. But in Rous sarcoma virus-transformed rat cells, neither the RNA template nor the DNA polymerase appear to be related to the transforming virus.

Uninfected chicken cells and normal chicken embryos contain endogenous RNA-directed DNA polymerase activity, as shown by its ribonuclease sensitivity, the nature of its early product, and hybridization of the product. The RNA template of this activity is not related to the RNAs of Rous sarcoma or reticuloendotheliosis viruses, and the DNA polymerase of this activity is not related to the DNA polymerases of avian leukosis or reticulonedotheliosis viruses or to the large soluble DNA polymerases of normal chicken embryos. Thus, this endogenous RNA-directed DNA polymerase activity from uninfected chicken cells is not related to RNA tumor viruses and may represent the activity suggested by the protovirus hypothesis and play an important role in normal development.

REFERENCES

1. Temin, H. M.: Mechanism of transformation by RNA tumor viruses. Ann. Rev. Microbiol., 25:609, 1971.
2. Temin, H. M., and Baltimore, D.: RNA-directed DNA synthesis and RNA tumor viruses. Adv. Virus Res., 17:129, 1972.
3. Temin, H. M.: Malignant transformation of cells by viruses. Perspectives Biol. Med., 14:11, 1970.
4. Temin, H. M.: The protovirus hypothesis. J. Natl. Cancer Inst., 46:III, 1971.
5. Coffin, J. M., and Temin, H. M.: Comparison of Rous sarcoma virus-specific deoxyribonucleic acid polymerases in virions of Rous sarcoma virus and in Rous sarcoma virus-infected chicken cells. J. Virol., 7: 625, 1971.
6. Coffin, J. M., and Temin, H. M.: Ribonuclease-sensitive deoxyribonucleic acid polymerase activity in uninfected rat cells and rat cells infected with Rous sarcoma virus. J. Virol., 8:630, 1971.
7. Nowinski, R. C., Watson, K. F., Yaniv, A., and Spiegelman, S: Serological analysis of the deoxyribonucleic acid polymerase of avian uncornaviruses. II. Comparison of avian deoxyribonucleic acid polymerases. J. Virol., 10:959, 1972.
8. Kang, C.-Y., and Temin, H. M.: Endogenous RNA-directed DNA polymerase activity in normal and uninfected chicken embryos. Proc. Natl. Acad. Sci. USA, 69:1550, 1972.
9. Temin, H. M., and Mizutani, S.: RNA-dependent Dna polymerase in virions of Rous sarcoma virus. Nature, 226:1211, 1970.
10. Hausen, P., and Stein, H.: Ribonuclease H.: An enzyme degrading the RNA moiety of DNA-RNA hybrids. Europ. J. Biochem., 14:278, 1970.
11. Molling, K. Bolognesi, D. P., Bauer, H., Busen, W., Plassman, H. W., and Hausen, P.: Association of viral reverse transcriptase with an enzyme degrading the RNA moiety of RNA-DNA hybrids. Nature New Biology, 234:240, 1971.
12. Peterson, D. A., Baxter-Gabbard, K. L., and Levine, A. S.: Avian reticuloendotheliosis virus (strain T): V. DNA polymerase. Virology, 47:251, 1972.
13. Mizutani, S.: in preparation.

12. STUDIES ON TRANSFORMING ACTIVITIES FROM HUMAN SOLID TUMOR CELLS FOLLOWING CO-CULTIVATION WITH HUMAN LEUKEMIC BONE MARROW CELLS[1]

Leon Dmochowski, Patton T. Allen, William A. Newton, Jr., J. Georgiades, Koshi Maruyama, James L. East and James M. Bowen

Department of Virology, The University of Texas at Houston
M.D. Anderson Hospital and Tumor Institute, Texas Medical Center
Houston, Texas 77025

INTRODUCTION

Both DNA and RNA viruses have now been shown to play an etiological role in neoplasia of different types in animals. However, the main thrust of research in the Department of Virology at The University of Texas at Houston M. D. Anderson Hospital and Tumor Institute has been directed toward studies of the RNA tumor viruses, and it is with this group of agents that this presentation will be concerned. RNA tumor viruses which bear the morphological designation "type C" have been shown to be the causative agents of various types of murine and feline leukemia and solid tumors, including: lymphoid, myeloid, and erythroid leukemia; lymphoid leukemia and autoimmune disease of mice of certain strains; reticulum cell neoplasms and Hodgkin's-like tumors of mice; sarcomas of mice and rats; bone tumors of mice, rats, and hamsters; and leukemia, lymphosarcoma, and fibrosarcoma of cats. These agents have also been isolated from and/or implicated in guinea pig leukemia, canine leukemia, and lymphosarcoma, in bovine leukemia, in fibrosarcoma of the wooly monkey, and in lymphosarcoma of the gibbon ape (1).

The past few years have seen the accumulation of very strong evidence that viruses may also play an etiological role in the development of human leukemia and of at least some types of solid tumors in humans. In fact, since the pioneering work of Gross (2), which showed the viral etiology of spontaneous leukemia of mice, our experimental approach has turned to the concept of a viral etiology for many human neoplasms. The demonstration, by Dmochowski and his co-workers, of virus particles of similar morphology in murine leukemia and in the tissues of some patients with leukemia and lymphoma (3 - 10) lent considerable impetus to this concept. Human cancers in which viruses have been or may be implicated include the same diverse types as those known to be associated with viruses in animals, including leukemia, lymphoma, and a variety of solid tumors, particularly breast cancer, different types of sarcomas of soft tissue, and malignant bone lesions. In addition, some other types of tumors not classified as particularly malignant, such as fibrosarcoma desmoides and the giant cell tumor of bone should be mentioned in the framework of viral studies.

[1] Supported, in part, by U.S.P.H.S. Contract PH 43-NCI-E-65-604 within the Special Virus Cancer Program of the National Cancer Institute, and, in part, by Grants CA-05831 and RR-05511 from the National Cancer Institute, National Institutes of Health, U.S.P.H.S.

Since the human being is obviously not an experimental animal and cannot be manipulated in biological studies, the establishment of a viral etiology for any human tumor is not a simple task. A variety of disciplines and experimental techniques must be brought to bear on the question, and the data obtained must of necessity be largely inferential. Our studies on the relationship of viruses to leukemia and the solid tumors of man have been based on a multifaceted approach utilizing morphological, immunological, biological, and biochemical methods and comprising the simultaneous application of appropriate animal models with studies on human clinical material.

EXPERIMENTAL STUDIES

As previously mentioned, the results of morphological studies have demonstrated the presence of virus-like particles in cells derived from animal tumors and in cells derived from the normal tissues of tumor-bearing animals. After biological studies had confirmed that these characteristic structures possessed the tumor-inducing activity, attention was turned to the search for similar particles in cells derived from human neoplasms. Morphological studies of human tumor materials have demonstrated the presence of type C virus particles in the lymph node tissues of patients with lymphocytic leukemia, lymphoma, and lymphosarcoma (5,6,8,10). Furthermore, type C virus particles have been found in cultured cells derived from human osteogenic sarcoma (10) and American-type Burkitt lymphoma (11). In addition, type C particles have recently been found in transformed cultured cells obtained by co-culturing cells of a fibrosarcoma desmoides with cells of a bone marrow aspirate from a patient with lymphocytic leukemia. The type C virus particles have been found after the transformed cells were grown in mice and then reestablished in tissue culture (12). Type C virus particles have also been demonstrated in cultures of human rhabdomyosarcoma cells passages in fetal kittens (13) and in cultures of rhabdomyosarcoma cells treated with chemical activators (14,15). The type C virus particles have been found in giant cell tumors, osteosarcomas, and liposarcomas (16 - 18).

As an over-all observation, it may be stated that there exists an essential similarity in the ultrastructure and sites of replication of type C virus particles in all types of leukemia, lymphoma, bone tumors, and other tumors of animals of all species and of type C virus particles observed in leukemia, lymphoma, soft tissue tumors, and bone tumors of man.

A point of interest and apparently of great importance is the difference in the number of type C particles encountered in spontaneous and induced leukemia and solid tumors of animals of different species. In most induced animal tumors, type C particles are encountered in great numbers and without difficulty, while in spontaneous animal tumors they occur in small numbers, infrequently, and can be found only after arduous search. Occasionally, however, cultures derived from spontaneous tumors yield type C particles in large numbers. In this respect the occurrence of virus particles in leukemia and solid tumors of man resembles that of virus particles in spontaneous animal tumors.

It is, therefore, necessary to resort to other methods and approaches in an attempt to clarify the relationship of viruses to some tumors of animals and to the neoplasia of man.

The discovery and subsequent characterization of mammalian sarcoma virus or viruses have established the relationship between leukemogenic and sarcomogenic viruses of animals and have led to further studies on the relationship of viruses to solid tumors of other animals and of man (1).

The murine sarcoma virus has been shown to induce foci of altered cells after infection of appropriate embryo cell cultures. The sarcoma viruses are defective and require coinfection of a helper leukemia virus for virus replication. Cells derived from virus-induced sarcomas of animals may not readily yield infectious sarcoma virus unless they contain a leukemia virus as a helper virus. The sarcoma virus can be rescued from the sarcoma cells by super-infection with an appropriate leukemia virus or co-cultivation of cells replicating leukemia viruses (19). The results of studies on animal model systems led to the application of similar retrieval techniques to the etiology of human leukemia and solid tumors.

At this point, it was important to bring to bear still another aspect of the multifaceted experimental approach to the study of the possible role of RNA viruses in animal and human enoplasia. The discovery, made independently by Temin and Mizutani (20), and by Baltimore (21), of the presence of RNA-dependent DNA polymerase in the virions of RNA tumor viruses has led to the development of specific and highly sensitive probes by which the possible role of oncogenic viruses in the origin of human tumors can be investigated in the laboratory. The sensitivity and potency of these biochemical probes in certain types of human neoplasia are already reflected in the recent work of Spiegelman and his co-workers (22,23), of Gallo and his co-workers (24,25), and in the work of other investigators (26).

We have applied the techniques and concepts gained from molecular studies of the RNA tumor viruses of animals to a study of cells from selected human neoplasms. These biochemical techniques have been combined with biological and morphological approaches in an effort to correlate different observations into a composite picture which would indicate the presence or involvement of a virus in the origin of the human tumors under study. The design of the investigation and the application of the various experimental approaches have been based directly on the findings in the animal model systems studied concomitantly with the study of human tumor material.

In experiments patterned after the animal sarcoma virus-helper leukemia virus model, cells derived from different types of human solid tumors (fibrosarcoma desmoides, giant cell tumor, osteosarcoma) were established in tissue culture and inoculated with cells derived from fresh bone marrow aspirates from patients with acute lymphocytic leukemia. Studies of this type were carried out independently by two different groups of investigators in our department. Some of the resulting cultures exhibited foci of morphologically altered cells. Rapidly growing epithelioid cells from foci which appeared in inoculated cultures of fibrosarcoma desmoides and giant cell tumor were established in tissue culture. The cells of the parent lines grew more slowly in a fibroblastic pattern and never exhibited any foci (27).

The appearanace of the foci in a culture derived from a fibrosarcoma desmoides after inoculation with the bone marrow cells of a patient with acute lymphocytic leukemia is shown in Figure 1.

When the cells derived from human osteosarcoma were inoculated with bone marrow cells from patients with acute lymphocytic leukemia, the cells of the co-

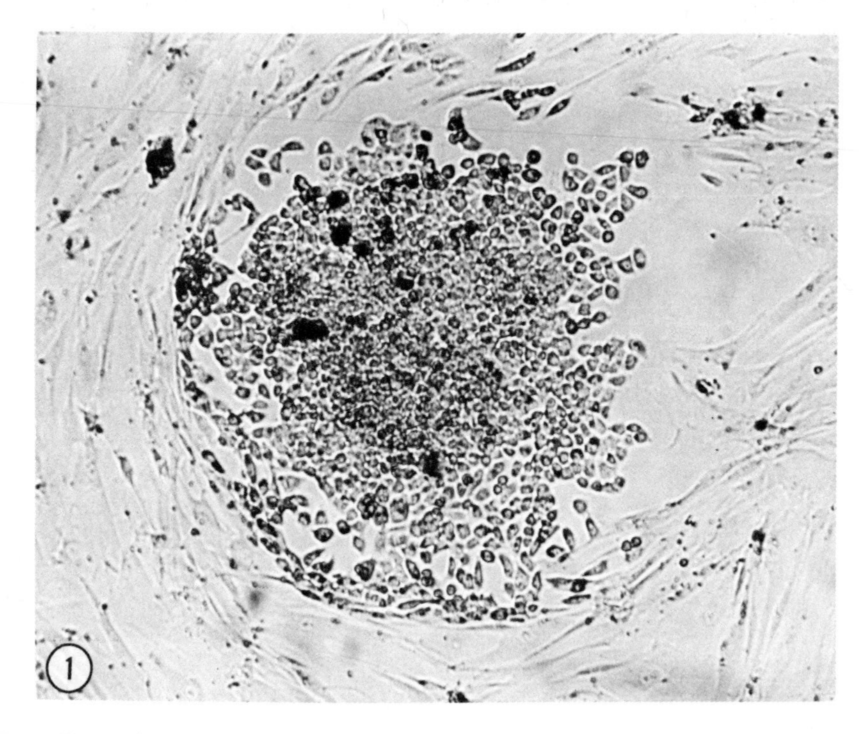

Fig 1. Focus of morphologically altered cells typical of those found in a culture derived from a fibrosarcoma desmoides (B) which appeared following inoculation with bone marrow cells from a patient (T) with acute lymphocytic leukemia. The cells may be distinguished from the surrounding fibroblast-like cells by their epithelioid morphology. The cell line derived from the cells of these foci was designated B-T. X 75.

culture exhibited growth characteristics distinct from cultures of the parental cells. Foci of morphologically altered cells appeared in these co-cultures, but could not be successfully isolated in culture. The appearance of foci in a culture derived from an osteosarcoma following inoculation with bone marrow cells from a patient with acute lyphocytic leukemia is shown in Figure 2. The cells of these foci usually remained unchanged for two to three weeks. Then their growth pattern changed, either forming more dense areas of pile-up cells, or the rounded cells gradually disappearing, and the remaining cells in the areas previously occupied by foci forming criss-cross patterns (Fig 3). The morphology of the cells derived from these co-cultures after long-term cultivation was fibroblast-like in contrast to those cells obtained by similar methods with the cultures of fibrosarcoma desmoides and giant cell tumor, which were epithelioid.

The cells from co-cultures grew more rapidly than the cells of either of the parental cell types (Fig 4). The cells of both the parental osteosarcoma line and the co-cultures were fibroblast-like in morphology but differed from each other. When cell-free supernatant fluid (0.45 um Millipore filtrates) from long-term (80 - 100 days) co-cultures was inoculated into cultures of human embryo cells (WHED169), foci of transformed cells appeared (Figs 5a-5e). These foci appeared after 5 - 8 subcultures. Noninoculated recipient cells and those inoculated with cell-free supernatant fluid from cultures of leukemic bone marrow or osteosarcoma did not show formation of foci.

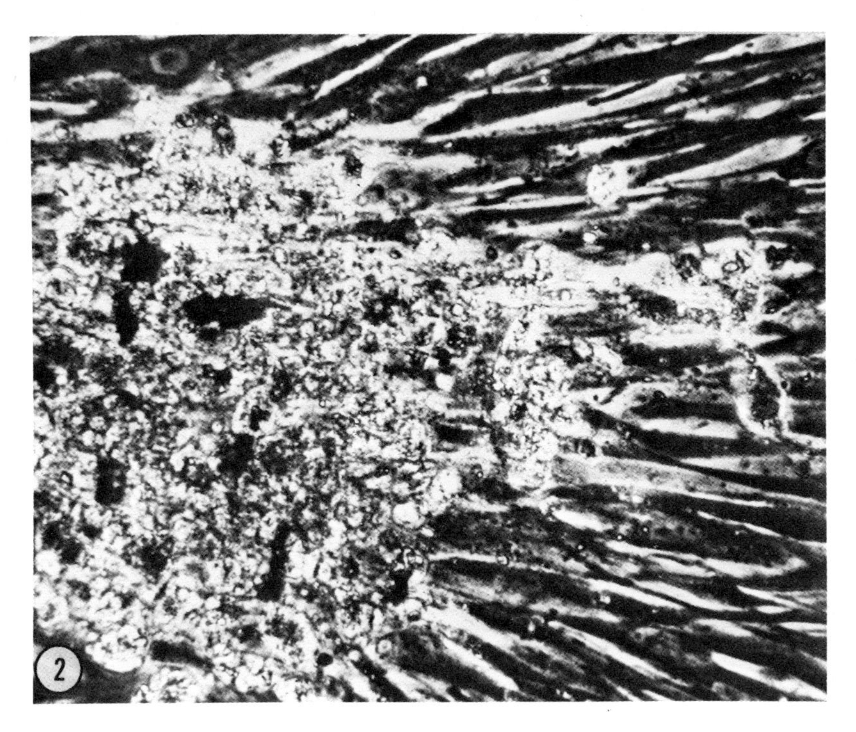

Fig 2. Phase contrast photomicrograph of a focus of rounded cells on top of the monolayer of fibroblastic cells. The rounded cells appeared after four weeks of co-cultivation of osteosarcoma biopsy specimens with leukemic bone marrow cells. X 190.

Cultures showing these foci were used for establishing suspension cultures. Filtrates (0.45 um) taken from these suspension cultures were tested again in human embryo recipient cells. After 1 - 3 months, colonies of fibroblast-like cells appeared in the recipient cultures. The cells from these colonies grew much faster than the nontransformed cells, and cultures consisting primarily of these transformed cells were readily obtained. In addition to the changes in morphology and growth dynamics, these cells exhibited the presence of a new antigen.

The results of indirect immunofluorescence tests of different human sera with parental human embryo cell cultures and with cells derived from morphologically altered human embryo cultures (WHE/"BCC") may be seen in Table I. The sera were tested before and after absorption for heterophile antibodies. None of the tested sera, even before absorption, gave positive immunofluorescence reaction with nontransformed human embryo cells. Only 2 our of 41 sera of apparently normal blood bank donors were positive. Fifty-two percent of sera of osteosarcoma patients gave positive reaction, and 55% of sera of patients with different types of leukemia were positive. The sera were positive in dilutions of 1:5 - 1:20. The results of absorption of the different sera with guinea pig kidney powder, sheep red blood cells, whole human embryo cells, and mycoplasma indicate a lower incidence of nonspecific (including heterophile and Forssman-like) antibodies in the sera of apparently normal donors than in the sera of patients with osteosarcoma or leukemia (Table I). The observed presence of heterophile and Forssman-like antibodies in the sera of patients with osteosarcoma and leukemia is in agreement with our previous observations based on a mixed

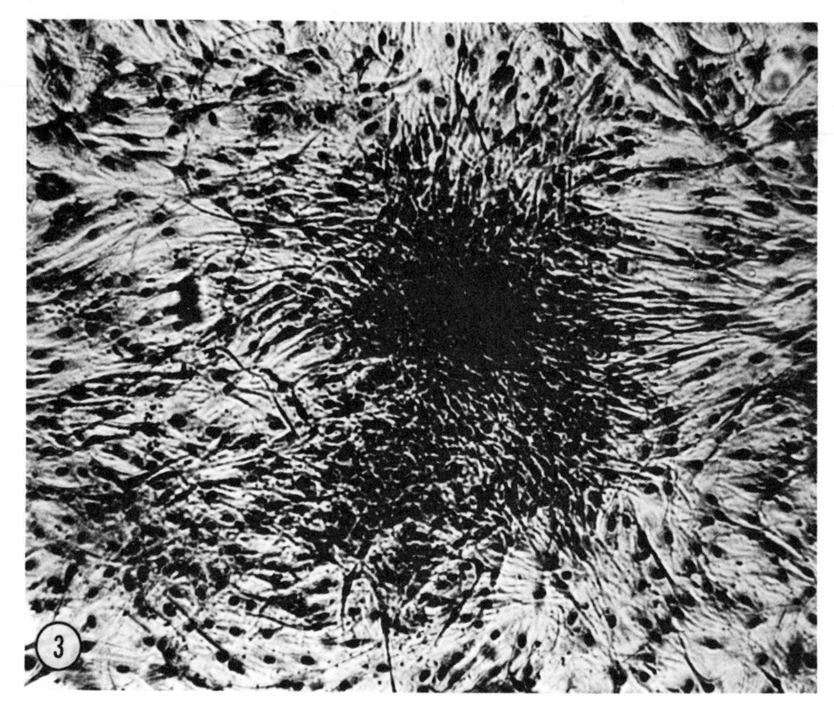

Fig 3. Giemsa-stained preparation of the culture shown in Figure 2 after 80 days of co-cultivation. The fibroblastic character of the co-culture has changed. The cells have lost their regular orientation, exhibiting unorganized growth patterns and, though still fibroblast-like, have changed to a criss-cross configuration. X 75.

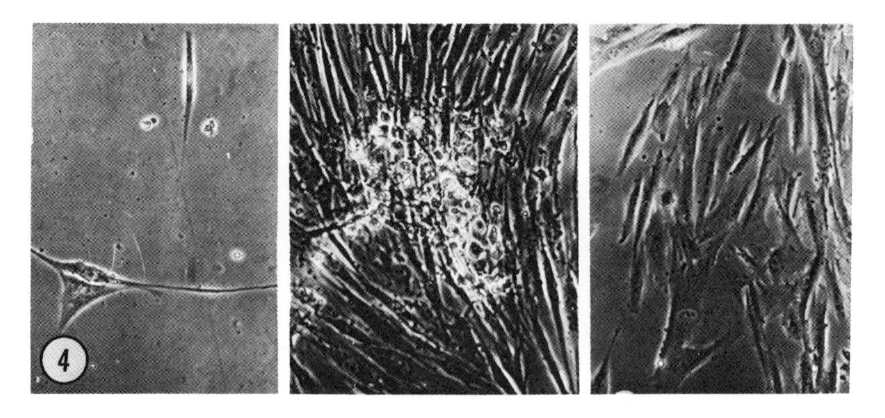

Fig 4. Phase contrast photomicrograph showing the appearance of cells of the two parental cell cultures and of the co-culture obtained from these two parental lines. The cells were photographed after 80 days of cultivation. A bone marrow culture is shown in the right panel, an osteosarcoma culture is shown in the left panel, and in the center panel the appearance of the co-culture with a focus of rounded cells is shown. X 175.

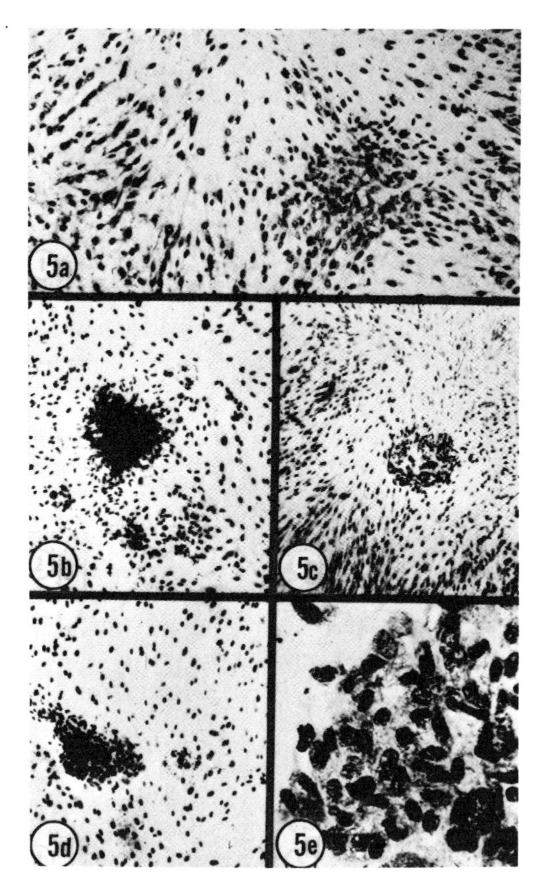

Fig 5. Photomicrograph of Giemsa-stained preparations of tissue cultures showing morphology of foci which appeared following treatment of whole human embryo cell cultures with cell-free supernatant medium from co-cultures of osteosarcoma and leukemic bone marrow cells.

Fig 5a. An early stage of focus formation characterized by a disorganized growth pattern may be seen. X 25.

Figs 5b, 5c, and 5d. Foci at later stages of development may be seen. X 25.

Fig 5e. The same focus as in Figure 5d shown at higher magnification. X 150.

hemadsorption test with sera of patients with different types of malignant disease (28). The appearance of cells showing positive cytoplasmic fluorescence with sera from patients with osteosarcoma may be seen in Figs 6a,b,c.

Electron microscopy of the parent cell lines and of cells from the foci of the different cultures failed to reveal the presence of type C virus particles. Biochemical studies were, therefore, carried out using reverse transcriptase as a probe in an attempt to apply a more sensitive means of detecting virus or viral gene expression in these cells. Fluids from the tissue cultures were concentrated by single-phase polyethylene glycol precipitation (25). The concentrates were centrifuged onto sucrose or ficoll cushions designed to retain particulate components with densities similar to those of members of the RNA tumor virus group. These preparations were tested for RNA-

TABLE I

Results of Fixed Immunofluorescence Test of Normal and "Transformed" Human Embryo Cells with Human Sera

Sera of	Normal cells[a] before absorption	"Transformed" cells[b] before absorption	"Transformed" cells[b] after absorption
Apparently normal blood bank donors	0/41	10/41	2/41
Patients with osteosarcoma	0/32	31/32	17/32
Patients with different types of leukemia and lymphoma	0/20	20/20	11/20

[a]WHED 169 cells
[b]WHED 169/0210/BCC
Numerators = positive sera
Denominators = number of sera tested

Note: Results of indirect immunofluorescence tests of selected sera of patients having neoplasms of various types and of apparently normal donors with normal and transformed human embryo cells. Cells were fixed in acetone. Sera were diluted 1:5 or higher with buffered saline (PBS). Sera in different dilutions were incubated with cells for 30 minutes. The cells were then washed three times with PBS and incubated with goat anti-human gamma globulin serum (Hyland). The slides were washed again and counterstained with Evans blue. After further washing with PBS and drying in air, the preparations were mounted in glycerol-buffered saline and examined with a Reichert fluorescence microscope, using a BG-12 exciter filter and a BG-9 barrier filter.

dependent DNA polymerase activity essentially according to the method of Gallo and his associates (25).

The results of DNA polymerase analyses of fractions obtained after isopycnic sucrose density gradient centrifugation of particulate components from one of the "transformed" cultures are shown in Figure 7 (26). This culture was established with cells derived from foci which appeared in a culture of fibrosarcoma desmoides inoculated with cells from the bone marrow of a patient with acute lymphocytic leukemia. This culture was designated as the B-T culture. When fractions from this gradient were analyzed, the peak of DNA polymerase activity was found at a density of 1.17 gm/ml. This polymerase activity was highly sensitive to RNAse and corresponds to that found in similar studies of RNA tumor virus particles released from productively infected animal cells (26).

The results obtained when the particles from this culture were subjected to rate zonal centrifugation analysis in a sucrose gradient are shown in Figure 8 (26). As may be seen, the position of ^{3}H-uridine labeled SD-MSV (SD-MSV-M) in an equivalent gradient is indicated by an arrow. The sedimentation behavior of the major peak of DNA polymerase activity from the transformed B-T cells closely approximated that of the reference tumor virus.

The results of the study of the kinetics of incorporation of tritiated thymidine monophosphate into acid insoluble material by the particulate components released

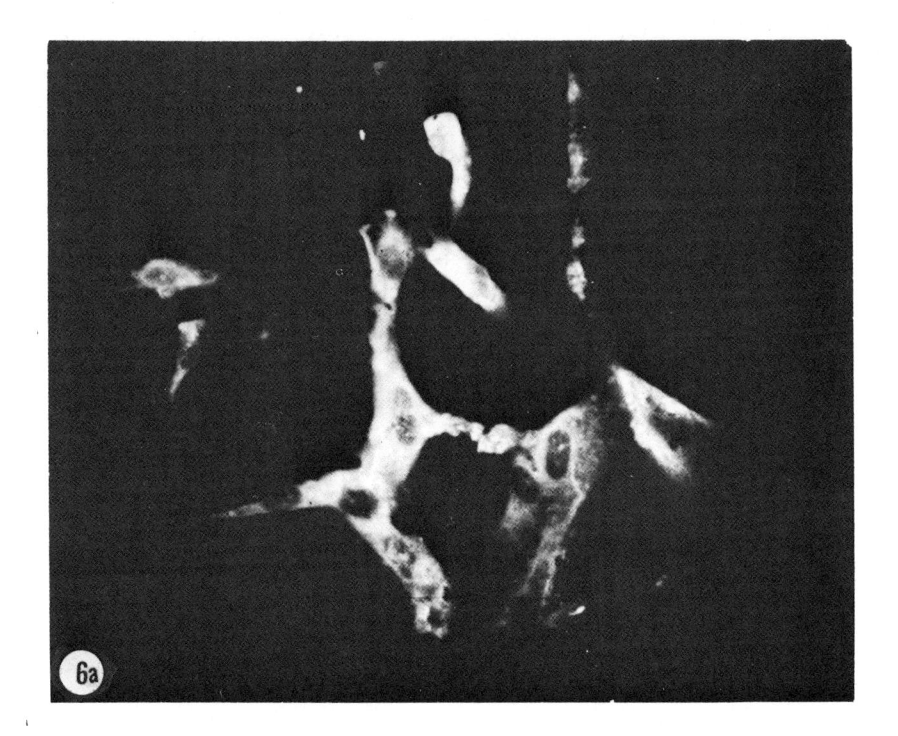

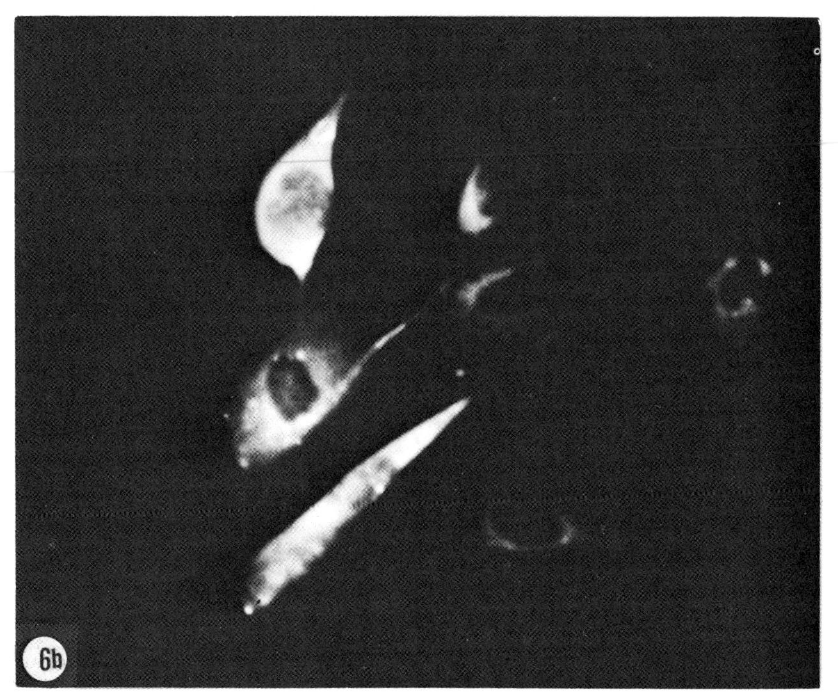

Fig 6. Appearance of fixed immunofluorescence reaction observed with sera of selected patients with osteosarcoma tested against cells from a transformed cell line (WHE/BCC-E).

Figs 6a and 6b. Positive cytoplasmic immunofluorescence with two of the sera may be seen. X 190.

Fig 6c. Negative reaction. X 190.

into the culture fluid by cells derived from foci appearing in cultures of fibrosarcoma desmoides cells inoculated with acute lymphocytic leukemia (ALL) bone marrow cells (B-T) confirm the susceptibility of the DNA-polymerase activity to ribonuclease (26).

Omission of magnesium or dATP from the reaction mixture resulted in a significant decrease in DNA synthesis. This demonstrated that the enzymatic activity observed is a true DNA polymerase. DNA polymerase activity released from B-T cells was sensitive to ribonuclease in the presence of 200 mM NaCl. In the presence of this high salt concentration ribonuclease sensitivity of the DNA polymerase reaction suggests that the enzyme utilizes an endogenous single-stranded RNA template. In this connection, it should be pointed out that all assays to be discussed in the present study were carried out with endogenous templates.

The results of $CsSO_4$ density gradient analysis of the DNA made by the B-T polymerase are shown in Figure 9 (26). As may be seen, the main peak of DNA banded at 1.52 gm/ml., a density characteristic of RNA:DNA hybrid molecules consisting of approximately equal moities of the two components. After RNAse treatment, the DNA banded at a density of 1.42 gm/ml, characteristic of free DNA. These data indicate that the product DNA from the B-T polymerase is synthesized along an RNA template with which it remains associated under these conditions. These results are similar to those obtained from studies of reverse transcriptase of the RNA tumor viruses of animals (29).

Since the activity of the particles released from B-T cells appeared to resemble that of RNA tumor viruses in several respects, it was important to determine if the presence of high molecular weight RNA characteristic of the genomic RNA of RNA tumor viruses could be detected in the B-T particles. The B-T cells were incubated in

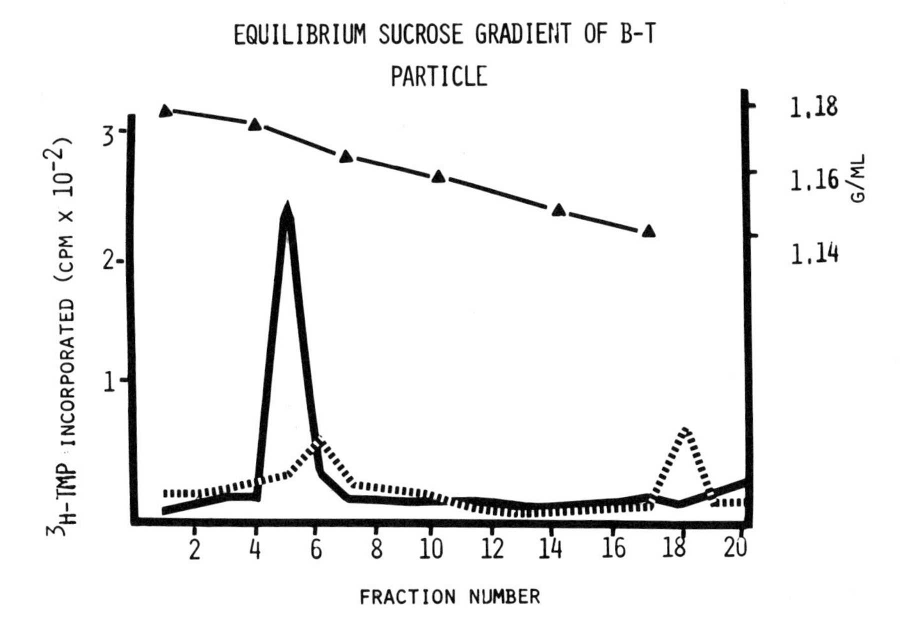

Fig 7. Sucrose density gradient analysis of the DNA polymerase activity found in the fluid of transformed cell cultures (B-T). DNA polymerase activity was concentrated from 1 liter of clarified culture fluid by single-phase polyethylene glycol precipitation as described earlier (Gallo et al 1971) (25). The resulting pellet was resuspended in NET buffer (100 mM NaCl, 10 mM Tris-HCl [pH 7.4] and 1 mM EDTA) at 0.01 times the original fluid volume. The activity was centrifuged through a sucrose/NET solution of 1.14 gm/ml onto a cushion of 1.18 gm/ml by centrifugation at 23,000 rpm in a Spinco SW 25.1 rotor for 90 min at 5°. The band of light-scattering material from above the cushion was diluted to 1.12 gm/ml with NET and layered onto a linear gradient of sucrose/NET. After centrifugation at 35,000 rpm in a Spinco SW 50 L rotor for 90 min at 5°, fractions were collected by bottom puncture and aliquots of each fraction assayed for ribonuclease-sensitive DNA polymerase activity. The standard 125 μl reaction mixture contained 25 μl of polymerase sample (in sucrose/NET), 40 mM Tris-HCl (pH 8.3), 14 mM magnesium acetate, 8 mM dithiothreitol, 0.64 mM each dATP, dGTP, and dCTP, and 12.5 μCi ^{3}H-dTTP (50 to 62 Ci/m mole). Prior to the addition of the other reaction components shown above, each sample aliquot was incubated for 15 min at 37° with Triton X-100 (0.08%) either alone or combined with pancreatic ribonuclease A (67 μg/ml) (Worthington Biochemical Corp.). DNA polymerase reactions were done at 37° and were terminated by adding 100 μg of yeast RNA and 2.0 ml of ice-cold 5% trichloroacetic acid (TCA) containing 0.02 M sodium pyrophosphate. After an additional 30 min at 0° the acid-insoluble material was collected on glass fiber filters (Reeve Angel), washed with 5% TCA and ethanol, and dried under heat lamps. Radioactivity was determined in a nuclear Chicago scintillation counter using a counting cocktail of toluene and omnifluor (New England Nuclear Corp.). The solid line indicates DNA polymerase activity without ribonuclease treatment, and the broken line indicates DNA polymerase activity remaining after treatment with ribonuclease (26, Fig. 5).

the presence of high levels of titriated uridine and the supernatant fluid was examined for the presence of particles containing high molecular weight RNA.

The results of this analysis are shown in Figure 10 (26). A high-speed pellet from the supernatant fluid of labeled B-T cultures was treated with sodium dodecyl sulfate and analyzed on a 15 – 30% sucrose gradient. The sedimentation profile of the RNA from the B-T particles is shown in the left panel of the figure. Feline leukemia virus RNA prepared identically and analyzed simultaneously is shown in the right

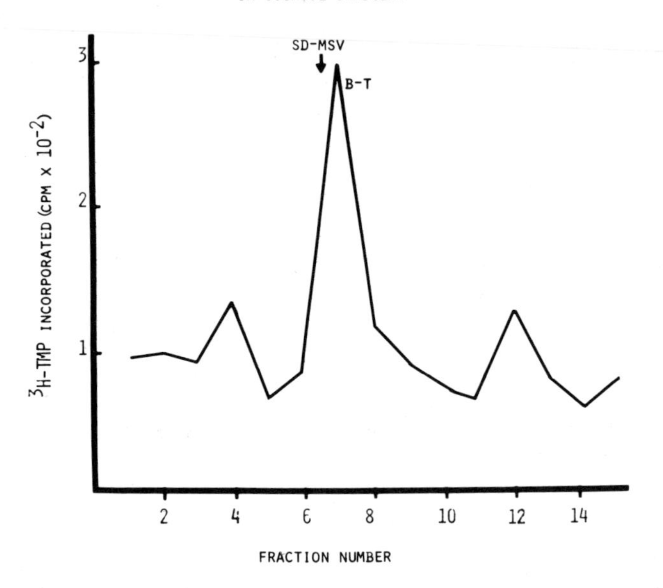

Fig 8. Rate zonal analysis of the DNA polymerase activity from the fluid of transformed cell cultures (B-T). The particles containing DNA polymerase activity were concentrated from 250 ml of culture fluid and partially purified by the sucrose cushion method as described in Figure 7. These particles were layered onto a linear gradient of 0.5 to 1.0 M sucrose in NET and centrifuged at 20,000 rpm for 30 min at 5° in a Spinco SW 50 L rotor. Fractions were collected by bottom puncture and assayed for DNA polymerase activity as described in Figure 7. The direction of migration was from right to left. The arrow indicates the position at which ^{3}H-uridine labeled SD-MSV was found under these conditions of centrifugation (26, Fig. 6).

panel. The arrows indicate the positions of 18 and 28 S cellular RNA and the 50 S RNA of Newcastle disease virus. The sedimentation coefficient of the high molecular weight RNA from the B-T particles was calculated to be 70 S. The feline leukemia virus RNA exhibited a sedimentation coefficient of 60 S (26).

Similar experiments were carried out on a cell line designated L-G, and derived from foci which appeared in a culture of cells from a giant cell tumor inoculated with bone marrow cells from a patient with acute lymphocytic leukemia. The results of kinetic analysis of the DNA polymerase released by L-G cells indicated the presence of RNAse sensitive DNA polymerase activity in the tissue culture fluid of L-G cells.

The product synthesized by the DNA polymerase released in L-G cell cultures was examined by rate zonal analysis in glycerol gradients. The major component in the L-G DNA polymerase reaction product sedimented at 60 S with minor components sedimenting more slowly (26). In addition, the results of uridine labeling experiments have shown the presence of high molecular weight RNA in the supernatant fluid from L-G cultures. The results are in agreement with those of Schlom and Spiegelman (3) and Schlom et al (22).

Several cell lines established from foci which appeared in human embryo cells inoculated with cell-free fluids from osteosarcoma and leukemia co-cultures have been

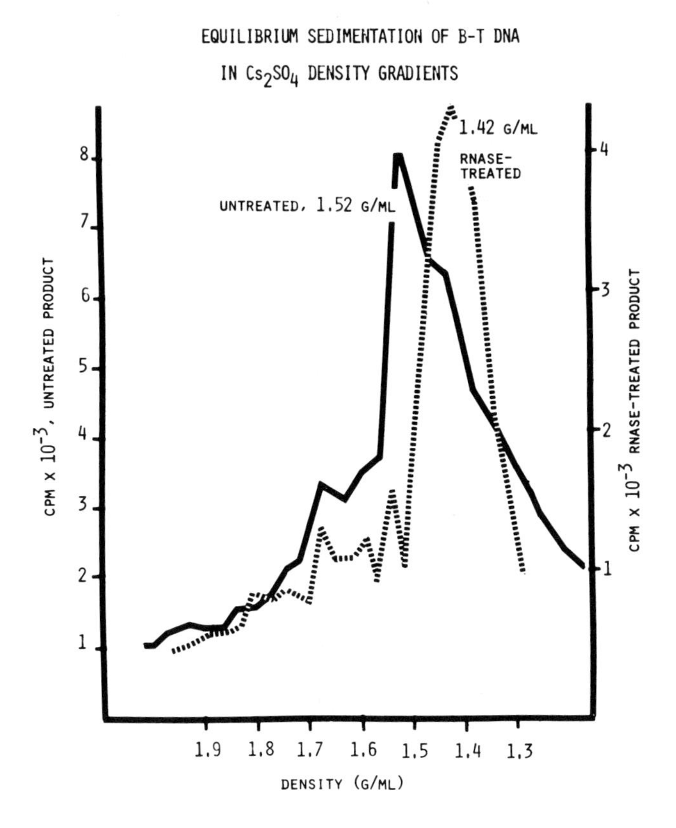

Fig 9. Cs_2SO_4 density gradient analysis of the product of the DNA polymerase activity found in the culture fluid of transformed cells (B-T). The particles containing DNA polymerase activity were concentrated and purified on a sucrose cushion as described in Figure 7. The product of a scaled-up reaction mixture was deproteinized by incubation with 2% SDS and pronase (500 μg/ml) for 30 min at 37°, followed by extraction with phenol and chloroform. After ethanol precipitation (2.5 vol at -20° for 18 hrs) the nucleic acid was dissolved in a buffer containing 10 mM NaCl, 10 mM Tris-HCl (pH 7.4), and 10 mM EDTA. Ribonuclease (pancreatic A, Worthington) (50 μg/ml) was added to one-third of the purified nucleic acid and incubated at 37° for 30 min. The ribonuclease-treated material and the untreated control were adjusted to 1.55 gm/ml with Cs_2SO_4 in NET and centrifuged for 72 hr at 37,500 rpm in a Spinco SW 50 L rotor at 5°. Radioactivity of each fraction was determined as described in Figure 7 (26, Fig. 4).

studied as described above using the presence of reverse transcriptase as a criterion for viral gene expression.

In these studies the methods of Schlom and Spiegelman (30) were utilized to test simultaneously for the presence of RNA-dependent DNA polymerase and of high molecular weight RNA in particles released into the culture medium by the transformed human embryo cells.

The products of the DNA polymerase reaction from each of the two transformed cultures designated as WHE/BCC and WHE/RLL were extensively deproteinized with SDS-, phenol, cresol, hydroxyquinoline and chloroform. The products were examined by rate zonal analysis in glycerol gradients in a SW41 rotor. The results are shown in Figures 11, 12. Analysis of the product made by the Soehner-Dmochowski

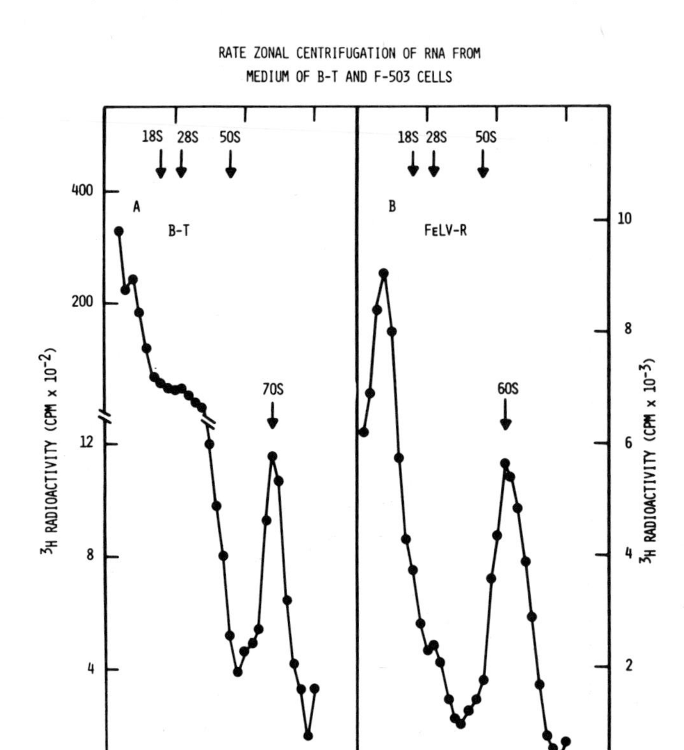

Fig 10. Sedimentation profile of RNA from particulate components in the fluid of transformed cell cultures (B-T) and of feline leukemia virus (Rickard) (FeLV-R) RNA. Fluid from cultures incubated 48 hrs with ^{3}H-uridine (100 μCi/ml) was clarified by centrifugation at 1,000 times g for 15 min. The supernatant fluid was centrifuged for 45 min at 30,000 rpm in a Spinco type 30 rotor. The resulting pellet was resuspended in NET buffer (100 mM NaCl, 1 mM EDTA, and 5 mM Tris-HCl, pH 7.4). Sodium dodecyl sulfate (SDS) was added to a concentration of 0.5%. The mixture was layered onto a gradient of 15 to 30% (W/W) sucrose in NET buffer containing 0.5% SDS and centrifuged for 16 hrs at 20° in a Spinco SW 25.1 rotor at 17,000 rpm. Fractions were collected from the top with an Isco fractionator and assayed for acid-insoluble radioactivity. The arrows indicate the locations of 18S and 28S cellular RNA and the 50S RNA of Newcastle disease virus (26, Fig. 7).

strain of murine sarcoma virus is shown in the right hand panel of Figure 11 for comparison. The arrows indicate the positions of 50 S Newcastle disease virus RNA and of 16 S and 28 S RNA of mouse embryo cells. The DNA made by the particles released by the cells of the transformed culture WHE/BCC (Fig 11, left panel) had a major component which sedimented at 31 S. This component is similar in size to the subunits of oncorna virus genomic RNA (31,32) and to one of the DNA:RNA hybrid components found by other investigators in the early reaction product of RNA tumor viruses (30,33). A smaller peak of similar molecular size is seen in the reaction product of murine sarcoma virus in the right panel. Schlom, Spiegelman, and Moore (22) reported that the product made by several preparations of human milk particles had only 35 S hybrid component.

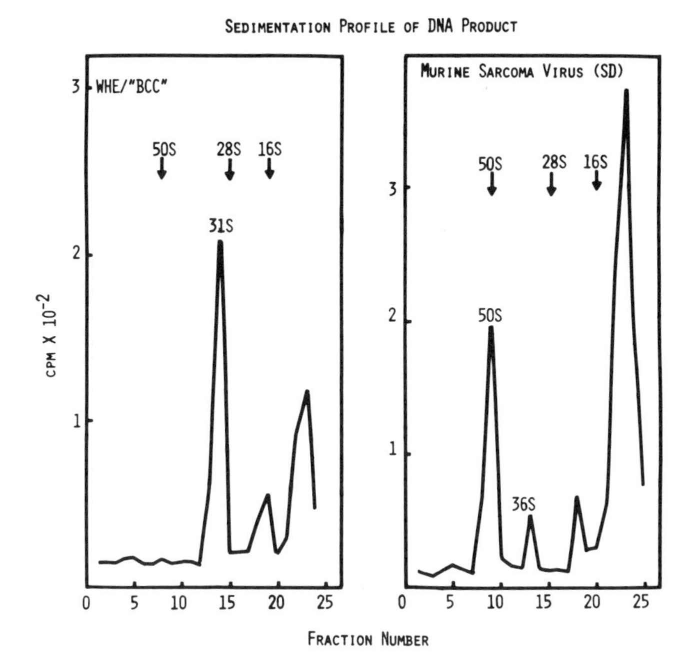

Fig 11. Sedimentation profile of the product of the DNA polymerase activity from the fluid of human embryo cell cultures (WHE/"BCC") transformed as described in the text. Clarified culture fluid was concentrated with polyethylene glycol as in Figure 7. Particulate components in the concentrate were centrifuged through a solution of 15% ficoll in NET buffer onto a cushion of 40% ficoll in NET buffer (90 min at 25,000 rpm in a Spinco SW 41 rotor at 5°). The band of light-scattering material on the cushion was diluted with 1.5 volumes of NET buffer and centrifuged for 60 min at 35,000 rpm in a Spinco type 40 rotor at 5°. The resulting pellet was resuspended in TDN buffer (50 mM Tris-HCl, pH 8.3, 20 mM dithiothreitol, 30 mM NaCl) and stored at -70°. The suspension containing 625 μg of protein was incubated 10 min at 30° with Triton times 100 (0.034%). Then product DNA was synthesized at 30° in a reaction mixture containing the following components: 50 mM Tris-HCl (pH 8.3), 10 mM magnesium acetate, 20mM dithiothreitol, oligo dT (dT_9, P & L Biochemicals) 10 μg/ml, 0.53 mM dGTP, dCTP and dATP, and ^{3}H-dTTP (33 μCi/ml). The reaction was terminated after 30 min by adding EDTA to 10 mM, SDS to 2%, and NaCl to 400 mM. The DNA product was extracted first with 1 volume of a mixture of phenol, m-cresol, and 8-hydroxyquinoline (30), then with a volume of chloroform. After low-speed centrifugation, the nucleic acid in the aqueous phase was precipitated at -20° with 2.5 volumes of ethanol in the presence of yeast RNA. The precipitate was resuspended in NET buffer and analyzed by rate zonal centrifugation in a gradient of 10 to 30% (w/w) glycerol in NET buffer containing 0.1% SDS for 2½ hrs at 41,000 rpm in a Spinco SW 41 rotor at 15° C. The acid-insoluble radioactivity in each fraction was determined (see Fig 7). The arrows show the positions of 16S and 28S mouse embryo cell RNA and the 50S RNA of Newcastle disease virus under these conditions of centrifugation. The right hand panel shows the sedimentation profile of the DNA product of purified SD-MSV prepared and analyzed identically in the same experiment.

The results of rate zonal analysis of the DNA product made by the polymerase released by transformed cells of another culture (WHE/RRL) is shown in Figure 12.

As may be seen (left and center panels of Fig 12) a fast sedimenting product component of 46 to 50 S was made under standard reaction conditions. However, this

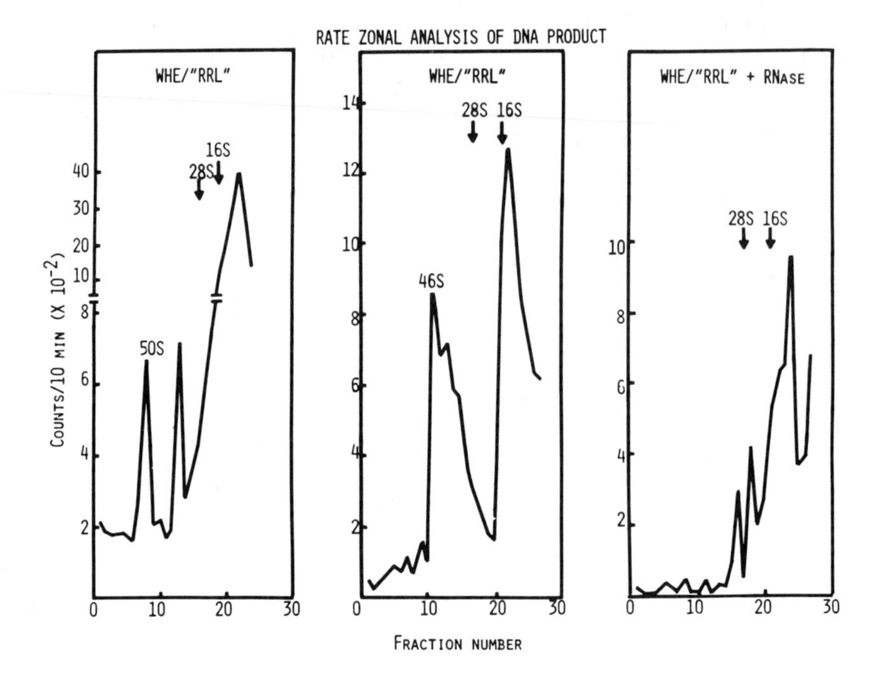

Fig 12. Sedimentation profile of the product of the DNA polymerase activity from the fluid of human embryo cell cultures (WHE/"RRL") transformed as described in the text. The DNA product of the left and center panels was made and analyzed as described in Figure 11. That of the right panel was made in a reaction mixture containing ribonuclease (14 μg/ml) in addition to the other components.

heavy component was absent in the product made in the presence of ribonuclease at a concentration of 14 ug/ml (right panel of Fig 12). This indicates that the heavy component is a DNA:RNA hybrid composed of small pieces of DNA associated with a large RNA molecule.

The presence of particles with reverse transcriptase activity and high molecular weight RNA in the transformed cells might be explained by activation of a covert or repressed sarcoma virus genome. Should this take place, it would be reasonable to expect occasional or minimal activity in the parent cultures. Careful examination of concentrates of the culture fluid of the human embryonic cells found susceptible to transformation has occasionally demonstrated reverse transcriptase activity in these cultures.

The results of two experiments in which the DNA product made by the polymerase present in the fluid of uninoculated human embryo cultures was analyzed in rate zonal glycerol gradients are shown in Fig 13. In one experiment, a 28 S component was seen. However, when the experiment was repeated with the fluid of the culture from a later passage this component was absent. The 28 S component is similar in size to the viral genomic RNA subunits found after mild denaturation in preparations of oncorna viruses incubated at 37° for several hours (31), and in newly released virions of SD-MSV (34). It can be concluded from the above results that a DNA:RNA hybrid molecule similar in size to oncorna virus genomic RNA subunits

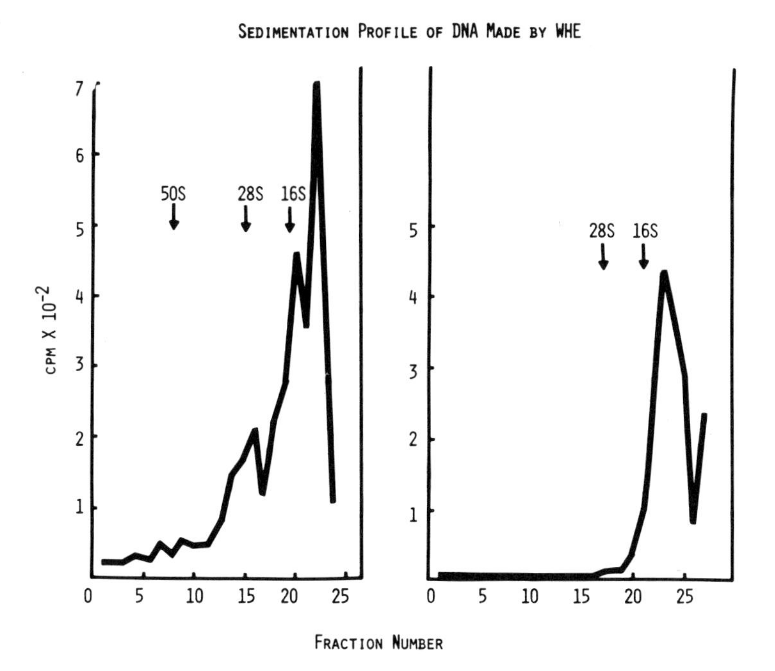

Fig 13. Sedimentation profile of the product of the DNA polymerase activity from the fluid of nontransformed human embryo cell cultures (WHE). The DNA product was made and analyzed as described in Figure 11. The results of two different experiments are shown.

may be present in the reaction product of the polymerase from fluid of the uninoculated cultures, but it is more difficult to demonstrate than that produced by the transformed cells. The observed difference in results of the two experiments and in other experiments with the nontransformed human embryo cultures may, at least in part, be due to the behavior of active cells in different culture passages.

The parent fibrosarcoma desmoides (B) and giant cell tumor (L) cell lines have recently been shown to possess ribonuclease-sensitive DNA polymerase activity, although at levels considerably lower than those of the DNA polymerase activity of the transformed cell lines (Fig 14).

The buoyant density of the particles exhibiting DNA polymerase activity, which were produced by the transformed human embryo cell culture designated WHE/"RLL," was analyzed by equilibrium centrifugation in 15 – 40% ficoll gradients (Fig 15). The main DNA polymerase component had a buoyant density of 1.14 gm/ml in ficoll. Pretreatment with ribonuclease inhibited the reaction by over 40%. The peak fraction containing this activity coincided with a peak in protein concentration. The addition of oligo dT to the assay mixture increased the incoporation of ^{3}H-TMP by this fraction to 2500 cpm, a 50% stimulation. Ribonuclease inhibited that reaction by 30% (data not shown). A minor component having DNA polymerase activity was seen in the WHE/"RRL" concentrate at a density of 1.12 gm/ml, but it was not inhibited by ribonuclease treatment. The density in ficoll of mouse mammary tumor virus particles was reported to be 1.14 gm/ml (35), while that of Rauscher murine leukemia virus was reported to be 1.12 gm/ml (36).

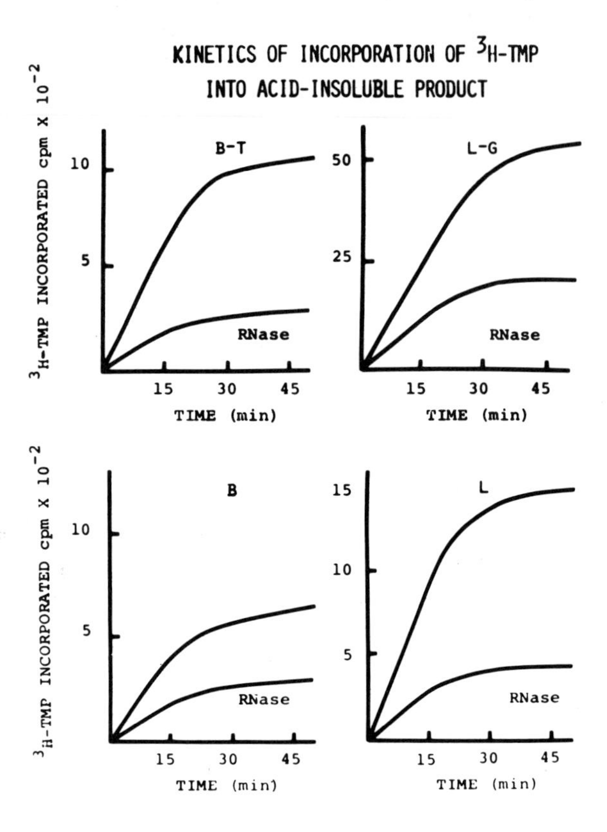

Fig 14. Reaction kinetics of the DNA polymerase activity from fluids of transformed and nontransformed human tumor-derived cells. The polymerases were prepared and assayed as described in Figure 7.

The buoyant density of these particles in sucrose density gradients has also been examined (Fig 16). Concentrates of the fluid from the transformed cell cultures WHE/"BCC" and WHE/"RRL" and from the uninoculated WHE culture were first centrifuged through 15% ficoll solutions onto cushions of 40% ficoll. They were then centrifuged to approach equilibrium in linear sucrose gradients and fractionated by bottom puncture. All preparations including the one from uninoculated WHE cultures contained DNA polymerase activity which banded near 1.2 gm/ml. The DNA polymerase from the uninoculated WHE cultures was not inhibited by pretreatment with ribonuclease, while that from transformed cells exhibited sensitivity to ribonuclease of up to 40%. The peak of activity of 1.16 gm/ml in the WHE/RRL" gradient (center panel, Fig 16) was also seen in other gradients of this material. However, it was not consistently sensitive to ribonuclease, while that at 1.2 gm/ml repeatedly exhibited partial ribonuclease sensitivity.

These data indicate that the component containing ribonuclease-sensitive DNA polymerase activity almost coincides in density with particles which contain DNA-dependent DNA polymerase activity. Due to the inability to separate clearly these components on the basis of their buoyant densities, the degree of ribonuclease sensitivity was only partial and a distinct peak of RNA-dependent DNA polymerase activity was not resolved.

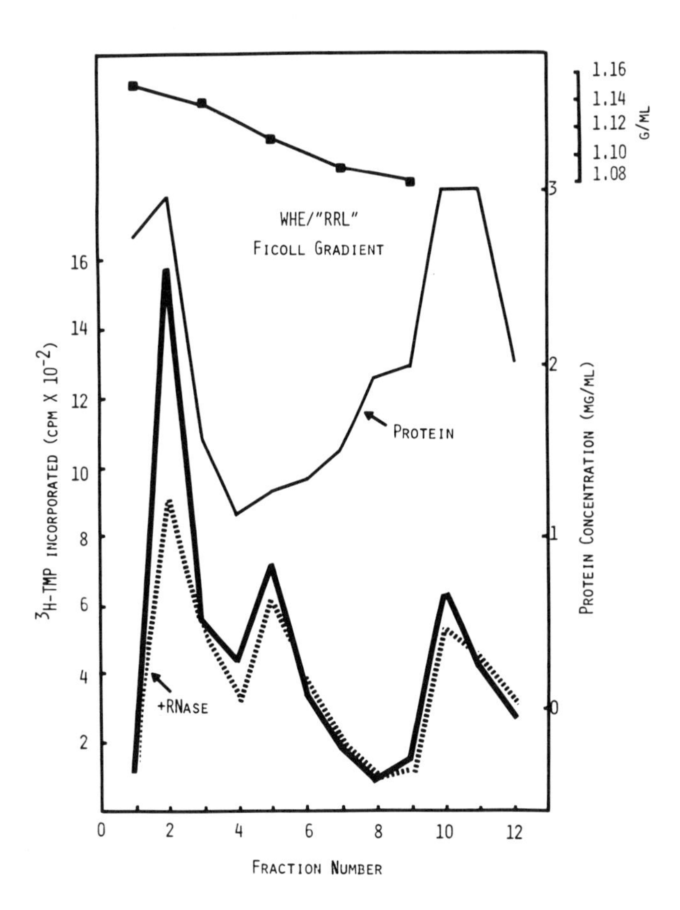

Fig 15. Ficoll density gradient analysis of the DNA polymerase activity from fluid of human embryo cells (WHE/"RRL") transformed as described in the text. The polyethylene glycol concentrate from 300 ml of medium was layered onto a gradient of 15 to 14% (w/w) ficoll in NET buffer and centrifuged 3½ hrs at 25,000 rpm in a Spinco SW 41 rotor at 5°. Fractions of 1 ml each were collected and stored at -70°. They were assayed for DNA polymerase activity under the reaction conditions described in Figure 11. Where present, ribonuclease concentration was 12 μg/ml.

The sedimentaion behavior in rate zonal sucrose gradients of the particles containing DNA polymerase activity which were found in human embryo cell cultures was studied. Concentrates of fluid from transformed WHE/"BCC" and uninoculated WHE cultures were analyzed by centrifugation for 30 min at 20,000 rpm in 0.5 to 1.0 M sucrose gradients in a Spinco SW 39L rotor. The resulting fractions were assayed for ribonuclease-sensitive DNA polymerase activity (Fig 17). The concentrates from both transformed and nontransformed human embryo cultures contained a major component exhibiting DNA polymerase activity which sedimented somewhat more slowly than did SD-MSV particles under these conditions. This polymerase activity was inhibited by ribonuclease. A third gradient of WHE/"BCC" (not shown) gave virtually identical results. From these results, it appears that the particles containing the DNA polymerase activity are somewhat smaller than type C virus particles.

The level of polymerase activity in the concentrate from nontransformed cells was about a third of that from the cultures of transformed cells. This was in agreement

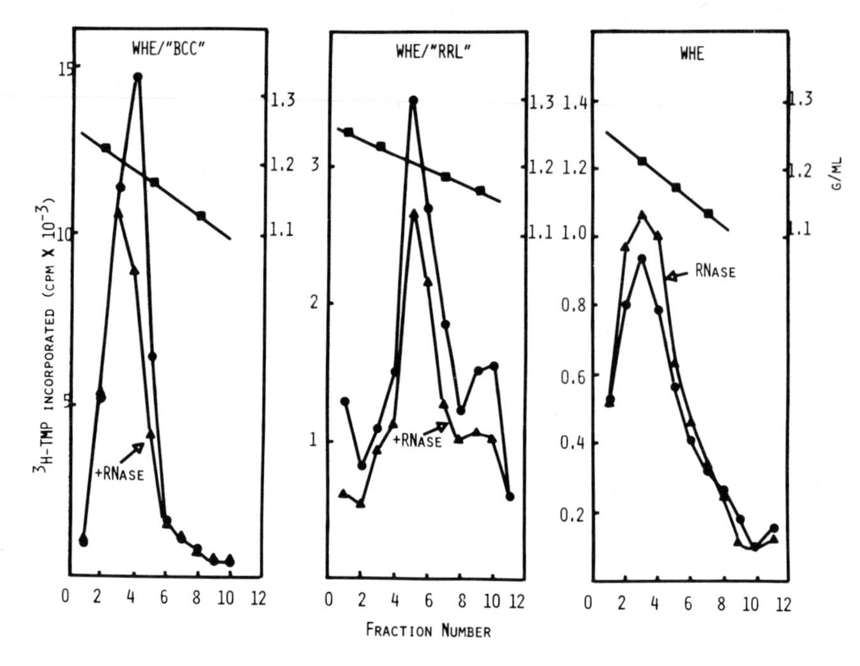

Fig 16. Sucrose density gradient analysis of the DNA polymerase activity from fluid of human embryo cell cultures. DNA polymerase preparations were prepared and centrifuged as described in the text. DNA polymerase activity was assayed as described in Figure 7, except the concentration of Triton times 100 was 0.1% during preincubation.

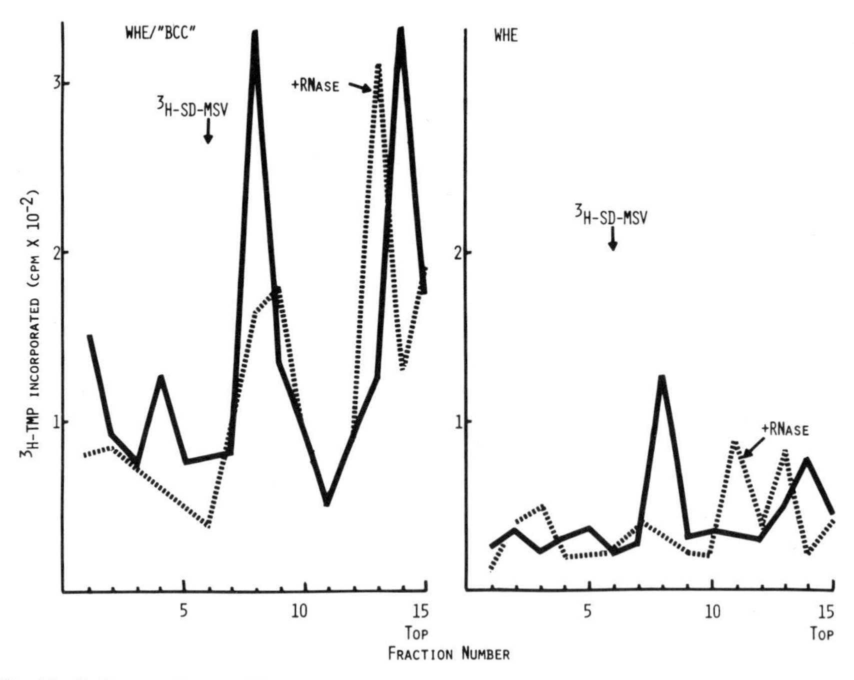

Fig 17. Sedimentation profiles of DNA polymerase activity from fluids of transformed (WHE/ "BCC") and nontransformed (WHE) human embryo cell cultures. Centrifugation conditions are described in the text. DNA polymerase activity in each fraction was assayed as described in Figure 16.

with earlier results, which indicated that a low level of ribonuclease sensitive DNA polymerase activity could occasionally be found in fluids from the uninoculated cultures.

These observations suggest that the origin of at least some solid tumors of man as well as that of human leukemia may be associated with particles with properties similar to those of viruses which cause tumors in animals.

SUMMARY

Cells derived from human tumors classified as giant cell tumor of bone, osteosarcoma, and fibrosarcoma desmoides were established in cell culture and then inoculated with bone marrow cells from patients with acute lymphocytic leukemia. Foci of transformed cells appeared in the inoculated giant cell tumor and fibrosarcoma desmoides cultures, and transient foci of rounded cells appeared in the inoculated osteosarcoma cultures. The cells obtained from these inoculated cultures when passaged in tissue culture exhibited morphology and growth characteristics distinct from the cells of cultures derived from any of the parental cell types. Long-term cultures (80 - 100 days) derived from the inoculated osteosarcoma cultures released a filterable (0.45 um Millipore) factor into the culture fluid which transformed cells in human embryo cultures. The factor in turn was released by the transformed human embryo cells. Several lines of the transformed cells were established.

The transformed human embryo cells revealed the presence of a new cytoplasmic antigen, as shown by the fixed immunofluorescence test with sera of patients with osteosarcoma, leukemia, and lymphoma.

Molecular studies demonstrated that the supernatant fluid of these transformed cell cultures contained particulate components exhibiting endogenously instructed reverse transcriptase activity. The fluid of uninoculated giant cell tumor, fibrosarcoma desmoides, and nontransformed human embryo cell cultures also contained reverse transcriptase activity, but the level of activity was lower than that from the transformed cells and was not detected at all in some of the assay attempts.

The particles with reverse transcriptase activity present in fluid of transformed cells derived from fibrosarcoma desmoides and giant cell tumor were similar in size and density to RNA tumor viruses, while those in fluid from transformed whole embryo cells were similar in density to viral nucleoids. The DNA synthesized by all particles was associated with high molecular weight RNA similar in size to the genomic RNA or RNA subunits of RNA tumor viruses. The role of these particles in cellular transformation remains to be determined.

REFERENCES

1. Proceedings of the Fourth International Symposium on Comparative Leukemia Research. Edited by W. Dameshek and R. M. Dutcher. Cherry Hill, New Jersey, 1969. S. Karger, Basel and New York, 1970.
2. Gross, L.: "Spontaneous leukemia" developing in C_3H mice following inoculation in infancy with AK leukemic extracts or AK embryos. Proc. Soc. Exp. Biol. Med., 109:27, 1951.

3. Dmochowski, L., and Grey, C. E.: Subcellular structures of possible viral origin in some mammalian tumors. Ann. N.Y. Acad. Sci., 68 (art. 2.):559, 1957.
4. Dmochowski, L., and Grey, C. E.: Studies on submicroscopic structure of leukemias of known or suspected viral origin: a review. Blood, 13:1017, 1958.
5. Dmochowski, L.: Electron microscope studies of leukemia in animals and man. In: Subviral Carcinogenesis, First International Symposium on Tumor Viruses. Edited by Y. Ito. Nagoya, Japan, 1966, "Nissha" Printing Co., Kyota, p 362, 1967.
6. Dmochowski, L.: Recent studies on leukemia and solid tumors in mice and man. In: Proceedings of Third International Symposium on Comparative Leukemia Research. Edited by H. J. Bendixen. Paris, 1967. S. Karger, Basel and New York, p 285, 1968a.
7. Dmochowski, L.: Ultrastructural studies in leukemia. In: Perspectives in Leukemia. Symposium of the Leukemia Society of America, Inc. Edited by W. Dameshek and R. M. Dutcher. Grune and Stratton, London and New York, p 34, 1968b.
8. Dmochowski, L.: Comparison of leukemia and sarcomogenic viruses at the ultrastructural level. In: Proceedings of the Fourth International Symposium on Comparative Leukemia Research. Edited by W. Dameshek and R. M. Dutcher. Cherry Hill, New Jersey, 1969. S. Karger, Basel and New York, p 62, 1970a.
9. Dmochowski, L.: Current status of the relationship of viruses to leukemia, lymphoma and solid tumors. In: Leukemia-Lymphoma. A Collection of Papers Presented at the Fourteenth Annual Clinical Conference on Cancer, 1969, at the University of Texas M. D. Anderson Hospital and Tumor Institute, Houston. Yearbook Medical Publishers, Inc., Chicago, p 37, 1970b.
10. Dmochowski, L.: Studies on the relationship of viruses to leukemia and solid tumors in man. In: Oncology, 1970. The Proceedings of the Tenth International Cancer Congress. Edited by R. L. Clark, R. W. Cumley, J. E. McCay. Yearbook Medical Publishers, Inc., Chicago, p 134, 1971.
11. Priori, E. S., Dmochowski, L., Myers, B., and Wilbur, J. R.: Constant production of type C virus particles in a continuous tissue culture derived from pleural effusion cells of a lymphoma patient. Nature New Biology, 232:61, 1971.
12. Maruyama, K., and Dmochowski, L.: In preparation, 1971.
13. McAllister, R. M., Nicolson, M., Gardner, M. B., Rongey, R. W., Rasheed, S., Sarma, P. S., Huebner, R. J., Hatanaka, M., Oroszlan, S., Gilden, R. V., Kabigting, A., and Vernon, L.: C-type virus released from cultured human rhabdomyosarcoma cells. Nature New Biology, 235:3, 1972.
14. Stewart, S. E., Kasnic, G., Draycott, C., and Ben, T.: Activation of viruses in human tumors by 5-iododeoxyuridine and dimethyl sulfoxide. Science, 175:198, 1972a.
15. Stewart, S. E., Kasnic, G., Draycott, C., Feller, W., Golden, A., Mitchell, E., and Ben, T.: Activation in vitro, by 5-iododeoxyuridine, of a latent virus resembling C-type virus in a human sarcoma cell line. J. Natl. Cancer Inst., 48:273, 1972b.
16. Priori, E. S., Wilbur, J. R., and Dmochowski, L.: Studies of sera from patients with osteosarcoma by immunofluorescence test. Abstract. In: Proceedings of the 61st Annual Meeting of the American Association for Cancer Research. Philadelphia, April, 9:64, 1970.
17. Morton, D. L., Hall, W. T., and Malmgren, R. A.: Human liposarcomas: tissue cultures containing foci of transformed cells with virus particles. Science, 165:813, 1969.
18. Morton, D. L., and Malmgren, R. A.: Human osteosarcomas: immunological evidence suggesting an associated infectious agent. Science, 162:1279, 1968.
20. Huebner, R. J., Hartley, J. W., Rowe, W. P., Lane, N. T., and Capps, W. I.: Rescue of the defective genome of Moloney sarcoma virus from a noninfectious hamster tumor and the production of pseudotype sarcoma viruses with various leukemia viruses. Proceedings of the National Academy of Sciences, USA, 56:1164, 1966.
21. Temin, H. M., and Mizutani, S.: RNA dependent DNA polymerase in virions of Rous sarcoma virus. Nature, 226:1211, 1970.

22. Baltimore, D.: Viral RNA dependent DNA polymerase. Nature, 226:1209, 1970.

23. Axel, R., Schlom, J., and Spiegelman, S.: Presence in human breast cancer of RNA homologous to mouse mammary tumor virus RNA. Nature, 235:32, 1972.

24. Gallo, R. C., Yang, S. S., and Ting, R. C.: RNA dependent DNA polymerase of human acute leukemia cells. Nature, 228:927, 1970.

25. Gallo, R. C., Sarin, P. S., Allen, P. T., Newton, W. A., Priori, E. S., Bowen, J. M., and Dmochowski, L.: Reverse transcriptase in type C virus particles of human origin. Nature New Biology, 232:140, 1971.

26. Allen, P. T., Newton, W. A., East, J. L., Maruyama, K., Bowen, J. M., Georgiades, J., Priori E. S., and Dmochowski, L.: Molecular studies of cells derived from human solid tumors. In: Molecular Studies in Viral Neoplasia. A Collection of Papers Presented at the 25th Annual Symposium on Fundamental Cancer Research. University of Texas M. D. Anderson Hospital and Tumor Institute at Houston, in press, 1972.

27. Maruyama, K., Dmochowski, L., Romero, J. J., and Wagner, S. H.: Studies of human cells infected by leukemia viruses. In: Proceedings of the Fifth International Symposium on Comparative Leukemia Research, 1972, in press.

28. Maruyama, K., Dmochowski, L., Bowen, J. M., and Hales, R. L.: Studies on human leukemia and lymphoma cells by membrane immunofluorescence and mixed hemadsorption tests. Texas Reports on Biology and Medicine, 26:545, 1968.

29. Spiegelman, S., Burny, A., Das, M. R., Keydar, J., Schlom, J., Travnicek, M., and Watson, K.: Characterization of the products of RNA-directed DNA polymerases in oncogenic RNA viruses. Nature, 227:563, 1970.

30. Schlom, J., and Spiegelman, S.: Simultaneous detection of reverse transcriptase and high molecular weight RNA unique to oncogenic RNA viruses. Science, 174:840, 1971a.

31. Watson, J. D.: The structure and assembly of murine leukemia virus; intracellular viral RNA. Virology, 45:586, 1971.

32. Bader, J. P., and Steck, T. L.: Analysis of the ribonucleic acid of murine leukemia virus. J. Virology, 4:454, 1969.

33. Schlom, J., and Spiegelman, S.: DNA polymerase activities and nucleic acid components of virions isolated from a spontaneous mammary carcinoma from a rhesus monkey. Nat. Acad. Sci. USA, 68:1613, 1971b.

34. East, J. L., Allen, P. T., Bowen, J. M., and Dmochowski, L.: In preparation, 1972.

35. Lyons, M. J., and Moore, D. H.: Isolation of the mouse mammary tumor virus: chemical and morphological studies. J. Natl. Cancer Inst., 35:549, 1965.

36. Oroszlan, S., Johns, L. W., and Rich, M. A.: Ultracentrifugation of a murine leukemia virus in polymer density gradients. Virology, 26:638, 1965.

13. BIOCHEMICAL PROPERTIES OF REVERSE TRANSCRIPTASE ACTIVITIES FROM HUMAN CELLS AND RNA TUMOR VIRUSES

Robert C. Gallo,[1] Prem S. Sarin,[1] Mangalasseril G. Sarngadharan,[2] R. Graham Smith[1] and Marvin S. Reitz[2]

For the past few years our laboratory has been concerned with an examination of the DNA polymerases of fresh human cells. The reasons for our interest are several-fold: 1. We were stimulated by the discoveries of the DNA polymerase in RNA tumor viruses which can synthesize DNA from RNA (1,2). It was obvious that if this were a novel DNA polymerase which could be distinguished from other DNA polymerases, then its presence in cells might indicate the presence of information from viruses of this class. 2. Further, neoplastic transformation by exogenous RNA tumor viruses requires cell division which is preceded by DNA synthesis which requires the function of DNA polymerases of the cell (distinct from the viral polymerase). 3. Cellular DNA polymerases are probably one of the most important target sites for many chemotherapeutic agents. 4. Transfer of information by an RNA/DNA pathway may have an important role in cellular metabolism as well as in the life cycle of RNA tumor viruses, as originally proposed by Temin (3,4). Differences in function of such pathways might be a contributing factor to the production of differences in some RNAs in cells which, in turn, may affect the state of differentiation of the cell. 5. Finally, investigation on purified DNA polymerases from mammalian cells was only recently begun in a systematic fashion, and there were virtually no detailed reports with purified DNA polymerases from fresh human cells. Therefore, for several reasons it was timely to pursue this approach in humam cellular systems.

APPROACH

At present there is no simple way to detect DNA polymerase with reverse transcriptase activity in crude cell extracts, certainly nothing which proves a viral origin. Simply demonstrating an activity with a given template or inhibition with different agents is not enough, nor is the demonstration of an RNase-sensitive DNA polymerase activity in a particulate fraction. For this and reasons referred to above our approach has been to attempt to purify and characterize each DNA polymerase that could be isolated from cells. Since it was possible that important properties of the "reverse transcriptase" might even differ with viruses isolated from different species (and this appears now to be the case), it was also important to purify and characterize the polymerase from various RNA tumor viruses. We then might compare the properties of each cellular DNA polymerase with the polymerase isolated from different viruses.

[1] Laboratory of Tumor Cell Biology, Experimental Therapeutics, National Cancer Institute, National Institutes of Health, Bethesda, Maryland.
[2] Litton-Bionetics, Inc., 7300 Pearl Street, Bethesda, Maryland.

The cells we are using in these studies are leukemic blood "blast" cells obtained by leukophoresis from the blood of leukemic patients, and blood lymphocytes from normal donors stimulated by the mitogenic agent, phytohemagglutinin, for 72 hours (a time of maximum induction of DNA polymerase). The details of these systems and the reasons why we chose them are described elsewhere (5 - 9).

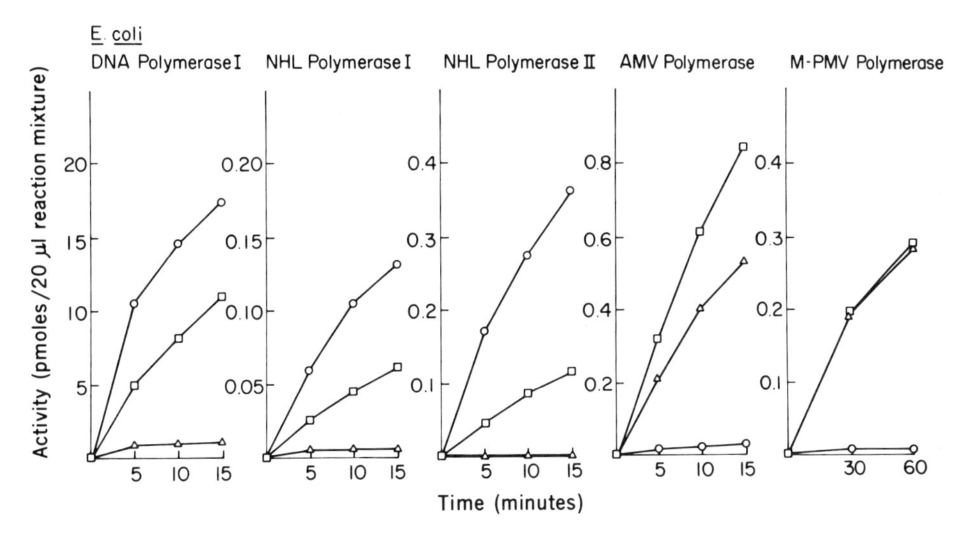

Fig 1. Kinetics of DNA polymerase activites with synthetic templates. DNA polymerase assays were carried out at 37°C in 100-μl standard reaction mixtures composed of 50 mM tris-HCl buffer, pH 8.3; 30 mM KCl; 10 mM $MgCl_2$; 5 mM dithiothreitol; 80 μM dATP; 5.6 μM [^{3}H] TTP (15,000 count/min per picomole); and 50 μg/ml template. Aliquots (20 μl) were taken at the indicated times, precipitated, and counted. The M-PMV was assayed as above, except that the template concentrations were 20 μg/ml, and the standard assay contained 5 mM $MgCl_2$. Results are expressed on the basis of 20 μl of reaction mixture, as the protein concentrations of the purified viral enzymes were not measured. Five microliters of AMV polymerase and 20 μl of M-PMV polymerase were routinely used in their respective reaction mixtures. The E. coli DNA polymerase was assayed at 2.8 ng of protein per 20 μl of reaction mixture. The NHL polymerases I and II had protein concentrations of 144 ng/20 μl and 36 ng/20μl, respectively. The poly (dA)•oligo $(dT)_{12-18}$ and poly (rA)•oligo $(dT)_{12-18}$ were prepared in ratios of 1:1 by annealing equimolar amounts of the polymer and oligomer in 0.01 M tris-HCl, pH 7.2; 0.1 M NaCl, at 70°C for 5 minutes, followed by slow cooling to room temperature over a period of 8 hours. The poly (rA)•poly (dT) was a product of Miles Laboratories. The other polymers and oligomers were obtained from Collaborative Research Inc. □--□, activity with poly (rA)•poly (dT); ○---○, activity with poly (dA)•oligo $(dT)_{12-18}$; △---△, activity with poly (rA)•oligo $(dT)_{12-18}$. (The figure is reproduced from reference 11 with permission of the publishers.)

NOMENCLATURE AND CRITERIA

Before reviewing our results on RNA-directed and DNA-directed DNA synthesis by cellular and viral polymerases, a definition of terms used in our laboratory will be useful. In view of the continued reports of "reverse transcriptase" in a variety of cells based on response to synthetic templates, it is perhaps necessary to clarify terms. Significant confusion has been generated by the use of synthetic DNA•RNA hybrids, such as poly dT•poly A. It is now well established that virtually any DNA polymerase can copy appropriately primed poly A (10 - 15) (Fig 1). Although more specific, the poly A strand of oligo dT·poly A is also transcribed by most DNA polymerases, particularly with Mn^{++} instead of Mg^{++} as the divalent cation. Further, we will show that cellular DNA polymerases (having nothing to do with a viral "reverse transcriptase") will copy poly A stretches contained within natural RNAs. Thus, the simple *demonstration* of *synthesis of poly dT (i.e., copying only poly A)* is *no evidence of anything* other than *detection of some DNA polymerase activity*.

In our laboratory the term "reverse transcriptase" is reserved for those DNA polymerases of RNA tumor viruses (or of cells) where: 1. an endogenous RNase sensitive DNA polymerase activity has been found in a particulate fraction; 2. the DNA product of this reaction is shown to be complementary to the RNA; and 3. the purified enzyme has been shown to copy *heteropolymeric* regions of exogenously added natural RNAs, a criterion which would seem to be a *sine qua non* of any "reverse transcriptase." So far this has only been demonstrated convincingly for the DNA polymerase of some RNA tumor viruses (see recent reviews, references 15, 16, and 17) and one from human leukemic cells (18, 19).

The possibility that RNA is a primer in DNA-directed DNA synthesis must be eliminated before one can assume that an endogenous RNase-sensitive DNA synthesis is an RNA-directed reaction.

TABLE I

Major Characteristics of Viral "Reverse Transcriptase"

1. Location: In virion core.
2. Endogenous DNA synthetic reaction: RNase sensitive synthesis of DNA and a significant portion of the endogenous reactions hybridizes to viral 70S RNA.
3. Purification: Requires high salt or detergent to solubilize.
4. Response to synthetic templates: Responds (crude or purified) to RNA, some RNA · DNA hybrids, and DNA templates (with appropriate primers). In general, the preferred synthetic templates is dT_{12}.rA.
5. Response to natural templates: to DNA: like all the cell DNA polymerases known, it can fill "gapped" double-stranded DNA (repair synthesis); to RNA: the avian enzyme will copy heteropolymeric regions of some exogenously supplied natural RNAs, for example, viral 70S RNA and cellular mRNA. *This has only been established with purified enzymes for the avian reverse transcriptase and the DNA polymerase isolated from leukemic cells.* (Some necessary factor[s] or component of the enzyme may be lost during the purification of mammalian virus reverse transcriptase, since the efficiency of transcription of RNA is less than with avian viral enzymes.
6. Distinguishing templates: The viral enzyme prefers dT_{12} · rA to dT_{12} · dA, copies the RNA strand of dG_{12} · rC, and at least some of the heteropolymeric region of some natural RNAs. The major cellular DNA polymerase (in presence of Mg^{++}) have strong preference for dT_{12} · dA over dT_{12} . rA, cannot transcribe the RNA strand of dG_{12} . rC, nor heteropolymeric regions of natural RNAs.
7. Association with RNase H (hybridase): (See Baltimore's discussion).
8. Species and virus-type heterogeneity: Immunological and biochemical studies indicate important differences between avian and mammalian enzymes, and even within the same species the enzyme appears to substantially differ between different virus types. The magnitude of this heterogeneity was not foreseen. It is clear, therefore, that everything that has been established for the avian virus DNA polymerase may not be generally applicable.

TABLE II

Criteria for Demonstration of RNA Tumor Virus DNA Polymerase ("Reverse Transcriptase")

1. Enzyme is in a particulate fraction.
2. Particulate enzyme (crude) catalyzes DNA synthesis which is RNase-sensivive ("RNA-dependent"), that is, there is a complex of RNA and polymerase.
3. The particulate enzyme requires detergent, divalent cation, and all four deoxynucleotides.
4. The DNA product synthesized by the particulate polymerase will (early in the reaction) band in Cs_2SO_4 gradients in the RNA region or in the RNA . DNA hybrid region.
5. The DNA product synthesized by the particulate polymerase should hybridize to RNA.
6. Solubilized and purified "reverse transcriptase" should respond to RNA, RNA . DNA hybrid, and DNA templates.
7. Purified "reverse transcriptase" shows strong preference for oligo dT . poly rA over oligo dT . poly dA as a template primer and transcribes the ribo strand of oligo dG . poly rC. The major DNA polymerases from normal lymphocytes in Mg^{++} show reverse preference, i.e., strong preference for oligo dT . poly dA over oligo dT . poly rA; in addition their utilization of oligo dG . poly rC is extremely poor.
8. The purified viral reverse transcriptase can transcribe heteropolymeric regions of exogenous 70S RNA. The reaction is RNase sensitive, the product is in part a hybrid, and the DNA of the hybrid can specifically hybridize back to the RNA template.
9. Immunological identification.

"REVERSE TRANSCRIPTASE" OF RNA TUMOR VIRUSES

The major properties of the tumor virus polymerase and criteria we utilize in looking in cells for this enzyme are summarized in Tables I and II.

DNA POLYMERASES OF HUMAN BLOOD LYMPHOCYTES

Normal Blood Lymphocytes Stimulated with PHA

Total DNA polymerase activity in these cells peaks approximately three days after treatment with PHA (6,20). Two major DNA polymerases have been isolated, purified, and characterized from these cells (9). We call these DNA polymerase I and II on the basis of their order of elution from a phosphocellulose column late in purification. This is illustrated in Figure 2. Bollum's (21,22), Weissbach's (23), and Baril's (24) laboratories have also found two major DNA polymerases in other mammalian cells. Bollum prefers the terminology "maxi" and "mini," referring to the large difference in molecular weight between the two. The major properties of these two enzymes are summarized in Table III. What we have termed DNA polymerase I is the high molecular weight enzyme (~150,000, which "aggregates" to 300,000 daltons) and polymerase II the smaller protein (~30,000 to 40,000 daltons). The major proportion of I is in the cytoplasm. However, it is possible that during S phase of the cell cycle the enzyme moves to the nucleus. This would appear likely, since most information indicates that this enzyme may be important in DNA replication in mammalian cells. For example, the activity of this enzyme increases in S phase, and we find that DNA polymerase I is much more sensitive to cytosine arabinoside (ara-CTP) than polymerase

II (Schrecker, Smith, and Gallo, unpublished results), and ara-CTP is a potent inhibitor of DNA replication. This has also been found recently by M. Goulian (personal communication).

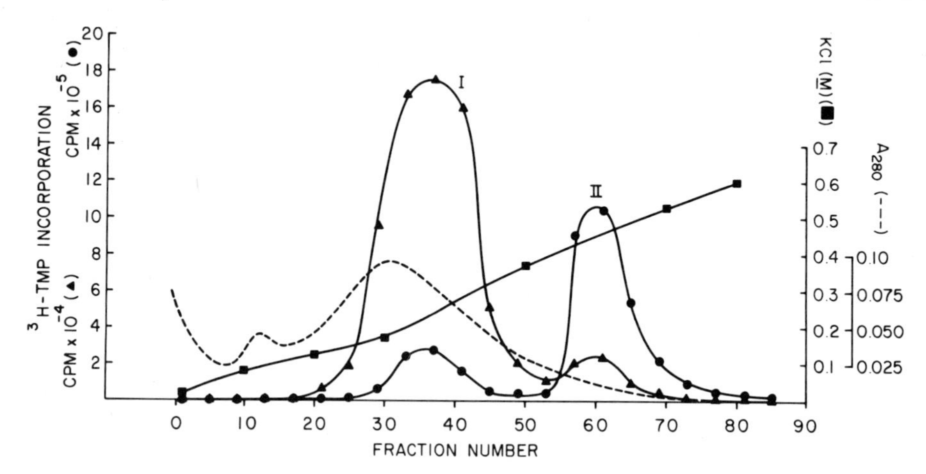

Fig 2. Separation on phosphocellulose of normal lymphocyte DNA polymerases I and II. Fraction 2 (DEAE eluate–see ref. 9) (156 ml) was absorbed to a 2.0 times 6.0 cm column of phosphocellulose (Whatman P-11) equilibrated with TDEG buffer. Following a 15 ml wash with TDEG buffer with 0.1 M KC1, 160 ml of a linear gradient of KC1 in buffer, extending between extremes of 0.1 M and 0.7M, was attached. Fractions of 1.6 ml were collected and assayed as described in Methods. (System A ▲---▲; System B o---o). Concentration of KCl (■---■) was measured with a conductivity meter. See reference 9 for details of buffer.

TABLE III

Summary of Characteristics of DNA Polymerase I and II of Normal Human Blood Lymphocytes

Characteristic	DNA polymerase	
	I	II
Size	7 - 10S (150,000-300,000 daltons)	3.3S (30,000 daltons)
pI	4–6	9.4
Template acceptance		
Activated DNA (Mg^{++})	28 pmoles/hr/μg	30 pmoles/hr/μg
Initiated homopolymeric DNA (Mn^{++}), e.g., $dT_{12} \cdot dA$	28 pmoles/hr/μg	880 pmoles/hr/μg
Inhibition by 0.05 mM NEM	+++	0
Inhibition by ara-(CTP)	+++ (Ki = 1.1 μM)	+
Subcellular localization	Cytoplasm	Nucleus and cytoplasm
Response to proliferation	Increases; absent in nonproliferating cells.	Increases; present in non-proliferating cells.

TABLE IV

Template (Primer) Responses of DNA Polymerases I and II of Normal Phytohemagglutinin Stimulated Human Blood Lymphocytes

Template (primer)	DNA polymerase (pmoles/hr/μg)	
	I	II
Activated DNA	23.4	12.6
Native DNA	0.81	0.72
Denatured DNA	1.00	0.73
Poly d(AT)	3.68	5.75
$dT_{10} \cdot dA$ (Mg^{++})	2.43	8.33
$dT_{10} \cdot dA$ (Mn^{++})	28	880
poly dT · poly rA (Mg^{++})	1.1	8.0
poly dT · poly rA (Mn^{++})	8.7	241
poly rA · rU (Mg^{++})	0	0
poly rA · rU (Mn^{++})	0.37	0
70S RNA from AMV	0	0.31[a]
+ RNase	–	0.31[a]
70S RNA + $dT_{12\text{-}18}$ (^{3}H-TTP) (Mg^{++})	0.17	0.22[a]
70S RNA + $dT_{12\text{-}18}$ (^{3}H-TTP) (Mn^{++})	0.55	9.78[a]
70S RNA + $dT_{12\text{-}18}$ (^{3}H-dATP) (Mg^{++})	0	0
70S RNA + $dT_{12\text{-}18}$ (^{3}H-dATP) (Mn^{++})	0	0
70S RNA + $dT_{12\text{-}18}$ (^{3}H-dCTP) (Mg^{++})	0	0
70S RNA + $dT_{12\text{-}18}$ (^{3}H-dGTP) (Mn^{++})	0	0

[a]See text for discussion.

As indicated in Table IV, these purified normal cellular DNA polymerases have great affinity for DNA, *and they can also copy the RNA strand of synthetic homopolymeric RNA•DNA hybrids*, such as poly dT•poly rA. However, they do not transcribe heteropolymeric stretches of natural RNAs and have relatively poor affinity for primer•RNA homopolymer complexes compared to primer •DNA homopolymer complexes. *The response to 70S RNA (or mRNA) is only to the poly A stretches of this RNA* as indicated by the lack of sensitivity to pretreatment with RNase and the incorporation *only* of TTP. They can be easily distinguished from the reverse transcriptase of RNA tumor viruses. A third DNA polymerase may be present in normal blood lymphocytes. This activity was isolated from a particulate fraction. However, we are not yet certain that it, in fact, represents a distinctly different DNA polymerase. This enzyme is described in more detail later.

Leukemic Leukocytes

We have also studied the DNA polymerase of human acute leukemic blast cells (acute lymphoblastic and acute myeloblastic leukemia). As in normal lymphocytes we find two major DNA polymerases by conventional purification techniques. These enzymes have been generally found to have properties identical to those in normal lymphocytes. In some cases, we have found that DNA polymerase I ("maxi") elutes at a significantly higher salt concentration from phosphocellulose columns compared to the enzyme from normal lymphocytes (Sarin, Sarngadharan, Smith, and Gallo, unpublished results).

TABLE V

Response of the Purified Human Leukemic "Pellet" DNA Polymerase to Distinguishing Templates[a]

Templates (primer)	^{3}H-TMP Incorporated (pmole/0.1 ml)
$dT_{12\text{-}18} \cdot dA$	2.4
$dT_{12\text{-}18} \cdot rA$	12.2
AMV 70S RNA (complete; ^{3}H-TTP)	2.5
70S RNA -Mg^{++}	<0.05
70S RNA +RNase	0.2
70S RNA (^{3}H-dCTP)	2.2

[a]These experiments were conducted with Mg^{++} as the divalent cation. In our experience no cellular DNA polymerases from normal lymphocytes have shown this preference for $dT_{12\text{-}18} \cdot rA$ over $dT_{12\text{-}18} \cdot dA$ in Mg^{++} nor incorporated ^{3}H-dCTP or ^{3}H-dGTP with AMV 70S RNA. (AMV = avian myeloblastosis virus.)

ISOLATION OF A DNA POLYMERASE FROM HUMAN LEUKEMIC CELLS WHICH COPIES NATURAL RNA

We have isolated a cytoplasmic pellet fraction from human leukemic cells that carries out endogenous synthesis of DNA, and this reaction is very sensitive to RNase, that is, an RNA-"dependent" DNA polymerase (18,19). This suggests but does not prove that the reaction is directed by RNA. This type of evidence is what was available in the initial reports with the viral reverse transcriptase (1,2). Like the viral enzyme, the activity is enhanced and the enzyme "solubilized" by non-ionic detergents. When the enxyme is purified it utilizes RNA, RNA•DNA hybrids, and DNA; transcribes the ribo strand of oligo dG•rC; has greater preference for $dT_{12} \cdot rA$ over $dT_{12} \cdot dA$ with Mg^{++}; and, most importantly, copies heteropolymeric portions of the 70S RNA isolated from avian myeloblastosis virus (AMV) (Table V). *Thus, it has the major biochemical characteristics of the reverse transcriptase of RNA tumor viruses and is distinct from the two main cellular DNA polymerases*

The evidence that the enzyme can copy heteropolymeric regions of a natural RNA is summarized below: 1. the endogenous DNA polymerase activity is completely sensitive to RNase; 2. the DNA product of the endogenous reaction is in part a hybrid; 3. the activity of the purified enzyme with AMV 70S RNA is completely destroyed by pretreatment of the RNA with pancreatic RNase (Table V); 4. the DNA product of the reaction of the purified enzyme with AMV 70S RNA as template is in part a hybrid,

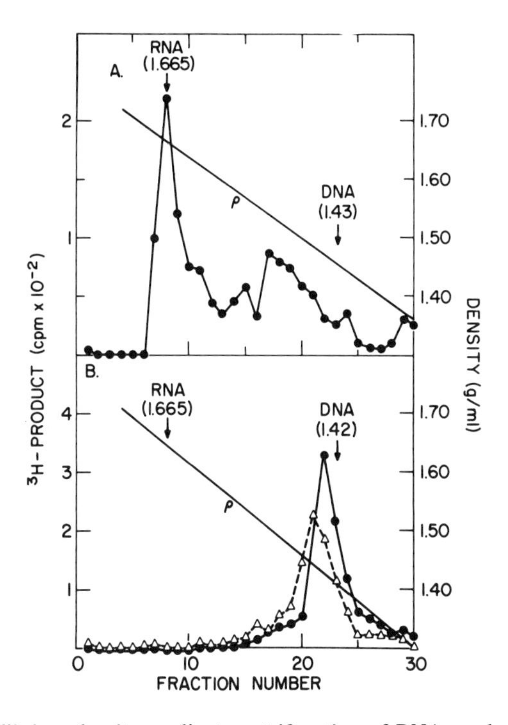

Fig 3. Cs_2SO_4 equilibrium density gradient centrifugation of DNA product from purified human leukemic pellet DNA polymerase with AMV 70S RNA as a template. A standard reaction mixture (0.05ml) as described in detail elsewhere (18-20) was incubated with AMV 70S RNA and oligo dT at 37°C for 60 minutes and then adjusted to 0.4 M NaCl and 1% SDS. This was extracted twice with phenol: m-cresol (9:1), 0.1% in 8-quinilinol, and twice with ether. These extracts were diluted to 1 ml with 10 mM Tris, 0.1 M NaCl and 1 mM EDTA, pH 7.24. 500 μg stripped yeast tRNA, 100 μg calf thymus DNA, and 25 μl 0.1 M cetyltrimethylammonium bromide (CTAB) were added to the final extracts. The CTA-nucleic acid salts were precipitated on ice for 30 minutes and collected by centrifugation (15,00 xg, 10 minutes). The pellet was dissolved in 1 ml lM NaCl and 3 ml EtOH was added. The nucleic acids were precipitated for 2 hours at -15°C and collected by centrifugation as above. The precipitate was then dissolved in NET (0.01 M Tris, pH 7; 0.1 M NaCl; 0.001 M EDTA) buffer and the NaCl concentration adjusted to 0.4 M. One ml of the solution was mixed in a 5 ml polyallomer tube with 4 ml of a solution of 1 ml NET per gm cesium sulfate (Schwartz-Mann optical grade). The tubes were centrifuged in an SW-50.1 rotor at 36,000 rpm for 65 hours at 20°C. Density was determined with Bausch and Lomb Abbe refractometer. Fractions were precipitated with 2 ml 10% TCA containing 0.02 M pyrophosphate and filtered on a millipore filter and counted in liquiflour-toluene scintillation fluid. A–Native DNA product B–Heated at 95°C for 10 minutes (●——●); treated with alkali (0.3 N; 95°, 10 minutes) (△---△).

that is, hydrogen bonded to the RNA template (Fig 3); and 5. the DNA product will hybridize back specifically to the RNA used as template. This demonstrates complementarity between product DNA and RNA (Figs 4 and 5) (18,19).

This is the first evidence that any "cellular" DNA polymerase can copy a natural RNA, and it is an extension of a preliminary communication from our laboratory two years ago (7). However, neither we nor any other laboratory can say anything about the cellular distribution of this enzyme. There is, as far as we know, no easy method for surveying a large series of patients or a wide variety of tissues.

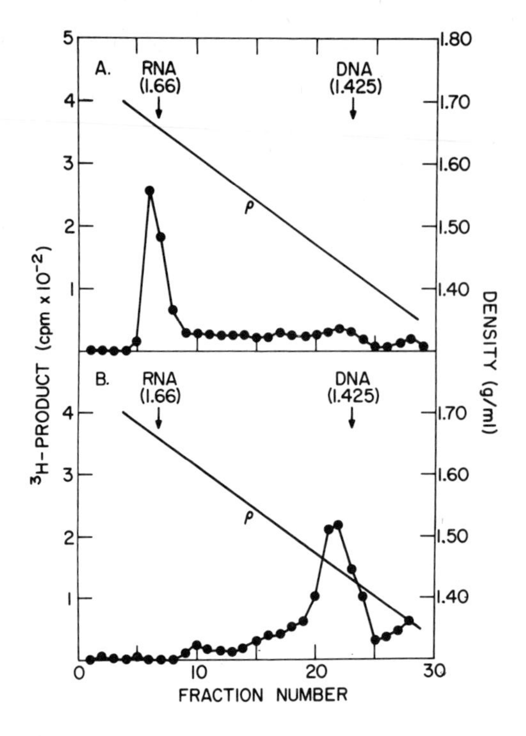

Fig 4. Specific hybridization of DNA product of purified human leukemic DNA polymerase (with AMV 70S RNA as template and with oligo dT stimulation) with AMV 70S RNA. Product was prepared as described in Figure 3, then alkali-treated (0.3 N KOH, 95°C, 10 minutes, followed by room temperature, 20 hours) and neutralized. This solution was then adjusted to 2XSSC and an equal volume of formamide was added. 2.5 μg AMV 70S RNA or 0.1 μg poly rA were added to 0.2 ml portions of this solution and annealed for 72 hours at room temperature. The product was analyzed as in Figure 3. A–The product annealed with AMV 70S RNA; B–The product annealed with MS_2 RNA.

RNA-DIRECTED DNA POLYMERASE ACTIVITY IN NORMAL CELLS?

Early reports suggested that this enzyme activity was present in a variety of normal tissue culture cells and lymphocytes. These and many subsequent reports were based on a nonspecific hybrid template, poly dT• poly rA, or with natural RNA known to be contaminated with DNA and for which neither product analysis nor RNase sensitivity was shown. As discussed in detail elsewhere (10,11,18,27), many DNA polymerases can copy the homopolymeric rA strand of poly dT• rA; therefore, activities based on response to this template only indicate that some DNA polymerase is present, but say nothing about the presence of viral reverse transcriptase or for that matter any enzyme which copies a natural RNA. Along these lines it is important to note that recently the 70S RNA of RNA tumor viruses was shown to contain poly rA (25 – 27). Since some DNA polymerases can copy poly A stretches to some degree with appropriate primers, small activities with 70S RNA as template do not prove the enzyme is acting as a "reverse transcriptase." The viral enzyme (and the enzyme from leukemic cells) copy heteropolymeric regions of some natural RNAs, including viral 70S RNA.

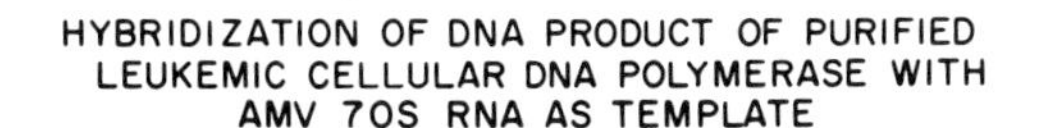

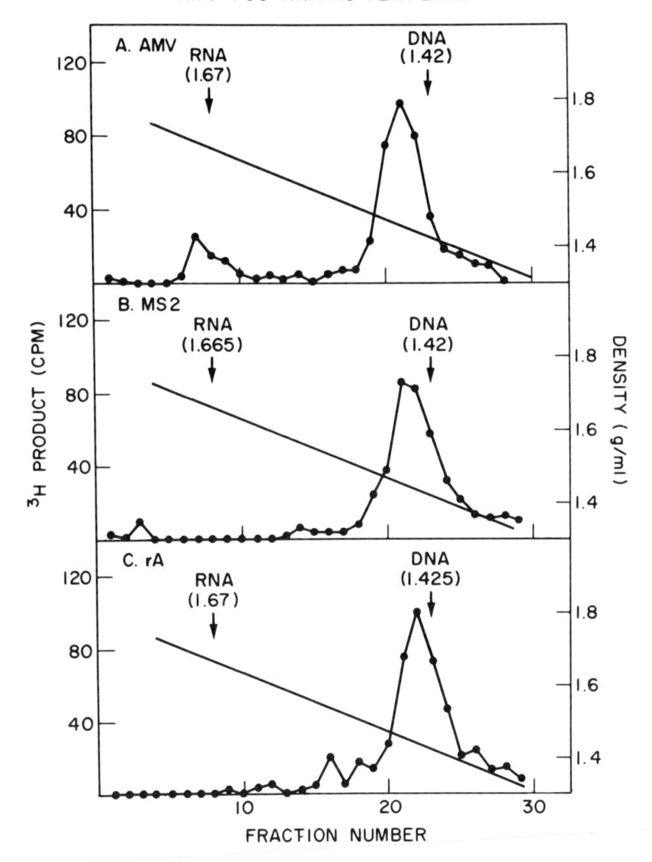

Fig 5. Specific hybridization of DNA product of purified human leukemic pellet DNA polymerase (with AMV 70S RNA as template and no oligo dT stimulation) with AMV 70S RNA. A–upper panel–the product was annealed with AMV 70S RNA; B–middle panel–with MS2 coliphage RNA; C–lower panel–with poly rA. Product was prepared as described in Figure 3, then alkali-treated (0.3 N KOH, 95°, 10 minutes, followed by room temperature, 20 hours) and neutralized. The product was then adjusted to 2XSSC and an equal volume of formamide was added. 5 μg AMV 70S RNA, or 5 μg MS2 RNA or 11 μg poly rA were added to 0.5 ml portions of this solution, annealed for 72 hours at room temperature, and then the Cs_2SO_4 density gradient centrifugation was carried out.

Recently two findings have been noted which indicate that some normal cells may have an enzyme capable of catalyzing true RNA-directed DNA synthesis. These are the findings of Kang and Temin reported in part very recently (28) and in this symposium with chick embryo, and Bobrow, Smith, Reitz, and Gallo in stimulated human blood lymphocytes (29). In both systems, endogenous RNase-sensitive DNA polymerase activities have been found in a particulate fraction, analogous to the leukemic enzyme and similar to the early observations made with RNA tumor viruses. Kang and Temin find that the chick embryo enzyme is not inhibited by antibody prepared against reverse transcriptase of AMV, and the purified enzyme apparently does not transcribe exogenous RNA, that is, the enzyme, free of its own endogenous nucleic acid, now behaves like the main DNA-dependent DNA polymerases of mammalian cells. Similarly, we find that the enzyme purified from the pellet of normal lymphocytes, although possibly a DNA polymerase distinct from the two main DNA polymerases, will not accept 70S RNA and has the characteristics which clearly indicate it is not like any of the known viral enzymes or the leukemic enzyme, but behaves like the main DNA polymerases (I and II). Temin has speculated that this may be due to a strict requirement of this enzyme for its endogenous RNA, that is, a specificty for RNA unlike viral reverse transcriptase, which utilizes a variety of heterologous as well as homologous RNAs. An alternative possibility that has not been rigorously eliminated is that these reactions in chick embryo and in the normal lymphocytes are initiated by an RNA molecule (RNA primed) but directed by DNA. If, however, these reactions are proven to be directed by RNA, then varying types of reverse transcriptase systems may exist. A derivation of frank RNA tumor virus from these systems would indeed be support for Temin's *protovirus* theory (3,4).

CONCLUSIONS AND SUMMARY

1. The DNA polymerase of RNA tumor viruses is not one of the main cellular DNA polymerases. It is almost certainly a novel and probably viral-coded enzyme.

2. Many DNA polymerases can copy some homopolymeric RNA stretches (for example, poly A) with appropriate primers. Therefore, responses to templates like dT•rA or copying only the poly A tracts of viral 70S RNA are not meaningful if one is attempting to identify the tumor virus polymerase, or for that matter any "reverse transcriptase" activity.

3. Normal PHA-stimulated human blood lymphocytes have at least two major DNA polymerases, this is probably true for all proliferating mammalian tissues. These enzymes, when purified, are easily distinguished from the viral polymerase.

4. Human leukemic blood cells contain a particulate cytoplasmic component which contains an endogenous RNase-sensitive DNA polymerase activity. The enzyme purified from this particle has the biochemical characteristics of the tumor virus reverse transcriptase, including the capacity to transcribe heteropolymeric regions of viral 70S RNA, greater response to oligo dT•rA compared to oligo dT•dA, and acceptance of oligo dG•rC.

5. Normal human blood lymphocytes stimulated with PHA also contain a particulate fraction with an endogenous RNase-sensitive DNA polymerase activity. However, the enzyme purified from this pellet fraction, unlike the leukemic and tumor virus polymerase, cannot transcribe heteropolymeric regions of exogenous natural RNAs, shows a marked preference for oligo dT•dA over oligo dT•rA, and does not utilize oligo dG•rC. Unstimulated blood lymphocytes do not have this particulate polymerase activity. The particulate enzyme from these normal cells and those reported by Temin in chick embryo could represent RNA-directed DNA synthetic systems with unique specificity for their endogenous RNA, while the viral (and leukemic) polymerase do not show this specificity. More evidence is still needed to prove these endogenous reactions in normal cells are RNA directed and not RNA primed and DNA directed. If they are directed by RNA, the findings will have obvious implications for Temin's protovirus theory.

REFERENCES

1. Baltimore, D.: RNA-dependent DNA polymerase in virions of RNA tumor viruses. Nature, 226:1209, 1970.
2. Temin, H. M., and Mizutani, S.: RNA-dependent DNA polymerase in virions of Rous sarcoma virus. Nature, 226:1211, 1970.
3. Temin, H. M.: The protovirus hypothesis: speculations on the significance of RNA-directed DNA synthesis for normal development and for carcinogenesis. J. Nat. Cancer Inst., 46:iii, 1971.
4. Temin, H. M.: Presented at the 25th Annual Symposium on Fundamental Cancer Research at The University of Texas M. D. Anderson Hospital and Tumor Institute at Houston, March 1972.
5. Gallo, R. C.: Synthesis and metabolism of DNA and DNA precursors by human normal and leukemic leukocytes: a summary of recent information. Acta Haemat., 45:136, 1971.
6. Gallo, R. C., and Whang-Peng, J.: Observations on the regulatory effects of the transfer RNA minor base, N^6-Δ^2-isopentenyladenosine, on human lymphocytes. In: Biological Effects of Polynucleotides. Edited by R. F. Beers and W. Braun. Springer-Verlag, New York, p 303, 1971.
7. Gallo, R. C., Yang, S. S., and Ting, R. C.: RNA dependent DNA polymerase of human acute leukaemic cells. Nature, 228:927, 1970.
8. Gallo, R. C., Yang, S. S., Smith, R. G., Herrera, F., Ting, R. C., and Fujioka, S.: Some observations on DNA polymerases of human normal and leukemic cells. In: Nucleic Acid-Protein Interactions and Nucleic Acid Synthesis of Viral Infection, Miami Winter Symposia 1971. Edited by D. W. Ribbons, J. F. Woessner, and J. Schultz. Holland Publishing Co., Amsterdam, p 353, 1971.
9. Smith, R. G., and Gallo, R. C.: DNA-dependent DNA polymerase I and II from normal human blood lymphocytes. Proc. Nat. Acad. Sci. USA, 69:2879, 1972.
10. Baltimore, D., and Smoler, D.: Primer requirement and template specificity of the DNA polymerase of RNA tumor viruses. Proc. Nat. Acad. Sci. USA, 68:1507, 1971.
11. Robert, M. S., Smith, R. G., Gallo, R. C., Sarin, P. S., and Abrell, J. W.: Viral and cellular DNA polymerase: comparison of activities with synthetic and natural RNA templates. Science, 176:798, 1972.
12. Goodman, N. C., and Spiegelman, S.: Distinguishing reverse transcriptase of an RNA tumor virus from other known DNA polymerases. Proc. Nat. Acad. Sci. USA, 68:2203, 1971.
13. Wells, R. D., Flugel, R. M., Larson, J. E., Schendel, P. F., and Sweet, R. W.: Comparison of some reactions catalyzed by deoxyribonucleic acid polymerase from avian myeloblastosis virus, Escherichia coli, and Micrococcus luteus. Biochem., 11:621, 1972.
14. Gallo, R. C., Abrell, J. W., Robert, M. S., Yang, S. S., and Smith, R. G.: Reverse transcriptase from Mason-Pfizer monkey tumor virus, avian myeloblastosis virus, and Rauscher leukemia and its response to rifampicin derivatives. J. Nat. Cancer Inst., 48:1185, 1972.
15. Gallo, R. C.: Reverse transcriptase–the DNA polymerase of oncogenic RNA viruses. Nature, 234:194, 1971.
16. Gallo, R. C.: RNA dependent DNA polymerase in viruses and cells–views on the current state. Blood, 39:117, 1972.
17. Temin, H. M., and Baltimore, D.: RNA-directed DNA synthesis and RNA tumor viruses. Adv. Virus Res., 17:129, 1972.
18. Gallo, R. C., Sarin, P. S., Sarngadharan, M. G., Reitz, M. S., Smith, R. G., and Bobrow, S. N.: Viral 70S RNA directed DNA synthesis with a purified DNA polymerase from human acute leukemic cells. In: Molecular Studies in Viral Neoplasia. 25th Annual Symposium on Fundamental Cancer Research, The University of Texas M. D. Anderson Hospital and Tumor Institute at Houston, 1972. In press.
19. Sarngadharan, M. G., Sarin, P. S., Reitz, M. S., and Gallo, R. C.: Reverse transcriptase activity of human acute leukemic cells: purification of the enzyme response to AMV 70S RNA, and characterization of the DNA product. Nature New Biology, 240:67, 1972.

20. Loeb, L. A., Agarwal, S. S., and Woodside A. M.: Induction of DNA polymerase in human lymphocytes by phytohemagglutinin. Proc. Nat. Acad. Sci. USA, 61:827, 1968.
21. Chang, L. M. S., and Bollum, F. J.: Low molecular weight deoxyribonucleic acid polymerase in mammalian cells. J. Biol. Chem., 246:5835, 1971.
22. Chang, L. M. S., and Bollum, F. J.: Low molecular weight deoxyribonucleic acid polymerase from rabbit bone marrow. Biochem., 11:1264, 1972.
23. Weissbach, A., Schlabach, A., Fridlender, B., and Boldon, A.: DNA polymerases from human cells. Nature New Biology, 231:167, 1971.
24. Baril, E. F., Brown, O. E., Jenkins, M. D., and Laszlo, J.: Deoxyribonucleic acid polymerase with rat liver ribosomes and smooth membranes–purification and properties of the enzyme. Biochem., 10:1981, 1971.
25. Gillespie, D., Marshall, S., and Gallo, R.: RNA of RNA tumour viruses contains poly A. Nature New Biology, 235:227, 1972.
26. Lai, M. M. C., and Duesberg, P. H.: Adenylic acid-rich sequence in RNAs of Rous sarcoma virus and Rauscher mouse leukaemia virus. Nature, 235:383, 1972.
27. Green, M., and Cartas, M.: The genome of RNA tumor viruses contains polyadenylic acid sequences. Proc. Nat. Acad. Sci. USA, 69:791, 1972.
28. Kang, C.-Y., and Temin, H. M.: Endogenous RNA-directed DNA polymerase activity in uninfected chicken embryos. Proc. Nat. Acad. Sci. USA, 69:1550, 1972.
29. Bobrow, S. N., Smith, R. G., Reitz, M. S., and Gallo, R. C.: Normal human lymphocytes contain a ribonuclease-sensitive DNA polymerase which is distinct from viral reverse transcriptase. Proc. Nat. Acad. Sci. USA, 69:3228, 1972.

14. AVIAN MYELOBLASTOSIS VIRUS DNA POLYMERASE: INITIATION OF DNA SYNTHESIS AND AN ASSOCIATED RIBONUCLEASE

David Baltimore, Inder M. Verma, Donna F. Smoler and Nora L. Meuth

Department of Biology
Massachusetts Institute of Technology
Cambridge, Massachusetts 02139

INTRODUCTION

In the two years since the first identification of a DNA polymerase in virions of the RNA tumor viruses much progress has been made in the understanding of the enzyme (reviewed in ref. 6). Initially an *endogenous* reaction was observed where the DNA polymerase was producing a copy of the 70S RNA of the tumor virus particles. It was observed, however, that a much more extensive synthesis of DNA could be stimulated by added DNA and RNA templates. Using such templates for assay, the soluble DNA polymerase has been purified from virion preparations. The purified enzyme can copy both DNA and RNA and no separation of the ability to copy these two templates has been observed. The enzyme is primer-dependent, that is, unable to initiate DNA synthesis without a preformed piece of DNA (or RNA) to initiate the reaction. Short oligomers of DNA, complementary in base sequence to the template, are very convenient primers. Aside from its polymerizing activity the enzyme purified from avian myeloblastosis virus (AMV) also has an associated ribonuclease activity.

Two topics will be discussed in this paper: the initiation of DNA synthesis when 70S viral RNA is the template and the nature of the ribonuclease activity associated with the DNA polymerase.

INITIATION OF DNA SYNTHESIS

A number of different methods have shown that the initial product formed when 70S RNA is used as a template for the DNA polymerase of AMV is a covalently bonded DNA-RNA molecule. The initial observations were made, using cesium sulfate gradients where the heat denatured product was shown to band as a covalently joined molecule of DNA and RNA (8). From the density and size of the product, it appeared that the initiator was a short RNA molecule (about 50 - 100 bases). More recently we have been studying the bond which joins DNA to the RNA primer using as an assay the transfer of [^{32}P] from an α-^{32}P-deoxyribonucleoside triphosphate to a ribonucleotide after alkaline hydrolysis of the product (7). Figure 1 shows the rationale for this type of experiment.

The electropherograms of alkali-hydrolized material are shown in Figure 2. It is evident that of the four deoxyribonucleotide precursors, only α-^{32}P-dATP is able to transfer a phosphorus to a ribonucleotide, and the only ribonucleotide which accepts the ^{32}P is 2′(3′)AMP. Table I quantitates the transfer data and shows that about 15%

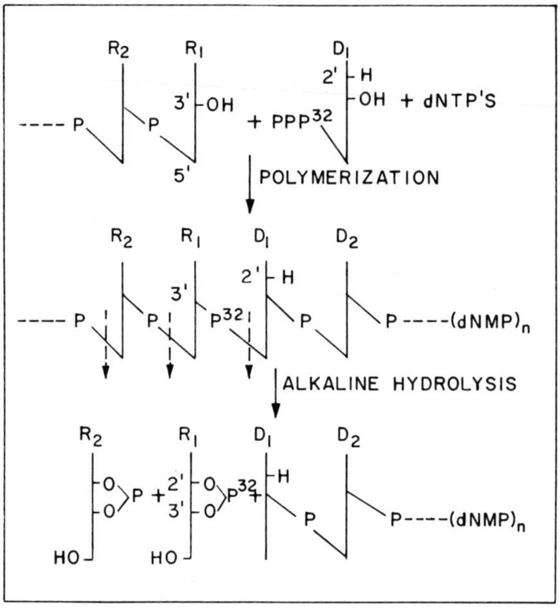

Fig 1. Schematic representation of the transfer of a [^{32}P] atom from the α-position of a deoxyribonucleotide to the 2′(3′)-position on a ribonucleotide. R_2 and R_1 represent penultimate and ultimate ribonucleotides of the presumed RNA primer. D_1 and D_2 represent the initial two deoxyribonucleotides incorporated.

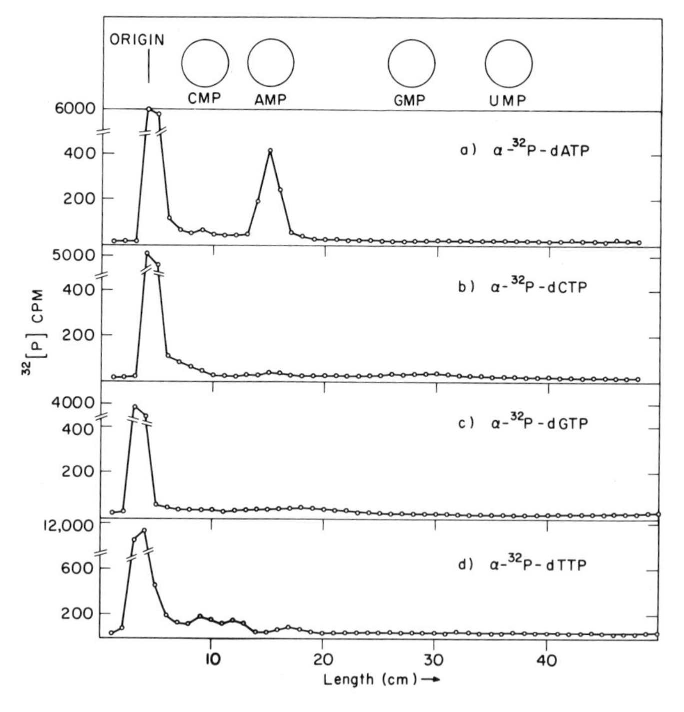

Fig 2. α– ^{32}P-transfer with individual deoxyribonucleoside triphosphates in the reconstructed system.

TABLE I

Transfer of ^{32}P to Ribonucleotides

Nature of reaction	Substrate α ^{32}P-dNTP	Percent recovered in			
		CMP	AMP	GMP	UMP
AMV virions	dCTP	<1	<1	<1	<1
(Endogenous system)	dATP	<1	19-23	<1	<1
	dGTP	<1	<1	<1	<1
	dTTP	<1	<1	<1	<1
AMV poly-merase plus	dCTP	<2	<1	<1	<1
60-70S AMV RNA	dATP	<1	11-15	<1	<1
(Reconstructed	dGTP	<1	<1	<1	<1
system)	dTTP	<1	<1	<1	<1

Note: Data is taken from a series of experiments like that depicted in Figure 2. All $\alpha^{32}P$-dNTP's were examined at least twice. The data for dCTP excludes two experiments where transfer to 2′(3′)-CMP and 2′(3′)-AMP was observed. This was not reproducible.

of the incorporated label can be recovered as ribonucleotide after a 60-minute reaction. Identical results were obtained whether purified RNA and enzyme were used or if detergent-disrupted virions were assayed. This data indicates that the DNA product is very small (50 - 100 nucleotides) and is in agreement with analysis of the product by sucrose gradient centrifugation.

The fact that of the sixteen possible DNA-RNA joints the only one found is ApdA indicates that the specificity of the initiation reaction is very great. Where on the 70S RNA molecule this initiation occurs, how many initiators there are per 70S RNA, etc., are questions for the future.

THE RIBONUCLEASE ASSOCIATED WITH THE AVIAN MYELOBLASTOSIS VIRUS DNA POLYMERASE

A ribonuclease associated with the AMV DNA polymerase was first observed by Mölling, Bolognesi, Bauer, Büsen, Plassmann, and Hausen (5). They found that the ribonuclease had the properties of a ribonuclease H (3), that is, it would degrade the RNA moiety of a DNA-RNA hybrid, but would not degrade either free RNA or double-stranded RNA. We have confirmed their data, using a somewhat different assay for the ribonuclease (2).

We initially observed the ribonuclease H when we were studying the fate of poly(A) after it was copied by the AMV DNA polymerase to form poly(A)·poly (dT). We found that when [^3H]poly(A) was used as template, the [^3H] label was rapidly converted to an acid-soluble form. Table II compares the synthesis of [^{32}P]poly(dT)

TABLE II

Degradation of Poly(A) as a Consequence of Its Use as a Template for the AMV DNA Polymerase

	% Poly(A) degraded	Poly(dT) synthesis (pmol)
Complete	71	31
Minus oligo(dT)	<5	<2
Minus dTTP	<5	-
Minus polymerase	<5	<2

Note: Standard reaction conditions for AMV polymerase were used with (per 0.1ml) 82 pmol of $[^3H]$poly(A) (22 cpm/pmol), 12 pmol of $(dT)_{14\text{-}18}$, 3.8 nmol of $[^{32}P]$dTTP (18 cpm/pmol) and 15 units of AMV DNA polymerase. Incubation was for 60 minutes at 37°C.

and the degradation of $[^3H]$poly(A) during this reaction. Both synthesis and degradation require the oligo(dT) primer (1) and degradation is dependent on the presence of TTP and the reaction mixture.

This coupled synthesis of poly(dT) and degradation of poly(A) is difficult to use as an assay for ribonuclease H, and we therefore turned to using preformed poly(A)·poly(dT) as a substrate. Table III shows that the degradation of poly(A) requires poly(dT) and magnesium and that the poly(dT) cannot be replaced by poly(U). This confirms the identification of the nuclease as a ribonuclease H. Using this assay we have followed the ribonuclease activity through our standard purification for the AMV DNA polymerase (2). We have found that polymerization and nuclease activity go hand-in-hand and that even glycerol gradient centrifugation of the purified material does not separate the two activities (Fig 3).

Studies on the nature of the polymerase-associated ribonuclease H indicate that it is an endonuclease which produces 3′-OH and 5′-phosphoryl termini at the

TABLE III

Catalysis of Poly(A) Degradation by Poly(dT)

	% Poly(A) degraded
Complete	70
Minus Mg acetate	<5
Minus poly(dT)	<5
Minus poly(dT); plus poly(U)	<5

Note: Reaction conditions were (in 0.1ml) 0.05M Tris HCl, pH 8.3; 20mM Mg acetate; 10mM dithiothreitol; 340 pmol $[^3H]$poly(A) (10 cpm/pmol); 186 pmol poly(dT) and, where indicated, 168 pmol poly(U) with 20 units of AMV DNA polymerase. Incubation was for 60 minutes at 37°C.

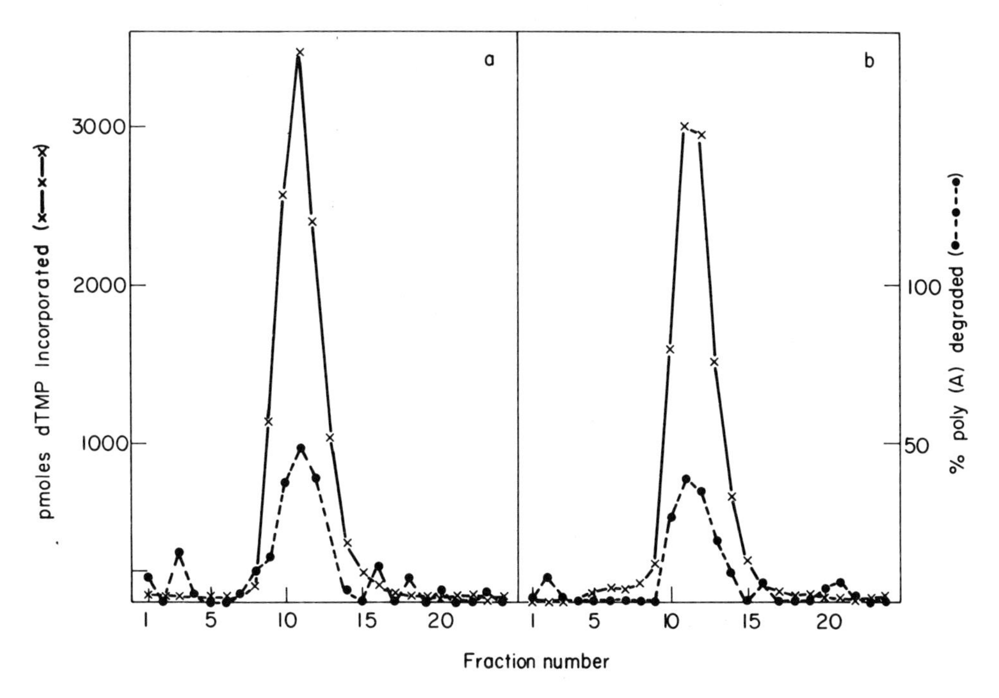

Fig 3. Glycerol gradient centrifugation of AMV DNA polymerase and associated nuclease. *a*, 0.2M KC1; *b*, 0.5M KC1.

cleavage site (2). Its ability to generate 3′-OH groups means that RNA which has been cleaved by ribonuclease H has primer sites for the DNA polymerase. This suggests that one of the functions of ribonuclease H might be to produce such primer sites.

The exact function of the ribonuclease H and the way that it participates in the DNA polymerase reaction are topics for future research. At present, we only know that it has not been separable from the polymerase, although future work might show that it is separable. The observation of two polypeptide chains in the purified AMV DNA polymerase (4) suggests that the nuclease and polymerase might be on different polynucleotide chains.

SUMMARY

Initiation of DNA synthesis by the avian myeloblastosis virus DNA polymerase requires a preformed primer. When 70S viral RNA is used as a template, a small RNA associated with the major species of viral RNA acts as primer. This has been proved both by analysis of the density of the initial product and by identification of the RNA-DNA joint in the product. This joint is ApdA.

Associated with the DNA polymerase is a ribonuclease which specifically degrades the RNA of a DNA•RNA hybrid. This enzyme has been inseparable from the DNA polymerase. It is readily assayed by the conversion of [^{3}H] in [^{3}H]poly(A)• poly(dT) to an acid-soluble form. The cleavage reaction leaves 3′-OH and 5′-phosphoryl termini and the enzyme appears to be an endonuclease.

REFERENCES

1. Baltimore, D., Smoler, D.: Primer requirement and template specificty of the RNA tumor virus DNA polymerase. Proc. Nat. Acad. Sci. USA, 68:1507, 1971.
2. Baltimore, D., and Smoler, D.: Association of an endoribonuclease with the avian myeloblastosis virus DNA polymerase. J. Biol. Chem., 247:7282, 1972.
3. Hausen, P., and Stein, H.: Ribonuclease H. An enzyme degrading the RNA moiety of DNA-RNA hybrids. Eur. J. Biochem., 14:278, 1970.
4. Kacian, D. L., Watson, K. F., Burny, A., and Spiegelman, S.: Purification of the DNA polymerase of avian myeloblastosis virus. Biochem. Biophys. Acta, 246:365, 1971.
5. Mölling, K., Bolognesi, D. P., Bauer, H., Plassmann, H. W., and Hausen, P.: Association of the viral reverse transcriptase with an enzyme degrading the RNA moiety of RNA-DNA hybrids. Nature New Biology, 234:240, 1971.
6. Temin, H., and Baltimore, D.: RNA-directed synthesis and RNA tumor viruses. Adv. in Virus Res., Vol. 17. New York and London: Academic Press, Inc., pp 129-186, 1972.
7. Verma, I. M., Meuth, N. L., and Baltimore, D.: The covalent linkage between RNA primer and DNA product of the avian myeloblastosis virus DNA polymerase. J. Virol., 10:622, 1972.
8. Verma, I. M., Meuth, N. L., Bromfeld, E., Manly, K. F., and Baltimore, D.: A covalently-linked RNA-DNA molecule as the initial product of the RNA tumor virus DNA polymerase. Nature New Biology, 233:131, 1971.

15. TRANSCRIPTION AND TRANSLATION OF VIRAL RNA IN CELLS TRANSFORMED BY RNA TUMOR VIRUSES[1]

Maurice Green, Nobuo Tsuchida, Giancarlo Vecchio, G. Shanmugam, Domenica Attardi, Martin S. Robin, Samuel Salzberg and Sauyma Bhaduri

St. Louis University School of Medicine
Institute for Molecular Virology
3681 Park Avenue
St. Louis, Missouri 63110

INTRODUCTION

The mechanism by which RNA tumor viruses, i.e., leukemia and sarcoma-producing viruses of animals, replicate and transform cells was shrouded in mystery two years ago (1). At that time, two puzzling requirements for virus replication, (i) DNA replication and (ii) actinomycin-sensitive RNA transcription, indicated that RNA tumor viruses differed from other animal and bacterial viruses with RNA as their genome (1). Moreover, the inability to detect the viral 70S RNA genome within infected cells suggested a unique mechanism for viral RNA synthesis. The discovery in 1970 of an RNA-directed DNA polymerase (RDP) in RNA tumor viruses (2,3) offered an explanation for these unique requirements and stimulated a burst of activity on the biochemistry of these viruses.

The analysis of the molecular events of virus replication and cell transformation is much more difficult for RNA-containing tumor viruses than for other RNA viruses. Detection of virus-specific RNA and protein molecules in cells replicating nononcogenic RNA viruses is facilitated by the use of actinomycin D to inhibit host cell RNA synthesis and by the virus-induced shutdown of cellular RNA and protein synthesis. But this strategy cannot be used to analyze virus-specific molecules in cells infected by RNA tumor viruses (1). To detect and quantitate viral RNA sequences in cells infected and transformed by RNA tumor viruses, it was necessary to develop a hybridization procedure which utilized the DNA product of the virion RDP (4). To measure the intracellular synthesis of viral proteins, an immunoprecipitation procedure which detected virion proteins was established (5). Using these methodologies we performed the following studies: 1. the characterization of virus-specific RNA molecules in virus-producing and in cryptic nonvirus-producing cells transformed by MSV (murine sarcoma virus); 2. the analysis of viral protein synthesis in transformed, virus-producing cells; 3. the demonstration of viral mRNA and nascent viral polypeptides in polyribosomes of transformed, virus-producing cells; 4. the synthesis of viral proteins in vitro, using polyribosomes containing viral mRNA; 5. the analysis of the molecular events during the rapid transformation of cells upon infection with MSV/MLV.

[1] This study was conducted under contract PH43-67-692 within the Special Virus-Cancer Program of the National Cancer Institute, National Institutes of Health, Public Health Service, Bethesda, Maryland.

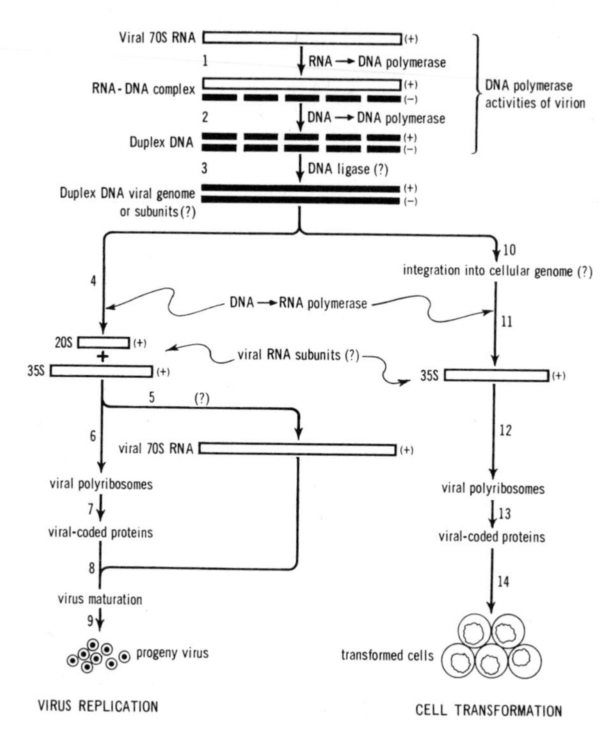

Fig 1. Scheme of RNA tumor-virus DNA polymerase activities and their possible functions in virus replication and cell transformation.

Figure 1 describes our current working model. The entire 70S RNA viral genome is transcribed by the virion RDP to small DNA molecules [1]; they may form duplex DNA copies of the entire viral genome or viral subunits [3] that are integrated into the cell genome of transformed [10] and perhaps of nontransformed, virus-replicating cells. During virus replication, viral DNA is transcribed to 20 and 35S RNA species [4] which may function as precursors of viral 70S RNA [5] and as viral mRNA on polyribosomes [6,7]. The assembly of virion proteins and viral RNA to form progeny virus probably occurs at the cell membrane [9]. In cells cryptically transformed by MSV (HT-1 cells), 35S viral RNA but not 20S RNA was found [11].

Viral RNA Transcripts in Mouse, Rat, and Hamster Cells Expressing Murine RNA Tumor Virus Information

The RNA genome of RNA tumor viruses sediments at 70S, but little or no 70S RNA is found in infected cells (1). Yet Green, Rokutanda, and Rokutanda (4) detected large quantities of virus-specific RNA sequences in the nucleus and cytoplasm of virus-producing rat and mouse cells transformed by the murine sarcoma virus and much smaller quantities in the cryptic, nonvirus-producing hamster cells (HT-1) transformed by the same virus. To understand this discrepancy, we analyzed the size of virus-specific RNA molecules in infected and transformed cells. Viral RNA molecules were detected by molecular hybridization with virus-specific DNA (Table I). The

TABLE I

Detection of Viral RNA in Cells Transformed by RNA Tumor Viruses

1. Preparation of viral ^{3}H-DNA copy of RNA genome

^{3}H-dTTP + dATP + dCTP + dGTP	70S viral RNA, Mg^{++} → RNA-directed DNA polymerase	viral ^{3}H-DNA(-)

2. Hybridization of viral ^{3}H-DNA(-) with cell RNA

Viral ^{3}H-DNA(-) + cell RNA → ^{3}H-DNA-RNA hybrid

formation of DNA-RNA hybrids was quantitated by differential elution from hydroxylapatite (4). As shown in Figure 2A, the hybridization reaction is specific for viral RNA, since the addition of normal cell RNA neither increased nor decreased the fraction of hybrid formed; 0.002 μg of virus-specific RNA was readily detected. ^{3}H-DNA specific for M-MSV (M is Moloney strain) was annealed with increasing amounts of RNA from 1. M-MSV(MLV) (MLV, murine leukemia virus); 2. M-MSV(MLV)-producing, transformed rat cells (78A1); 3. H-MSV(MLV) (H is Harvey strain) producing, transformed mouse cells (MEH); 4. "cryptic" nonvirus-producing M-MSV transformed hamster cells (HT-1), 5. nonvirus-producing mouse cells (NIH-X) derived from the leukemic spleen of an X-irradiated mouse; 6. adenovirus transformed rat cells (8617); and 7. normal rat cells (F-1853). As shown in Figure 2B, virus-specific RNA was detected readily in virus-producing 78A1 and MEH cells (70 - 80% hybridization) and in nonvirus-producing HT-1 and NIH-X cells (30 - 40% hybridization), but not in control 8617 and F1853 cell lines (6). Nonvirus-producing cells contain 1/15 as much virus-specific RNA as do virus-producing cells, as estimated from the half-saturation values.

Total cellular RNA was isolated by the hot phenol method, treated with dimethyl sulfoxide to dissociate RNA aggregates and fractionated by electrophoresis on

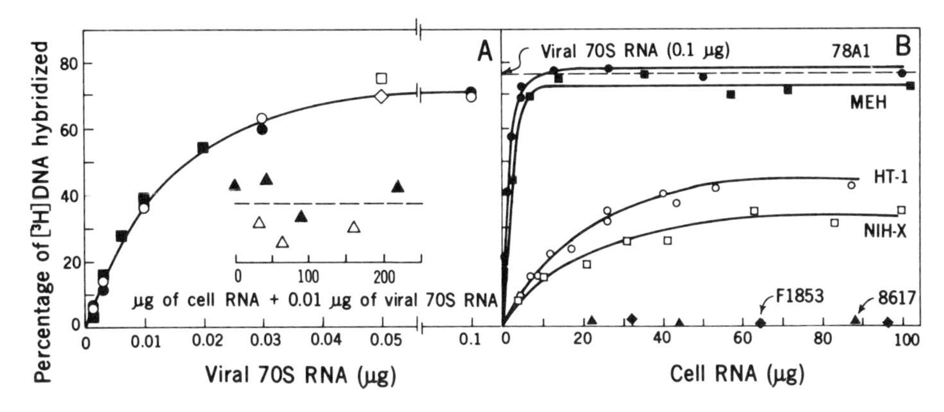

Fig 2. (A) Quantitation of virus-specific RNA. (B) Hybridization-saturation of M-MSV ^{3}H-DNA with RNA from virus-producing and nonvirus-producing transformed cells. See Tsuchida, Robin, and Green (6).

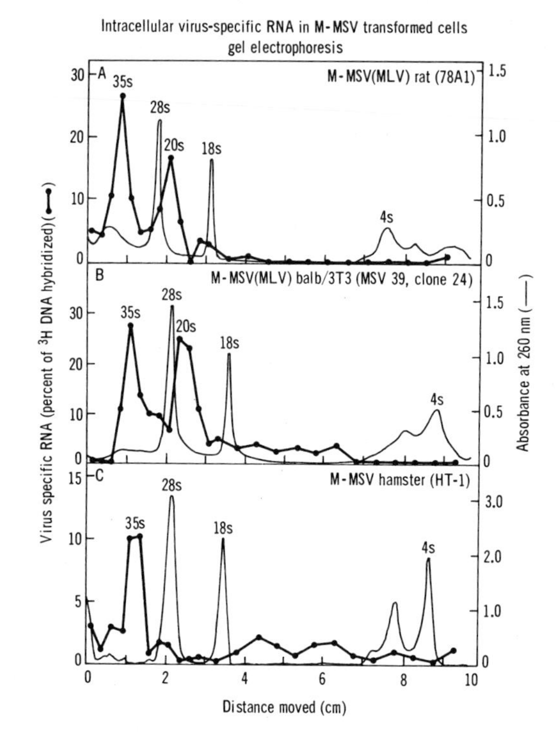

Fig 3. Resolution of intracellular virus-specific RNA in cells transformed by M-MSV. Total cell RNA was extracted from the following transformed cell lines:

(A): M-MSV transformed rat (78A1)

(B): M-MSV transformed balb/3T3 (MSV 39, clone 24)

(C): M-MSV transformed hamster (HT-1)

RNA was fractioned by electrophoresis on *2.5% polyacrylamide gels* (Tsuchida and Green, in manuscript).

polyacrylamide gels. Each gel fraction was annealed with M-MSV(MLV) ^{3}H-DNA. As shown in Figure 3, two distinct peaks of virus-specific RNA with sedimentation coefficients of 35S and 20S (determined independently on sucrose density gradients) were isolated from MSV(MLV) virus-producing cells. A single 35S peak was found in cryptic HT-1 cells. Both 35S and 20S viral RNA species are single-stranded RNA molecules, since they are digested by RNase but not by DNase.

The 35S and 20S viral RNA species may serve two functions: 1. precursor to the 70S viral RNA genome, and 2. mRNA for viral protein synthesis. In support of the precursor role, we have found that the MSV 70S RNA genome is irreversibly denatured to relatively homogenous 35S RNA (Tsuchida and Green, unpublished data). Whether 20S RNA is a precursor to 35S viral RNA or serves only as mRNA is not known.

The cryptic state of HT-1 cells is characterized by two defects: a level of viral RNA only 2 - 6% of that of virus-producing cells, and a specific deficiency in viral 20S RNA. To determine whether these two characteristics are related to the cryptic state of MSV transformed cells, we have analyzed the virus-specific RNA of three other nonvirus-producing mouse cell lines transformed by M-MSV. In each case we found 1. smaller amounts of virus-specific RNA; 2. a deficiency in some viral gene sequences;

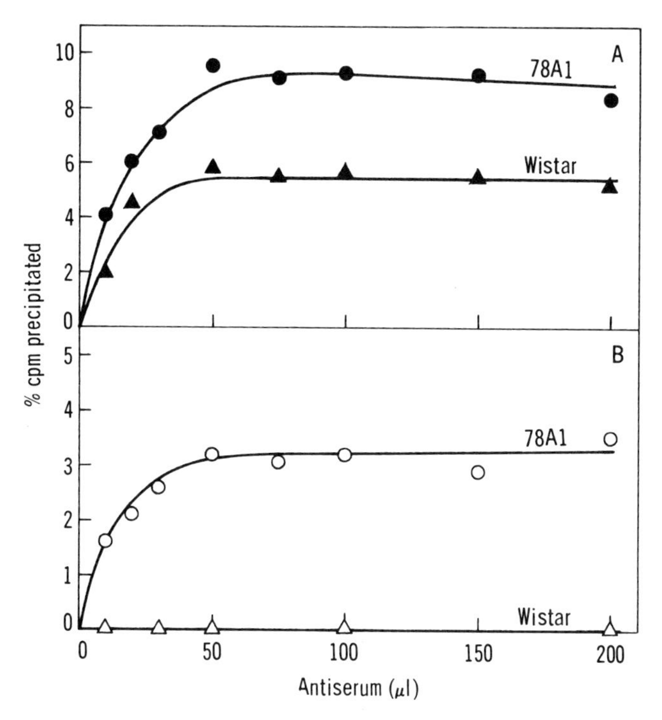

Fig 4. Immunoprecipitation curves of ^{3}H-amino acid labeled 78A1 and Wistar cell extracts by unabsorbed (A) and absorbed (B) anti-MSV(MLV) serum.

and 3. no 20S RNA, but instead a 26S RNA species (Tsuchida and Green, unpublished data).

Synthesis of Viral Polypeptides in Cells Replicating Murine Sarcoma-Leukemia Virus

Antibodies to disrupted M-MSV(MLV) were used to analyze the synthesis of viral polypeptides in the transformed, virus-producing 78A1 rat cell line. Antiserum prepared against purified virus contained antibodies directed against both virion and cellular proteins. Unabsorbed anti-MSV serum precipiated proteins of normal Wistar rat cells as well as of 78A1 cells (Fig 4A), while a serum absorbed with normal Wistar cells precipitated proteins only from 78A1 cells (Fig 4B). The specificity of the absorbed antibodies was demonstrated by electrophoretic analysis of the immunoprecipitate obtained with purified virus. The electrophoretic profile of purified virus before immunoprecipitation (Fig 5) shows seven labeled peaks (I–VII). Peaks II, III, and IV are the three major virion polypeptides; peak (IV), molecular weight 31,000, is the murine gs antigen. The same virus preparation was precipitated with absorbed anti-MSV serum or with anti-gs serum and analyzed by SDS polyacrylmaide gel electrophoresis (Fig 6). The absorbed antiserum precipitated the major virion proteins (II, III, and IV). The anti-gs serum precipitated mainly the gs antigen (peak VII may contain an aggregate of gs antigen).

Viral proteins that were synthesized by 78A1 cells in the presence of ^{3}H-amino acids were isolated from various subcellular fractions and analyzed by electro-

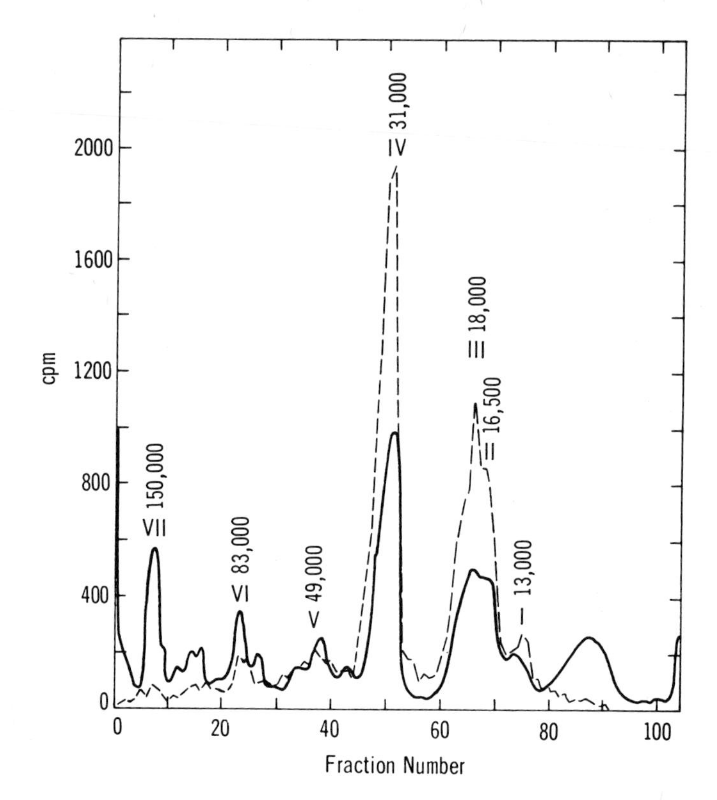

Fig 5. Electrophoretic pattern of M-MSV(MLV) polypeptides on polyacrylamide-SDS gels. MSV(MLV) (28,000 counts/min) labeled for 24 hr with ^{14}C-amino acids was disrupted with SDS and mercaptoethanol and subjected to electrophoresis (5). (——) densitometer scanning of stained bands; (---) ^{14}C radioactivity.

phoresis on SDS polyacrylamide gels, together with ^{14}C-labeled virion polypeptides added as markers. A comparison of gel patterns obtained with unabsorbed (Fig 7A) and absorbed (Fig 7B) sera show that both immunoprecipitates contained the major virion polypeptides (II, III, and IV) in proportions similar to those present in the assembled virion. However, components V and VI were present at much higher levels in the immunoprecipitate obtained with the unabsorbed serum and most likely are cell-specific proteins. This was further indicated by electrophoretic analysis of labeled proteins immunoprecipitated from the membrane fraction and from the cytoplasmic fraction of 78A1 cells (5). Polypeptides V and VI are present in the membrane fraction, but not in the cytoplasmic fraction. Polypeptides V and VI may become associated with RNA tumor viruses during the budding process of virus maturation. Whether they are structural components of the virion or tightly bound contaminants is not known. The origin of polypeptides I and VII is not established.

As described below we may utilize these immunological techniques to detect and identify virus-specific polypeptides in polyribosomes and those that are synthesized in cell-free protein synthesizing systems.

Detection of Virus-Specific RNA and Nascent Viral Polypeptides in Free and Membrane-Bound Polyribosomes of Virus-Producing Cells Transformed by M-MSV

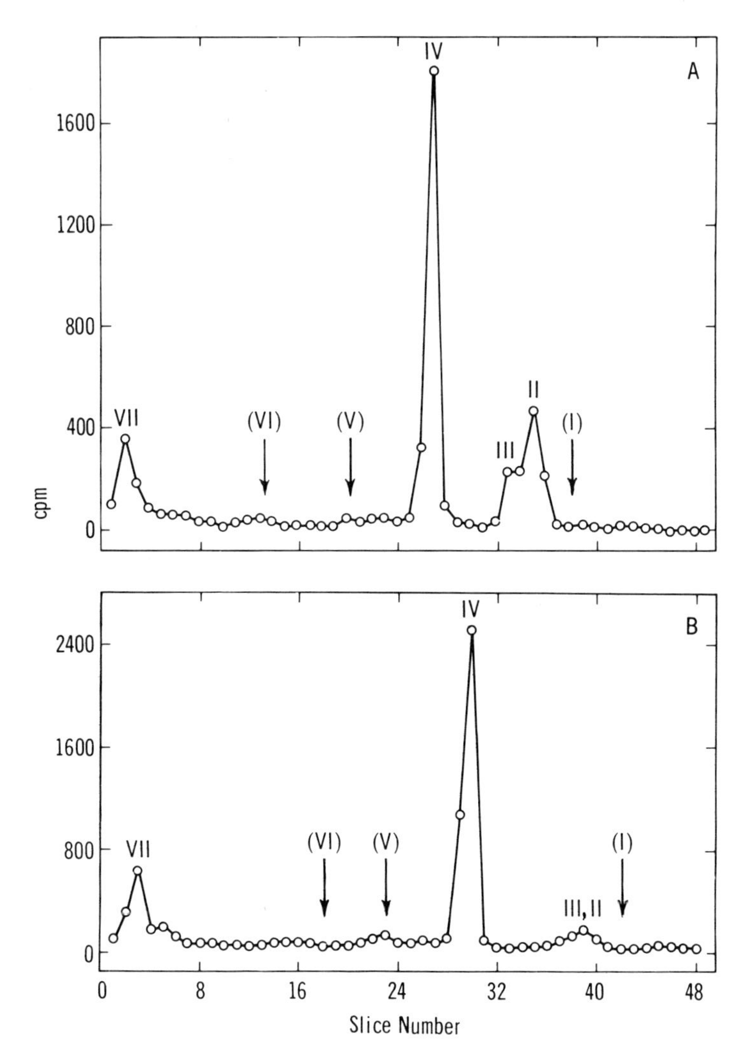

Fig 6. Electrophoretic analysis on SDS-polyacrylamide gels of immunoprecipitated MSV(MLV) polypeptides. MSV(MLV) (40 μg of protein) was disrupted with 0.5% Triton X-100 and precipitated with 400 μl of absorbed anti MSV(MLV) serum (A) or with 300 μl of anti gs antigen serum (B) (5).

Free (Fig 8A) and membrane-bound (Fig 8B) polyribosomes isolated from 78A1 cells were fractionated by sedimentation on sucrose gradients. RNA was extracted from each fraction and annealed with MSV-^{3}H-DNA to detect viral RNA. As shown in Figure 8, viral RNA was present on all classes of polyribosomes with a peak in the 350S region of sedimentation. Analysis by hybridization of RNA from free and membrane-bound polyribosomes showed that the latter contained approximately a four-fold higher concentration of virus-specific RNA.

Evidence that the viral RNA in polyribosomes is messenger was shown in three types of experiments. 1. When free and membrane-bound polyribosomes were treated with 25 mM EDTA, most of the virus-specific RNA was released. This is typical for mRNA. 2. Different size classes of free and membrane-bound polyribosomes

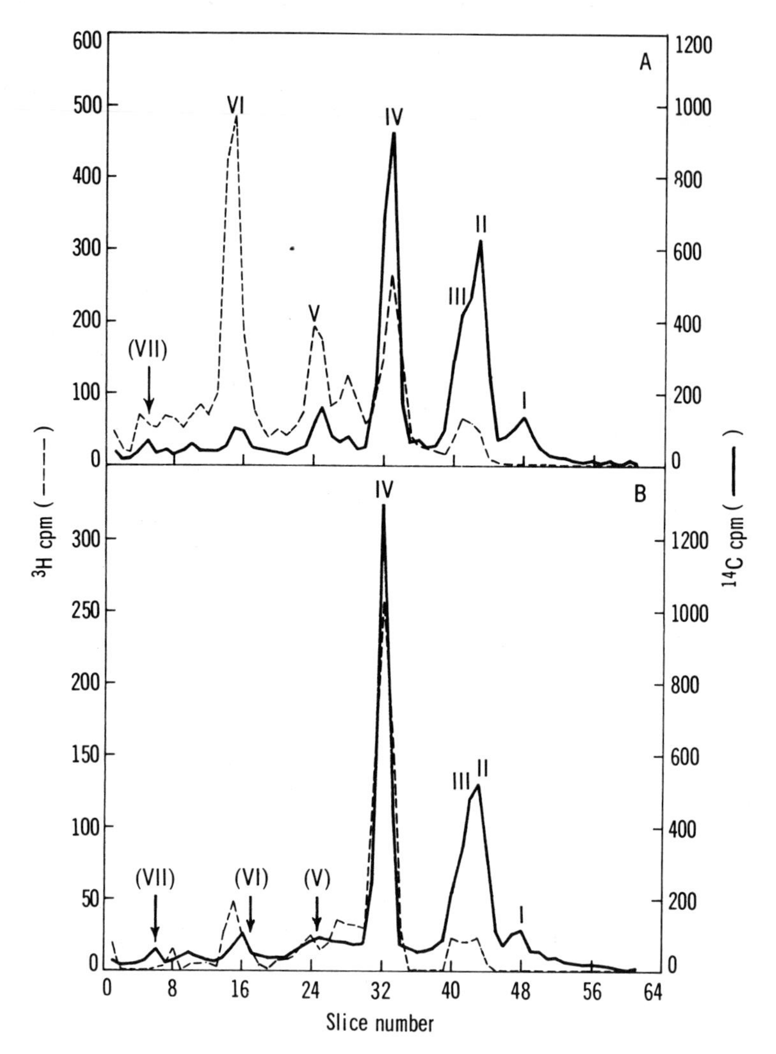

Fig 7. Electrophoretic patterns on SDS-polyacrylamide gels of ^{3}H-amino acid-labeled proteins immunoprecipitated from 78A1 cell extracts. The immunoprecipitates obtained with unabsorbed (A) and absorbed (B) anti-MSV(MLV) serum were subjected to co-electrophoresis with ^{14}C-amino acid labeled virus (5). (——) ^{14}C amino acid labeled marker; (– – –) ^{3}H amino acid intracellular proteins immunoprecipitated.

were treated with RNase and EDTA. Released nascent polypeptides contained virion-specific proteins as shown by immunoprecipitation with the anti-MSV serum. 3. Polyribosomes containing virus-specific RNA synthesize virus-specific proteins in vitro (see below).

In Vitro Synthesis of M-MSV(MLV) Proteins

In vitro synthesis of RNA tumor virus proteins was studied using pelleted polyribosomes prepared from 78A1 cells. All the components necessary for protein synthesis were present in the polyribosomes (Table II). Protein synthesis was optimal

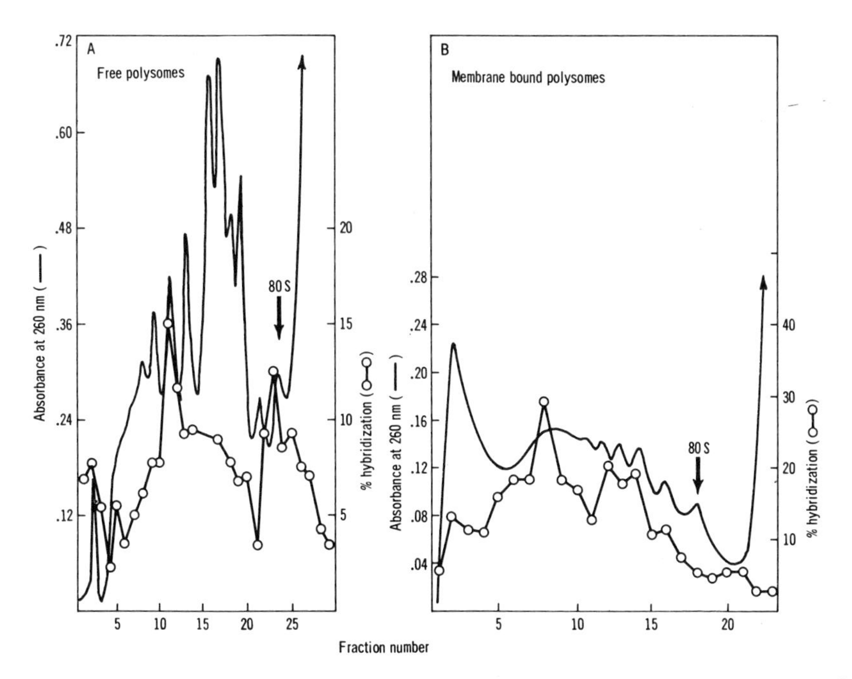

Fig 8. Virus-specific RNA in free and membrane-bound 78A1 cell polysomes. Sucrose density gradient profiles of (A) free and (B) membrane-bound polysomes after centrifugation on 15–45% sucrose gradients for 75 min at 36,000 rpm in the Spinco SW41 rotor. The absorbance was continuously recorded (——) and virus-specific RNA in each fraction was determined by hybridization with the ^{3}H-DNA product (o–o) (see Vecchio, Tsuchida, Shanmugam, and Green, in manuscript).

TABLE II

Characteristics of Cell-free Protein Synthesizing System[a] from MSV Transformed 78A1 Cells

	^{14}C-cpm incorporated/ 100 μl incubation mix
Complete	8,107
-Polysomes	88
-tRNA	8,100
+S-100[b]	6,300

[a]The complete system contained per ml: 26.6 μmoles Tris-HCl, pH 7.6; 2 μmoles $MgCl_2$; 10.6 μmoles mercaptoethanol; 57 μmoles KCl; 0.665 μmoles ATP; 1.67 μmoles GTP; 3.3 μmoles phosphenol pyruvate; 11.6 μmg phosphenol pyruvate kinase; 80 μmoles each of 1-cysteine, 1-glutamic acid, 1-asparagine, 1-tryptophane, and 1-methionine; 4 μC of ^{14}C-amino acid mixture (NEN); 19.5 A_{260} units of 78A1 polysomes; 160 μg tRNA from KB cells. Identical results were obtained by using 78A1 tRNA instead of tRNA from KB cells. Incubation was carried out at 37°C for 20 min.

[b]S-100 = 100,000 times g supernatant from 78Al disrupted cells added at a concentration of 0.94 A_{280} units/ml.

TABLE III

Immunoprecipitation of Virus Specific Proteins Synthesized in the 78A1 Cell-free System

	Input cpm ^{14}C in vitro product	^{14}C cpm in precipitate	^{14}C cpm immunoprecipitated
Rabbit control antiserum	6,592	21	0.31%
Rabbit anti-MSV serum	139,757	2,333	1.60%

at 2 mM $MgCl_2$ and was independent of exogenous tRNA prepared from 78A1 cells. Pyruvate kinase was about two times more efficient than creatine kinase as an energy source. Amino acid incorporation was linear up to 20 min at 37°C and was inhibited by 10^{-2}M NaF (38%), 100 μg/ml puromycin (74%), and 10^{-5}M pactamycin (32%), 400 μg/ml cycloheximide (77%), and 5 μg/ml RNase (96%). Of the proteins synthesized during a 20-minute incubation, 50 - 60% were released from polysomes.

The released proteins include both viral and cellular polypeptides. Viral polypeptides were precipitated specifically by anti-MSV serum (Table III). As shown by gel electrophoresis, two major polypeptide peaks (P1 and P2) were present in the immunoprecipitate. P1 and P2 have similar migration properties as the two major virion polypeptides and are present in the same relative proportions as in the virions. Further studies involve the characterization of these proteins by tryptic peptide mapping.

Analysis of the Molecular Events during the Rapid Transformation of Cells by MSV

Few systems are available for studying the molecular events during the transformation of mammalian cells. Transformation by chemicals and viruses is generally a very inefficient process which occurs over a prolonged period of time and generally involves a small fraction of the cells exposed. We have developed a system consisting of a clonal line of mouse 3T6 cells which can be rapidly transformed by H-MSV(MLV). At an input multiplicity of 1 - 2 ffu/cell, over 80% of the cells are infected within 1 - 2 hours and all cells are morphologically transformed within 2 - 4 days. As shown in Figure 9, virus-specific RNA synthesis begins as early as 7 hours after infection, as determined by hybridization of total cellular RNA with the ^{3}H-DNA product of the virion RDP. This suggests that the intracellular synthesis of viral DNA by the virion RDP, the integration of viral DNA, and the transcription of viral DNA have occurred within 7 hours after infection. The synthesis of the first virus particles was detected at 14 - 16 hours after infection, as determined by three independent procedures: formation of focus-forming units, synthesis of ^{3}H-uridine-labeled particles, and synthesis of particles containing RDP. This system presents many advantages for further analysis of transcription and translation during transformation by the viral genome.

It was shown recently that cells transformed by and continuously producing the MSV-MLV complex are strongly agglutinated in the presence of the plant lectin, concanavalin A (con A) (7) (see Fig 10). In order to determine whether this is a direct consequence of cell transformation, the agglutinability of cells by 250 μg/ml of con A

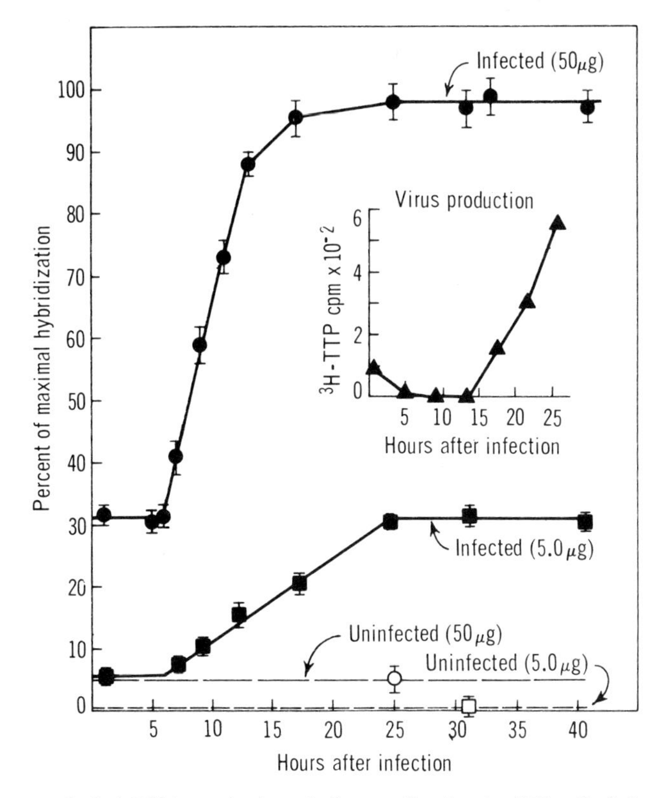

Fig 9. Kinetcs of viral RNA synthesis and virus replication in 3T6 cells infected with H-MSV (MLV). Viral RNA was determined by hybridization with the viral DNA product. Extracellular virus was analyzed by assay for RNA-directed DNA polymerase.

was determined at various times after infection. A small increase in agglutinability was observed at 22 hours after infection and a major increase occurred at 22 - 68 hours, at which time over 70% of cells were agglutinated. These results may suggest that a viral-coded function is responsible for the surface alterations associated with increased agglutinability and with the process of cell transformation.

SUMMARY

Several virus-cell systems expressing RNA tumor virus genetic information were studicd in order to understand the mechanism of transcription and translation of viral RNA during virus replication and during cell transformation with the following results: 1. single-stranded 35S and 20S viral RNA species are synthesized in virus-producing mouse and rat cells transformed by MSV (murine sarcoma virus). Transformed hamster and mouse cells that do not produce virus do not synthesize 20S viral RNA but do synthesize 35S or 26S RNA species. 2. Antibodies to disrupted MSV (MLV) (murine sarcoma-leukemia virus complex) were used to study the synthesis of viral polypeptides in the MSV transformed virus-producing rat cell line, 78A1. Intracellular viral proteins labeled with radioactive amino acids were isolated by immunoprecipitation and analyzed by electrophoresis in sodium dodecyl sulfate-polyacrylamide gels. The mobilities of intracellular viral polypeptides were identical to those of

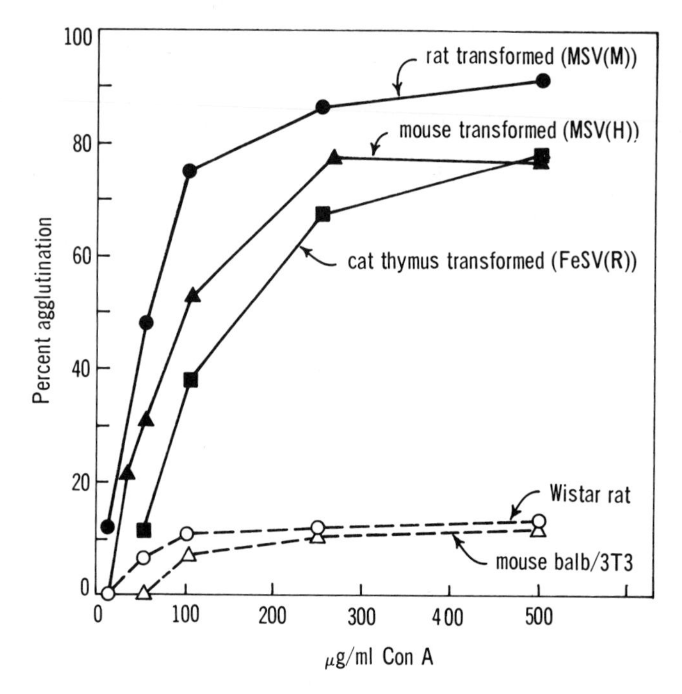

Fig 10. Agglutinability of sarcoma virus-producing transformed cells by concanavalin A. Agglutination assays were performed at 3–4 days after seeding, as described by Inbar and Sachs (8).

purified virions. However, labeled polypeptides having electrophoretic mobility lower than that of the major virion polypeptide, the group-specific antigen of molecular weight 31,000, were present in higher proportions in the total cell extract and in the membrane fraction than in the virion. These polypeptides appear to be of cellular origin and are associated with the cellular membrane. 3. Utilizing molecular hybridization and immunoprecipitation, it was demonstrated that membrane-bound and free polyribosomes of 78A1 cells contain virus-specific RNA and nascent viral polypeptides. 4. Polyribosomes isolated from MSV-transformed cells synthesize virion polypeptides in vitro. 5. A system for studying the molecular events during the rapid transformation of cells by MSV was developed. Virus-specific RNA is formed at 7 hours and virus synthesis begins at 14 - 15 hours after infection. An alteration in the cell surface was detected by agglutination with concanavalin A and was associated with the morphological expression of the transformed morphology and the production of MSV(MLV).

REFERENCES

1. Green, M.: Oncogenic virus. Ann. Rev. Biochem. 39:701, 1970.
2. Temin, H. M., and Mizutani, S.: RNA dependent DNA polymerase in virions of Rous sarcoma virus. Nature, 226:1211, 1970.
3. Baltimore, D.: Viral RNA dependent DNA polymerase. Nature, 226:1209, 1970.
4. Green, M., Rokutanda, H., and Rokutanda, M.: Virus specific RNA in cells transformed by RNA tumor viruses. Nature New Biology, 230:229, 1971.

5. Shanmugam, G., Vecchio, G., Attardi, D., and Green, M.: Immunological studies on viral polypeptide synthesis in cells replicating murine sarcoma-leukemia virus. J. Virol., 10:447, 1972.
6. Tsuchida, N., Robin, M. S., and Green, M.: Viral RNA subunits in cells transformed by RNA tumor viruses. Science, 30:1418, 1972.
7. Salzberg, S., and Green, M.: Surface alterations of cells carrying RNA tumour virus genetic information. New Biol., 240:116, 1972.
8. Inbar, M., and Sachs, L.: Interaction of the carbohydrate-binding protein concanavalin A with normal and transformed cells. Proc. Nat. Acad. Sci., 63:1418, 1969.

16. INFECTIOUS DNA INTERMEDIATE OF AN ONCOGENIC RNA VIRUS

L. Montagnier and P. Vigier

Foundation Curie–Institut de Radium, Biologie
Batiment 110, 91–ORSAY (France)

INTRODUCTION

Several kinds of evidence already suggest that a DNA intermediate is formed during the replication of oncogenic RNA viruses. In particular:

1. The action of inhibitors of DNA synthesis at the early phase of infection (1).

2. The existence of some homology between sequences of viral RNA and DNA of transformed cells (2,3).

3. The isolation of an RNA-directed DNA polymerase constantly associated with virions (4,5) which makes in vitro sequences complementary to at least some sequences of viral RNA.

However, a direct proof that the whole length of incoming viral RNA is transcribed in infected cells into a DNA carrying all the genetic information of the virus was still lacking. It could be argued notably that only a portion of the viral RNA is transcribed into DNA and that the remainder of viral RNA might be replicated by some other mechanisms or transcribed from cellular DNA sequences (6).

In contrast, isolation of infectious DNA from transformed virus-producing cells would be unequivocal proof that these cells do contain a proviral DNA, possibly integrated into cellular DNA, as has been postulated by Temin (7).

A preliminary report by Hill and Hillova (8) in 1971, indicated that a DNA fraction extracted from a permanent line of nonpermissive rat cells transformed by Prague strain Rous sarcoma virus, the XC line, could induce a C-type transforming virus in chicken cells, the virus being detected only after several passages of the treated cells.

However, proof that the virus obtained from DNA was identical to that of the donor virogenic cells was lacking. It could therefore be argued that DNA had derepressed an endogenous virus present in the chicken cells used.

EXPERIMENTS

In order to confirm or refute these results, we decided to undertake similar experiments using permissive cells with the following approach. We extract DNA from chicken cells transformed by the Schmidt-Ruppin strain of RSV, incubate uninfected chicken cells of the same origin with this DNA and look for foci of transformed cells, after one or several passages of the treated cells. A preliminary report of our first results has been published (9).

In a first series of 7 experiments, DNA was extracted from cells with sodium dodecylsulfate following the Marmur method (10), deproteinized with chloroform, treated with RNase, Pronase, and precipitated with alcohol. DNA fibers were fished out and dissolved in Tris-Hanks buffer (Tris 0.01 M in Hanks solution, pH 7.4).

For DNA infection, primary or secondary cultures of chick embryo fibroblasts, in 60 or 100 mm Falcon petri dishes were pretreated with DEAE-Dextran (30 - 100 μg/ml) in Tris-Hanks buffer, for 20 to 60 min at 37°C. The fluid was removed and the DNA solution at concentrations varying from 10 to 75 μg in 1.5 ml per 100 mm petri dish was added for 60 minutes. The DNA solution was removed and fresh culture medium was added. The treated cultures were incubated under liquid medium for three or four passages of about a week each (at least thirty days in all). The cells were divided in two or three lots and kept at each passage.

Each time a control was done with DNA pretreated with DNase. Of these first seven experiments, three were positive, that is new foci of transformed (Rous) cells and free virus appeared at the second or the third passage in some plates. No foci could be seen in cultures incubated with DNase-treated DNA.

In order to separate the infective fraction and to determine its intracellular localization, we used a technique devised by Hirt to separate chromosomal DNA from episomal polyoma DNA in cells infected with this virus (11): cell monolayers are lysed by SDS, NaCl in the lysate is raised to 1 M, and the lysate is kept at 4°C, 12 to 18 hours, then centrifuged at 14,000 g: the pellet contains the bulk of chromosomal DNA, the supernatant episomal or plasmidal circular DNA.

TABLE I

Recovery of RSV from CE Cells Infected with DNA from Rous Cells

Method of extraction	μg DNA tested[a]	DEAE-Dextran treatment (μg/ml, time)	N° of passages[b]	N° positive experiments	Rous cells and virus recovered at passage N°
Marmur	20/1D to 200/10D	30/60 mn to 100/60 mn	3-4	3/7	2(14d)-3(25d)
Hirt	10/1D to 300/24d	50/20 mn to 100/60 mn	3-4	8/10[c]	2(9d)-3(24d)

Note: Primary or secondary CE cell cultures (1-3 day-old), in 6 cm (d) or 10 cm dishes (D), were infected for 1 hr with DNA from Rous cells, following treatment with DEAE-Dextran. The treated cultures were incubated and subcultured in liquid medium at all passages (Marmur), or in liquid medium at first passage and under agar medium at subsequent passages (Hirt).

[a]Total amount on number of 10 cm (D), or 6 cm (d) dishes, in positive experiments.

[b]Cultures were kept at least for 30 days when negative.

[c]Including two experiments in which DNA was fractionated on a sucrose gradient, one in which it was fractionated on a CsCl gradient, and two in which it was recovered from Rous cells transformed by the FU-19 *ts* mutant (cf. Table III).

TABLE II

Detail of Positive Experiments with DNA Prepared by the Method of Marmur

Expt. N°	μg DNA tested[a]	CE cells[b]	DEAE-Dextran treatment (μg/ml, time)	N° of pas-sages	Rous cells and virus at passage N°	N° of positive dishes
1	200/10D	I, 1d	30/60 mn	4	3 (25d)	All (pooled)
2[c]	30/2D	II, 3d	50/20 mn	3	2 (14d)	1/2
3[c]	20/1D	I, 3d	100/20 mn	3	2 (17d)	1/1

Note: Primary (I) or secondary (II) CE cell cultures were infected for 1 hr with DNA from Rous cells, following treatment with DEAE-Dextran. The cultures were subsequently incubated in liquid medium and subcultured at 4-10 day intervals.

[a]Cf. Table I.
[b]CE cells were allowed to reach confluence before infection with DNA.
[c]Control cultures infected with DNA treated with DNase were negative.

Using this technique, we always found the infectivity in the pellet fraction (Table III): this means that RSV-DNA is associated with chromosomal DNA, although it does not prove that both are covalently linked. Moreover, the experiments using this pellet fraction were almost constantly positive (Tables I and III).

On the other hand, all foci induced by the DNA produced virus. The virus obtained was phenotypically the same as that used to transform the donor cells. It produced Rous cells and foci of the Schmidt-Ruppin type and it was neutralized by antiserum against type D avian leukosis virus, to which our SR-RSV strain belongs, and not by antiserum against other types.

This means that the DNA molecule carries at least three kinds of information: for making the envelope of the virions, for inducing or for making the DNA polymerase, and for transforming the host cell.

Owing to the very low sensitivity of the assay, it was not possible to quantitate accurately the system by making a dose-response curve. However, as far as we can see, the response seems to be nonlinear at high concentrations of DNA. Notably, foci were more readily obtained by using relatively low concentrations of DNA, such as 10 - 20 μg per plate of 1 - 2 times 10^6 cells. Higher concentrations were less active.

Finally, we attempted to fractionate DNA by sedimentation in sucrose gradient. The optimal density profile (Fig 1) is that of cellular DNA extracted from the pellet of the Hirt technique. One obtains a rather broad peak at about 20 S, as judged by ribosomal RNA markers sedimented in another gradient run under standard conditions of ionic strength. Fractions of the gradient were pooled two by two, and assayed for infectivity, each in 60 mm plates.

Three fractions were found to be infectious, as shown by the crosses in the Figure 1:

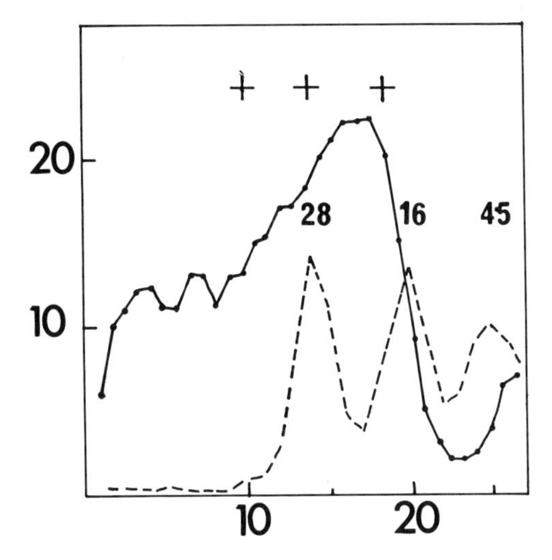

Fig 1. Sedimentation in a sucrose gradient of Roux cells DNA (full line). The crosses indicate the infective fractions. The dotted line represents the sedimentation of ribosomal RNA markers under the same conditions.

–fraction 10: 1/3 plates (1 plate was lost by contamination)(35 S)
–fraction 14: 1/4 plates (30 S)
–fraction 18: 1/4 (22 S)

As we have already pointed out, since we do not know the dose-response curve of the assay, these results should be taken as preliminary. Two further remarks may be made:

1. Infectivity is associated with several sizes of DNA. This should be expected if the RSV-DNA is integrated into chromosomal DNA. Alternatively, the DNA may correspond to tandem copies of RSV-DNA molecules in varying number.

2. The minimal size seems to be 6 million (fraction 18).

Assuming that DNA is double stranded (as it seems from preliminary studies with CsCl), this would mean transcription into a single strand of viral RNA of 3 million, which is the size of only one subunit of viral RNA.

This raises a number of problems, but since the determination of size is a preliminary estimate, it would not be wise to speculate further at the present time.

Anyhow, the fact remains: it is possible to extract in a reproducible manner from cells infected with Rous sarcoma virus a DNA which carries its genetic information.

Similar results have been obtained in three other laboratories. Svoboda and associates have confirmed the infectivity of DNA extracted from XC rat cells (12). F. Lacour and associates could induce myeloblastosis in chickens by inoculating chicken fibroblasts incubated with DNA from myeloblasts (13). Hill and Hillova have recently shown that DNA extracted from hamster cells transformed by a ts mutant of RSV (14) could induce in chicken cells the production of virus with the same characteristics.

TABLE III

Detail of Positive Experiments with DNA Prepared by the Method of Hirt

Expt. N°	Fraction tested	μg/ DNA tested[a]	CE cells[b]	DEAE-Dextran treatment (μg/ml, time)	N° of pas-sages	Rous cells and virus at passage N°	N° of positive dishes
1	Prec.[c]	45/2d	I, 1d	50/20 mn	3	2 (13d)	1/2
	Sup.	30/2D				-	
2	Prec./ sucrose gradient	ε-60d/ 4-8d	II, 1d	50/40 mn	3	2 (9-12d)	F2 1/6 F3 2/6
	Sup.	30/1D				-	0/1
3	Prec./ sucrose gradient	ε-50[d] 3-4d	II, 1d	50/30 mn	4	3(14-22d)	F10 1/3 F14 1/4 F18 1/4
4	Prec./ CsCl gr.	ε-40[d]/ 3-4d	II, 1d	50/40 mn	4	3 (22d)	F5 1/4
5	Prec.	290/24d	II, 1d	50/60 mn	3	2 (16d)	2/24
		50/21d				2 (16d)	4/21
6	Prec.:	150/5d	II, 1d	50/60 mn	3	3 (18d)	1/5
	65°C[c]	30/5d				-	0/5
	100°C	150/5d				-	0/5
		30/5d				-	0/5
7	Prec.	225/5d	II, 1d	100/60 mn	4	3 (24d)	1/5
	FU-19	45/5d				-	0/5
8	Prec.[c]	60/2D	I, 1d	100/40 mn	3	2 (16d)	1/2
	FU-19	12/2D				2 (16d)	1/2
	Phenol	50/2D				3 (23d)	1/2
		10/2D				3 (23d)	1/2

Note: CE cell cultures (cf. Table II) were infected with DNA from Rous cells (precipitate [prec.] and supernatant [sup.]), following treatment with DEAE-Dextran. Cultures were incubated 3-10 days in liquid medium, then subcultured and incubated for 2-3 more passages under agar medium.

[a,b,c]Cf. Table II.
[d]Amount per fraction tested.

Recently, we have been able to confirm this latter result in the permissive system and also to recover infectious DNA from RAV-infected chick cells.

REFERENCES

1. Bader, J. P.: The requirement for DNA synthesis in the growth of Rous sarcoma and Rous-associated viruses. Virology, 26:253, 1965.

TABLE IV

Some Properties of Infectious DNA from RSV-Transformed Chick Cells

- Induces foci of transformed, virus-producing cells.
- The virus produced is identical to that used for infection of donor cells.
- Activity destroyed by Dnase
  not destroyed by Rnase, Pronase
  not destroyed by heating at 65°C for 10 min
- Associated with chromosomal DNA
- Density in CsCl: that of cellular DNA, or slightly lower
- Sediments between 20 to 35 S (?)

2. Baluda, M. A., and Nayak, D. P.: DNA complementary to viral RNA in leukemic cells induced by avian myeloblastosis virus. Proc. Natl. Acad. Sci. USA, 66:329, 1970.
3. Rosenthal, P. N., Robinson, H. L., Robinson, W. S., Hanafusa, T., and Hanafusa, H.: DNA in uninfected and virus-infected cells complementary to avain tumour virus RNA. Proc. Natl. Acad. Sci. USA, 68:2336, 1971.
4. Mizutani, S., and Temin, H. M.: RNA-dependent DNA polymerase in virions of Rous sarcoma virus. Nature, 226:1211, 1970.
5. Baltimore, D.: RNA-depentent DNA polymerase in virions of RNA tumour viruses. Nature, 226:1209, 1970.
6. Montagnier, L.: Replication of oncornaviruses. In: RNA Viruses and Host Genome in Oncogenesis. Proceedings of a conference held in Amsterdam May 12 – 15, 1971. North-Holland Publishing Company, Amsterdam–London, p 49, 1972.
7. Temin, H. M.: Nature of the provirus of Rous sarcoma. Natl. Cancer Inst. Monogr., 17:557, 1964.
8. Hill, M., and Hillova, J.: Production virale dans les fibroblastes de poule traités par l'acide desoxyribonucleique de cellules XC de rat transformées par le virus de Rous. C. R. Acad. Sci. (D) (Paris), 272:3094, 1971.
9. Montagnier, L., and Vigier, P.: Un intermédiaire infectieux et transformant du virus du sarcome de Rous dans les cellules de Poule transformées par ce virus. C. R. Acad. Sci: (D) (Paris), 274:1977, 1972.
10. Marmur, J.: A procedure for the isolation of deoxyribonucleic acid from micro-organisms. J. Mol. Biol., 3:208, 1961.
11. Hirt, B.: Selective extraction of polyoma DNA from infected mouse cell cultures. J. Mol. Biol., 26:365, 1967.
12. Svoboda, J., Hlozanek, I., and Mach, O.: Detection of chicken sarcoma virus after transfection of chicken fibroblasts with DNA isolated from mammalian cells transformed with Rous virus. Folia biol. (Praha), 18:149, 1972.
13. Lacour, F., Fourcade, A., Merlin, E., and Huynh, T.: Detection de virus de la myéloblastose aviaire dans des cultures de fibroblastes de poule traités par l'ADN de cellules leucémiques productrices de virus. C. R. Acad. Sci. (D) (Paris), 274:2253, 1972.
14. Hill, M., and Hillowa, J.: Recovery of the temperature-sensitive mutant of Rous sarcoma virus from chicken cells exposed to DNA extracted from hamster cells transformed by the mutant. Virology, 49:309, nl, 1972.

17. DISCUSSION

Dr. Gillespie:
(Baltimore, Md.)

I would just like to mention a couple of experiments we have done in collaboration with Bob Gallo's group, because it bears on the total picture of the leukemic cell with respect to the presence of the reverse transcriptase and its possible templates, and also bears on some information that Dr. Spiegelman will present this afternoon.

We have been looking for a viral-like RNA in leukemic cells. We reported a few months ago an assay system for potentially doing this. The assay essentially is based on a search for a 70-S RNA which contains long Poly-A stretches.

In itself it is not enough to demonstrate a viral RNA but it certainly gives one a probe to screen potential viral RNAs. What we didn't have at that time was the appropriate compartment in the cell to look for, since the cell itself contains some 70-S Poly-A containing RNA.

It became clear from Bob's work that the reverse transcriptase was in a particular cytoplasmic pellet fraction at least of leukemic cells, and it was also becoming clear from work in Dr. Spiegelman's lab that the reverse transcriptase was apparently associated with a transcribable RNA molecule.

We, therefore, isolated the RNA from this pellet fraction, analyzed it on sucrose gradients, and looked for a 70-S Poly-A containing RNA. This was done with Sally Gillespie in my laboratory and Dr. Marjorie Roberts in Bob Gallo's laboratory. In eight out of ten leukemic cells we have looked at of different types they all have 70-S Poly-A containing RNA in the pellet fraction in the amount of 25 molecules per cell. That is what we recover. It is, moreover, in a low density particle, density of less than 1.2. It contains poly-ribo-A, at least. We are not certain at this point that it is pure RNA, but we know that the Poly-A entity is a poly-ribo-A and we know that it is very long.

This RNA component is not present in unstimulated lymphocytes or, at least if it is present, it is approximately 1/100th of the concentration that you find in leukemic cells, but it is present in PHA-stimulated normal lymphocytes.

We have looked at three different preparations. All three have shown the 70-S RNA, although it appears to be about 1/10th that you find in leukemic cells. It does not seem to be simply a function of rapidly proliferating cells, because we have not seen it in tissue culture cells.

From this we would like to suggest that there is a virus-like particle in these cells. It is not clear at what point you undergo the transition from particle to virus particle. Certainly one thing that would be very important in this regard would be some information on the nature of the genetic information in the RNA and I hope we hear some more about that this afternoon.

Dr. Temin: I would like to ask Dr. Green whether he thinks that the two-sized classes of RNA in the virus-producing cells could relate to the fact that you have a mixture of two viruses, MSV and MULV. Specifically, can you test it the way Stephanson and Aaronson did by getting a spot that had a majority of MSV particles and making two probes to see if one is MULV and one is MSV RNA?

Dr. Green: We would love to have a cell line producing MSV. I don't know of any, but it is not because it is a mixture. We have looked at the straight leukemia, murine leukemia, and malony viruses. They have both subunits. So, it is not because it is a mixture.

Dr. Montagnier: Dr. Green, when you are detecting the viral RNA, can you tell if it is minus or plus?

Dr. Green: It is single-stranded plus.

Dr. Montagnier: You are using only single-stranded DNA?

Dr. Green: Yes.

Dr. Baltimore: Dr. Montagnier, when you test the DNA from transformed cells, do you do it on GS-plus or GS-minus chicken cells, or is there any difference?

Dr. Montagnier: I forgot to mention this. Most of the cells we use are GS-plus, and in the positive experiments there were mixtures of three embryos, so the probability that only one, this one, is GS-plus is very high. This may raise a question whether there is some information coming from the cells.

Dr. Baltimore: Did you use DEAE dextran?

Dr. Montagnier: Yes.

Dr. Beers: Dr. Gallo, to what extent can you rule out the apparent specificity that you described on the basis of primer requirement?

Dr. Gallo: Well, if it is specificity, I can't rule out that it is a difference due to primer requirement, one enzyme versus another. I can't say what the difference is. I can only say the difference is there.

Dr. Beers: In your natural RNA with the strings of Poly-A, did you attempt to complex them at all with poly dT?

Dr. Gallo: Yes. When we were using purified enzymes from leukemic cells, we tested the RNA alone from AMV and in some experiments the RNA plus oligo dT. We get the same kind of stimulation that one obtains with the viral enzyme purified from virus, about a five-fold stimulation.

Dr. Green: Dr. Gillespie, on the 70-S RNA, how do you know that it is RNA and not a ribosomal subunit or a complex of RNA and protein? I don't know what sort of gradients were run.

Dr. Gillespie: It is conceivable that it is a ribonucleoprotein complex, but not conceivable that it is as gross as a ribosomal subunit. These are all phenol-chloroform extracted materials, pH-8, pH-9, or rather stringent conditions. It could still conceivably be some sort of protein that is difficult to get rid of, I agree.

Dr. Green: You don't have a double-label experiment.

Dr. Gillespie: We have not tried to label the RNA. This is all nonradioactive RNA which is detected with a radioactive probe.

Dr. Gallo: In the tissue culture cells one doesn't see this and in conditions where you would isolate cellular cytoplasmic RNA under the same phenol conditions you don't see 70-S RNA anymore.

Are these DMSO-treated prior to sedimentation so they are not aggregates?

Dr. Gillespie: In the case of the tumor virus RNA sedimentation runs on neutral sucrose gradients, we do see aggregates. We have been counting only the 70-S material; we see no 35-S subunits.

Dr. A. K. Bandyopadhyay: (Houston, Texas) Dr. Green, I would like to know in your in vitro protein synthesis what is the source of elongation factor, t-RNA synthesis and t-RNA?

Dr. Green: I am not sure myself. It is a polysome fraction. You don't need t-RNA. It has t-RNA bound to the polyribosomes, and I am not sure whether or not there is an S-100 added to it.

Dr. Cordaro: (Baltimore, Md.) Dr. Green, those cells that you showed were cryptically transformed. Could you recover the virus, much as is done with single and double lysogens of lambda? Can you recover the transforming virus out of those cryptically transformed cells?

Dr. Green: The one that I reported was the HT-1, the hamster cell line, and Heubner and his colleagues reported in 1966 that they could recover the virus by co-cultivation with mouse cells and superinfection with a marine leukemia virus.

Dr. Cordaro: In the last slide you presented you didn't talk much about the last column where you added alphamethylglucoside to the Con A agglutinating experiment and you didn't get any agglutination.

Dr. Green: That is a control. It's competing with Con A, yes.

Dr. Cordaro: Dr. Baltimore, could it possibly be that naively RNase-H in the formation of the hybrids takes away the second complementary DNA strand mate, so that it can be copied with the formation of two double-stranded DNA, which is then inserted into its pro-virus? If this were true, you should be able to mutaganize your virus and plate out for the non-transforming mutants, among which should be RNase-H-minus.

Dr. Baltimore: Yes, but it is not true. The ribonuclease-H activity does not completely degrade to acid solubility.

Dr. Cordova: You mentioned an excess of endonuclease.

Dr. Baltimore: The point is that although it will completely degrade Poly-A, it will not completely degrade the RNA template to acid solubility when a natural RNA is used.

We suspect that all it will do, although we don't know this, is make a limited number of endonucleolytic cuts in the other side, presumably to produce primer sets for DNA synthesis in the opposite direction. But we have no direct evidence at present that that is what is happening.

Dr. Green: I wonder if we could open up the question of what normal reverse transcriptase means in the cell and what the possibilities are that these are not significant structures, but just sort of artifacts that Dave Baltimore described a number of years ago with proteins and extracts. How can we attack that problem?

Dr. Temin: We tried to attack that problem in two ways; first, by attempting to make artificial reconstructs and not finding any effect and, second, by showing that the polymerase which was part of that structure was not found elsewhere in the cell. That still does not say it has a biological function, but it couldn't arise from that kind of thing. So the approach we are taking is to look at the distribution of this polymerase in different embryonic stages, different organs, and to do cross-hybridization with the templates and the products and see if there is any relationship in that way.

Dr. Gallo: Maybe I could add to that. When we don't use unstimulated normal lymphocytes, we don't see this. It is only with the stimulated cell or leukemic cell. The kind of DNA we get with the purified enzyme in AMV-70 S RNA is almost as good as you get with the viral system. Of course, this doesn't prove it is biologically functional in the cell in that way.

Dr. Green: Let me ask Howard again, how do you use two milligrams of RNA for hybridization? That should tell you that the actual template was present in very, very small amounts.

Dr. Temin: Yes. If we work with 200 micrograms per 0.4 milliliter, we get very low hybridization, about 12%. If we go up to 800 or 1,000 micrograms, we get about 40%. To show that this is explained by the hypothesis that you have suggested, we have reisolated DNA from the hybrid region of the gradient.

Now, if we do this with a virus system as a model, that is, if we take the viral endogenous DNA polymerase product and anneal that with viral RNA, 40% of the DNA product moves to the RNA region of the gradient. If we isolate that DNA and reanneal with the same concentration of viral RNA just at the break in the saturation, we get 80 to 90% moving with the RNA region, which shows high efficiency.

If we do the same experiment with the endogenous cellular DNA polymerase activity, that is, make DNA from the endogenous reaction, and hybridize that DNA with very large amounts of RNA from that fraction, isolate the DNA that hybridizes so that we know it is capable of hybridizing, and then reanneal it with fresh RNA, we find again low efficiency. So this shows that there is a low concentration in the template RNA.

Dr. Green: So the RNA fraction is very heterogenous. How would that fit in with the protovirus hypothesis? There are many sequences being continuously transcribed reversibly. Does it seem to be very functional?

Dr. Temin: No. The question that you are now asking has to be answered by a different kind of experiment. The first question you asked was answered by the experiment I described regarding the concentration of the sequences which were copied. Now you are asking about the complexity of this. This we would have to study with DNA-DNA reassociation kinetics, to see how complicated it is. And those experiments are just now being carried out. We have no idea yet how complex the template is. We can just say it is in low concentration.

I should mention that it takes about five dozen embryos to get enough to do one or two experiments. There is not an awful lot of the material around.

Dr. Gallo: Since you brought it up—maybe Dr. Baltimore could comment on it—I wasn't familiar with the artifact systems by reconstitution that you have. How would they compare to what he said? Is it analogous?

Dr. Temin: No, the analogy was something else that we found, I think a long time ago, that has really little to do with this, although it is formally the same thing.

Dr. Leon Dmochowski: (Houston, Texas) You mentioned this particular material or "particle" connected with the reverse transcriptase activity of normal cells. Have you examined what kind of particle you have?

Dr. Temin The point about saying it was "particle" was so that we wouldn't be asked that question. We did not call it a particle. It is a fraction with activity in its sediments. All we know is that it sediments and it has a low density. We have not looked at it because it is not very pure.

Dr. Baltimore: I am just interested in what kind of structure has a density as low as what, 1. 1?

Dr. Temin: These are membrane structures from animal cells and plasma reticulum. Sometimes one can actually see a visible band of membrane. So it is a fair amount of the cell.

Dr. Gallo: Ours bands at 1.1 also. We did look; there is nothing to see.

Dr. Dmochowski: I might add we looked, too, but didn't see anything.

Dr. Green: I didn't look and I didn't see anything.

Dr. Gallo: Protovirus is invisible.

Dr. Dmochowski: Dr. Montagnier, in your cells transformed by the infectious DNA, are full virus particles synthesized?

Dr. Montagnier: Yes.

Dr. Dmochowski: You mean morphologically or immunologically?

Dr. Montagnier: Immunologically.

Dr. Temin: Helen Hilova did publish electron micrographs of their covered virus, showing they were a C-type, if that is any help to you.

Dr. Morris Pollard: (South Bend, Ind.) I feel uncertain, listening to this entire discussion and the papers that preceded it, about the role of viruses in all oncogenic tissues. How solid is this evidence that viruses are involved in reverse transcriptase in all cases?

Dr. Temin: It is very certain that the Rous sarcoma virus can cause tumors in chickens.

Dr. Pollard: Do you have any virus-negative cells?

Dr. Temin: That was the first question, whether viruses have anything to do with the oncogenic process.

The second question, whether viruses are always responsible for RNA-directed DNA synthesis in cells, was examined by looking at the nature of the template, whether it was related to the RNA of known viruses, and looking at the nature of the polymerase to see if it was related to the polymerase of known viruses. In both cases the results were negative. So we can say two things: that the activity in normal cells is unrelated to known viruses or, if you prefer, that the activity in normal cells is related to unknown viruses.

Dr. Baltimore: There is a known covert virus in all chicken cells, a Group E virus. Do you know whether the polymerase in Group E viruses, Rous 60 or Rous 0, is neutralized by antiserum against AMV polymerase?

Dr. Temin: I guess it is known that the RNA cross-hybridizes. The polymerase antibody has been tested with A, B, C, and D, and it neutralizes all of them, so it probably does E; but I guess it has not been tested. But the RNAs all cross-hybridize.

Dr. Gallo: There has been no one, I think, who has said that all of these virus systems are reverse transcriptase systems and are playing a definite role in oncogenesis. One is trying to understand if reversal of information flow is present in cells, which cells, which kind of function it might have, and possibly use it as a probe to see if you can detect virus by a more sensitive technique or what might be something in the formation of a virus. But even if you know virus is there, you don't know for sure if it is causing disease, certainly not in man, because you can't do the experiment to prove it.

Dr. Howard Elford: (Durham, N.C.) I know, Dr. Gallo, that you know it was Dr. Beril of Duke University who first reported that DNA polymerase-I is associated with a membrane fraction in the cytoplasm in rat hepatomas and that it is related to cell proliferation rate.

My question is: What is it doing in the cytoplasm when you would expect that activity to be in the nucleus?

Dr. Gallo: It is obviously a very good question and one that perplexes everybody. Dr. Beril was right. Dr. Boland in his system has similar data. I think there are at least three, if not four, laboratories working with mammalian cells that have found these two major DNA polymerases. It seems to be generally agreed by almost every technique approaching it that DNA polymerase-I, or the high molecular weight enzyme, is in the cytoplasm. One could speculate that during replication of DNA this enzyme moves into the nucleus. I don't know of any proof of that, although I think Larry Lerb had some evidence in either synchronized cell systems or some other proliferative system that that might be the case, but I am not positive of that. Otherwise, I don't know. Like Boland says, maybe we ought to rethink about biology.

Dr. Baltimore: There was a paper about eight years ago by Marvin Gold, when he was in Toronto, in which he looked at DNA polymerase activity in the cytoplasm as a function of the cell cycle. Although he found a very small movement, apparent movement (at that time he didn't know there were two enzymes) of enzymes from cytoplasm to nucleus. In fact, the take-home lesson from that was that even in the S-phase, when the cells are quite nicely synchronized, most of the DNA polymerase is in the cytoplasm.

Dr. Temin: It is an enigma.

Dr. R. Graham Smith: (Bethesda, Md.) Dr. Montagnier, Hill and Hilova in their report showed that they had to treat these cell lines at least three times with DNA to get the transformation. Is your efficiency any better than that and do you have any ideas how this might be improved?

Dr. Montagnier: Yes. In our system, which is the permissive system, we need only do DNA once; we don't need to repeat incubation two or three times to get foci. The foci never appeared in the first passage of the cells. You have to pass the cells one or two more passages to see some foci.

So it seems that probably a few cells are infected at the first when they are treated with DNA. You don't see these cells before the system is amplified with this polyoma virus and then you can see foci.

So the efficiency is higher. It should be improved very much to get some quantitative results. We hoped by now to have foci of the first passage but we do not.

Dr. LeRoy Kuehl: (Salt Lake City, Utah) I would like to make a comment on the localization of DNA polymerase, apparently in the cytoplasm. There was some work done quite a few years ago which showed that DNA polymerase may indeed not be localized in the cytoplasm. This may be an artifact that arises when you isolate the nuclei. These workers isolated the nuclei under nonaqueous conditions, and when they did that they found considerable amounts of the polymerase in the nucleus, as you might more reasonably expect.

Dr. Gallo: I think I can answer that. That was before it was known that there were two DNA polymerases. There is a nuclear DNA polymerase. But interestingly enough, the nuclear DNA polymerase is not inhibited by cytosine arabinoside, which blocks DNA synthesis and does not increase in S-phase of the cell cycle nor in proliferative response. There is a nuclear

DNA polymerase, but it doesn't appear by biological studies to be the one that is relevant to DNA replication.

Dr. Kuehl: Has it been excluded that the polymerases which are supposedly cytoplasmic could also be nuclear because originally the nuclear polymerase was thought to be cytoplasmic?

Dr. Gallo: I think the people who have done the most competent work say it is not in the nucleus.

Dr. Kuehl: Could you briefly say on what evidence they say it is not in the nucleus?

Dr. Gallo: By isolating the nucleus and not finding it.

Dr. Kuehl: Did they isolate the nucleus in nonaqueous media or in an aqueous system, because things leak out?

Dr. Gallo: I can't tell you each of the techniques, but I would be fairly confident that Boland would be aware of that. You don't find liberated DNA or other nuclear proteins extracted out under the conditions he used.

Dr. Dmochowski: I would like to ask Dr. Gallo, with reference to localization of the DNA polymerase in the cytoplasm, about its relationship to mitochondria. Mitochondria do contain DNA. Have any studies of this type been done?

Dr. Gallo: That is obviously a good question. These polymerases may play a role with mitochondrial DNA, but it is surprising that it is the dominant activity, when there is so much more DNA in the nucleus.

Dr. Hattmore: (Rochester, N.Y.) Dr. Baltimore, I am somewhat confused by the initiation of DNA synthesis by reverse transcriptase in view of the co-valent linkage of the DNA to the 3′-adenine. What do you consider the role of the oligo dT to be?

Dr. Baltimore: The linkage to an A is only found in 60 or 70-S RNA when it is used as template without an added primer. You can superstimulate the activity of that RNA by addition of a primer. Under those conditions, presumably you don't make a linkage to an RNA, although I haven't looked at it directly. Presumably, in that case, the oligo dT is binding to the Poly-A regions which are ordinarily not used by the endogenous primer.

It raises serious questions because the place that one would like to put a primer, at least in a simple model, would be very near the Poly-A, which is somewhere near the 3′-end.

There are some theoretical problems that you can see in these models, but the only thing we can say is that we are not seeing a Poly-A primer, because you can remove the primer with ribonuclease.

Dr. Hattmore: If you use oligo dT with something like globin message—

Dr. Baltimore: In that case you don't get a linked molecule. You get a pure DNA.

Dr. Elford: I would like to respond to the problem of mitochondrial relationship to the cytoplasm of DNA polymerase. Mitochondrial marker enzymes are not found with this membrane particle. However, I have found that other enzymes related to DNA synthesis, like ribonucleotide reductase, thymidine kinase, thymidylic synthetase, are associated with this endoplasmic smooth membrane material.

Dr. Thompson: (Berkeley, Calif.) Dr. Temin, have you used detergent in obtaining your endogenous activity, and if so, does it affect the amount you can get?

Dr. Temin: Yes, we use nonadet and that is part of the fractionation scheme, so we can tell what happens without it. We haven't used a variety of isolation procedures.

Dr. Babrow: (Bethesda, Md.) Dr. Baltimore, have you investigated inhibitors of this ribonuclease-H, namely, specific inhibitors that would inhibit this ribonuclease and not the DNA polymerase?

Dr. Baltimore: No.

Dr. Ryser: (Baltimore, Md.) Dr. Montagnier, since you get three positive experiments out of seven, I wonder whether you might have any suggestions as to what the variables might be that are responsible for the inconsistency of the results?

Dr. Montagnier: Well, maybe this was not clear in my talk, but in the last series, using the Hirt technique, in seven experiments, six were positive. Now, we can reproduce nearly every time the results of infectious DNA. This could not be nuclear-capsid virus incorporated characteristic of DNA because heating at 65 degrees does not destroy the activity. The activity is destroyed by DNase and not by RNase in conditions where any double-strand RNA or RNA-DNA hybrid will be destroyed. So the only possibility is that the infectivity lies in DNA.

Dr. Aulakh: May I ask Dr. Gallo for the reasons why the DNA synthesized is a 5 to 6-S instead of closer to 70-S RNA?

Dr. Gallo: I wish one person knew that. Even with the viral enzyme that comes from what you know to be virus, this is what happens. No one has obtained a big size of DNA mate to my knowledge yet, either with our enzyme from the leukemic cells or when we do work with purified viral enzymes and add 70-S RNA. To my knowledge, no one has ever obtained synthesis of a large size DNA. Now, presumably factors may be missing when you go through purification.

Dr. Aulakh: You are not sure what factors are missing?

Dr. Gallo: No, we have no idea of the nature of the factors–is that what you asked?

Dr. Aulakh: Yes.

Dr. Gallo: Presumably we are missing factors that would allow for greater transcription. This is also true in the case of the enzyme purified from the virus. It is also possible that small sizes, multiple small DNA molecules, are made and linked later.

Dr. Meltz: (San Antonio, Texas) Dr. Montagnier, in my past experience, using a winding-out technique to isolate the DNA from mammalian cells, I succeeded in leaving at one point approximately 90% of the DNA behind from the expected yield. I am just curious to know if you have tried any other DNA isolation technique to obtain the DNA that you are isolating to use in these transformation experiments?

Dr. Montagnier: In some experiments we also precipitated DNA by ethanol and spun down the precipitate with no more efficiency. The infectivity is obtained only after, with 10 - 20 micrograms per plate. We don't need to increase the yield of DNA fraction.

Dr. Meltz: What I am mostly concerned about is possible differences in size which are obtained using a winding-out technique or possibly an ethanol precipitation technique. Do you get the same size distribution in your yield? That is the ultimate concern, because your infectivity and your measurements of sucrose are based also on the size of the DNA. If your DNA isolation technique is in low yield, you may be selecting against low molecular weight DNA species which are responsible for the infectivity.

Dr. Montagnier: I agree we may not be using the proper conditions.

Dr. Winkert: (Washington, D.C.) With regard to Dr. Gallo's puzzle about whether there was a high concentration of the polymerase in cytoplasm and what it might be doing there, you might look at the problems another way. Maybe the enzyme really is there but just doesn't have much primer. Consequently, there must be a high enzyme concentration to get any work done. Nonaqueous media might be the approach, to see whether the polymerase is really getting out or not. But one thinks of using nonaqueous media when you are worried about small molecules getting out. I wouldn't know whether this polymerase would be big enough.

Dr. Gallo: It is big.

Part IV

18. DETECTION OF TYPE C VIRUSES IN NORMAL AND TRANSFORMED CELLS

George J. Todaro, Edward M. Scolnick, Wade P. Parks, David M. Livingston and Stuart A. Aaronson

Viral Leukemia and Lymphoma Branch
National Cancer Institute
Bethesda, Maryland 20014

INTRODUCTION

Potential RNA containing tumor viruses have been recognized by a number of methods based on biological, biochemical, and immunological properties (1). More recently an enzyme has been found in all known RNA tumor viruses–the reverse transcriptase–which has provided another potentially extremely sensitive method for virus detection (2,3).

Following the initial reports of RNA-dependent DNA polymerase in the virions of certain RNA tumor viruses, it was important to see if the enzyme was specifically restricted to tumor viruses and whether it was specifically restricted to tumor cells (4). All of the oncogenic RNA viruses tested so far have been found to have DNA polymerase, as indicated both by endogenous reaction using the viral RNA and by synthetic polymer-stimulated reactions. The non-oncogenic RNA viruses have shown no evidence of this enzyme activity. Two apparent exceptions were found (Table I). The first, visna virus, produces a chronic, progressive, neurological disease of sheep but has, heretofore, not been associated with tumors in sheep. The second major exceptions are the group of "foamy" viruses. These RNA-containing viruses are frequently found in healthy as well as diseased monkeys, cattle, and cats, and they have not yet been associated with any disease. Visna and "foamy" viruses, then, are apparent exceptions to the rule that only tumor viruses contain DNA polymerase.

"FOAMY" VIRUSES: A GROUP OF REVERSE TRANSCRIPTASE-CONTAINING VIRUSES

The foamy viruses of monkeys, cats, and cattle are an interesting group of reverse transcriptase-containing viruses (5). These viruses are very common in these three species. Simian foamy virus antibody, for example, can be detected in most monkeys kept in captivity. Cat foamy viruses are almost invariably found as contaminants of cat cell cultures and cat leukemia cells. These viruses can be transmitted horizontally from animal to animal, but so far have not been incriminated in any disease. They are capable of remaining latent in animals for long periods without any overt expansion of disease. Whether these viruses are capable of producing in vitro cell transformation is as yet unclear. Their typical effect is to alter the membranes of cells

TABLE I

DNA Polymerase in RNA-containing Viruses

Virus	Endogenous	Polymer-stimulated
Oncogenic		
C-type (9 species)	+	+
B-type (MTV)	+	+
Non-oncogenic		
Sendai	–	–
Influenza	–	–
Respiratory syncytial	–	–
NDV	–	–
VSV	–	–
LCM	–	–
Measles	–	–
Mumps	–	–
Rubella	–	–
Possibly oncogenic		
Visna	+	+
"Foamy" viruses–monkey, cow, cat	+	+

they infect, so that the cell membranes will fuse with the membranes of adjacent cells to form syncytia. The viruses can be titered by their syncytium-forming ability. Their biologic properties are very much like the oncogenic viruses; replication is blocked by inhibitors of DNA synthesis, such as BrdU and by actinomycin-D (6). The expression of the transformed state, in this case an altered cell membrane that will fuse with other cells, again requires cell division, as does transformation by oncogenic viruses. There is no evidence that the viruses, in fact, are oncogenic; however, this point has not been adequately tested. So, two apparent exceptions (visna and foamy viruses) may really turn out not to be exceptions. They both may be viruses that, like the known tumor viruses, are capable of transforming cells. In this case, the enzyme would have been very valuable, for it would allow us to detect viruses that have oncogenic potential. The alternate explanation, however, is that RNA-containing viruses that have reverse transcriptase have the capacity to form stable association, perhaps by becoming integrated in cells they infect, and that this property is not necessarily linked to oncogenic capacity. The class of viruses that have reverse transcriptase, then, may be considerably larger than the class of viruses that also have oncogenic potential. Since the "foamy" viruses are so prevalent in nature, and since they can be transmitted horizontally, finding of reverse transcriptase in a virus does not necessarily prove that the virus is oncogenic. In spite of the diversity of species of origin, host range, and biological properties of viruses with reverse transcriptase (retraviruses), the enzymes are quite similar in their template characteristics and can be distinguished from cellular DNA polymerase.

SEPARATION OF REVERSE TRANSCRIPTASE FROM CELLULAR DNA POLYMERASE

A DNA polymerase that can utilize RNA-RNA and RNA-DNA synthetic templates has been found in a variety of normal mouse and normal human cells. The partially purified cellular enzyme has some properties that are similar to those of the mouse leukemia virus enzyme; however, the cellular enzyme can be separated from the viral enzyme, using Sephadex and phosphocellulose columns (7). The cellular enzymes have very different template properties and antigenic properties when compared to the viral enzymes. In order to distinguish the viral enzyme from the various contaminating cellular and microbial enzymes, the use of viral RNA has been suggested (8); this provides some potential specificity, but lacks sensitivity. Certain synthetic templates have been found that provide sensitivity and some specificity (9). Neither approach, however, distinguishes between type C viruses of leukemia and sarcoma from a variety of other viruses also having reverse transcriptase where there is no direct evidence that they are, in fact, tumor viruses.

ANTIGENIC RELATIONSHIPS BETWEEN REVERSE TRANSCRIPTASES

The polymerase, as an antigen, like the gs antigen has both species-specific and interspecies characteristics. Tumor-bearing animals can make antibody to the viral polymerase (10) and some sera appear to be more broadly reactive than others. The murine polymerase has been partially purified and used to produce an antibody in rabbits. The antibody, an IgG immunoglobulin, is directed against the enzyme and not against the template (11). The antibody to the mouse leukemia virus polymerase will also inhibit the enzymatic activity of hamster, rat, and cat leukemia virus polymerase (Table II). Thus, the polymerases from different mammalian tumor viruses are antigenically related. However, the crosses with other mammalian type C viruses are only partial crosses. The mouse mammary tumor virus and avian leukemia virus polymerases are not inhibited at all by this serum. The antibody to the avian virus polymerase inhibits all the major avian type C viruses, but not any mammalian type C (12).

TABLE II

Inhibition of Polymerase Activity by Antibody to the Mouse Leukemia Virus DNA Polymerase

Inhibited	Uninhibited
Mouse leukemia	Mouse mammary tumor
Mouse sarcoma	Monkey mammary tumor
Cat leukemia	Avian sarcoma
Cat sarcoma	Avian myeloblastosis
Rat leukemia	Avian leukosis
Hamster leukemia	Foamy viruses–simian, cat and bovine
	Visna

PURIFICATION OF TRANSCRIPTASE BY AFFINITY CHROMATOGRAPHY

Another approach to reverse transcriptase purification and identification utilizes immunoadsorbent columns. Solid-phase affinity chromatography of proteins is based on the principle that an insoluble compound or macromolecule with selective affinity for a given protein will specifically bind to that protein. The method has proven valuable for the rapid and selective purification of enzymes from partially purified mixtures and crude cell extracts. One application of this method involves the specific binding of an enzyme to antibody covalently coupled to Sepharose.

A solid-phase immunoadsorbent specific for murine and feline RNA tumor virus reverse transcriptase was prepared (13). Reverse transcriptases from murine and feline virus, but not avian virus, bind to columns of this material. Bound enzymes can be eluted in active form from the Sepharose antibody conjugate, using alkaline pH (10.6) in 0.20M NH_4OH, 0.30M KC1, 1% BSA. Presumably, at this pH and under these conditions, the structure of the immunoglobulin is so altered that its affinity for the enzyme is greatly reduced. This method has permitted selective purification of viral enzyme from crude extracts of virus-transformed cells, since the immunoadsorbent has no affinity for cellular DNA polymerases.

Since the procedure involves only two steps, the preparation of a crude extract and the affinity column procedure, it is rapid and simple. Thus, it should allow identification of viral enzyme from crude mixtures, where activity might not be readily detected or distinguished in the impure state, or even after conventional separation methods. It should be possible to obtain small quantities of enzyme in concentrated form from large volumes of extracts or tissue culture medium. These properties make affinity chromatography of the viral RNA-dependent DNA polymerase useful for the recognition of other members of this interesting class of enzymes.

NEW PRIMATE TYPE C VIRUSES

Two new type C viruses of primates have recently been described (14,15). They have a polymerase with the characteristic properties of tumor viruses and can be classified as type C based on morphology. They also have the biological properties of type C viruses and a cross-reacting gs antigen (16).

The polymerase antibody studies, however, show a very weak or absent cross reaction with antibody to rodent or cat virus polymerase. Both the murine and feline type C viruses can grow in primate cells without losing the immunologic specificity of their polymerase. These findings provide additional evidence that the polymerase is coded for, at least in part, by the viral genome. From the experiments on polymerase inhibition (16), the woolly monkey and gibbon ape viruses appear to have a polymerase that is distinguishable from that of the type C viruses of lower mammals.

By using a combination of methods—biological, biochemical, and immunological—it is concluded that there is a new class of primate type C viruses that have many properties comparable to the avian and the rodent type C viruses but also have immunological differences in two virus-coded proteins, the major group-specific antigen and the reverse transcriptase, that distinguish them from the previously described type C viruses.

The isolation of type C viruses from both old world and new world monkeys from naturally occurring tumors greatly strengthens the possibility that related viruses will be directly isolated from human tumors.

INDUCTION OF TYPE C VIRUSES FROM NORMAL CELL CLONES

In previous studies, we have shown that when cell lines derived from Balb/c and random-bred Swiss mouse embryos were established in cell culture it was found that early in their in vitro life the cells were free of detectable virus. After they had "spontaneously" transformed and acquired malignant properties in culture, some of the lines began to release type C virus (17). Since the cells have been viably frozen during the course of their development into established lines, they could be thawed and systematically tested for virus production. The pattern for a given subline was quite reproducible; virus production developed only in those lines that had lost normal cellular growth control. The above studies, along with many genetic and sero-epidemiological observations, raised the possibility that the genetic information for making a complete murine type C virus might be present in an unexpressed form in each and every mouse cell (17,18). It was clear that direct proof of this aspect of the hypothesis would require the specific manipulation in tissue culture of virus-free cell lines derived from single cell clones, with the resultant release of complete virus.

Subsequent studies revealed that clonal lines of Balb/3T3, after exposure to high doses of radiation, were virus-negative in tissue culture but became positive for virus after they had formed tumors in animals (19). These and similar studies with hamster embryo cells, however, could not resolve whether the virus came from the inoculated cells or from the animal in which the tumors had developed (20).

With the development of very rapid and sensitive methods to detect type C virus production, using the reverse transcriptase assay and an antibody that specifically inhibits the viral enzyme, and the finding that the thymidine analogs, BrdU and IdU, are efficient inducers of virus from high leukemic AKR cell clones (21), it has become possible to search for low levels of virus in tissue culture fluids of strains with a low leukemia incidence. The results showed that each and every Balb/c clone tested can be induced to release type C viruses (22). These findings provide strong evidence to support the concept that the genetic information for virus production is present in every mouse cell and that this information can be transmitted vertically from cell to progeny cell for many hundreds of generations without being expressed.

Since continuous cell lines offer a great advantage for the detailed study of virus cell interactions and their genetic control, it was important to look at other continuous lines. Complete type C viruses now have been induced from clonal lines of high and low leukemic strains of mice (21,22) and from clones of rat cells (23,24). We have recently found that type C virus is readily inducible from clones of Chinese hamster cells. Three to five days following exposure to 100μg/ml of IdU there is production of type C virus readily detectable by the reverse transcriptase assay and by electron microscopy (25). Table III summarizes the continuous cell lines that are now known to have inducible virus. The finding that Chinese hamster ovary cell lines, CHO, have inducible virus is of considerable interest, because chromosome number is small and the individual chromosomes are well characterized. This may prove to be a system where it is possible to specifically localize the oncogenic virus information and its control to particular chromosomes.

SPONTANEOUS RELEASE OF TYPE C VIRUS FROM SPONTANEOUSLY TRANSFORMED CELLS

A particular clone of Balb/3T3 (clone A31) was used to derive the transformed subclones. The radiation transformed cell clone, R4, has been previously described (19). Two other subclones, S1 and S2, are spontaneous transformants that

TABLE III

Continuous Cell Lines Whose Clones Have Inducible Type C Virus

Species	Line designation
Mouse (AKR)	Line 26 Line 32
Mouse (Balb/c)	Balb/3T3 Balb/3T12
Mouse (random bred Swiss)	3T3-SV40 3T6 3T12
Rat	NRK
Chinese hamster	CHO

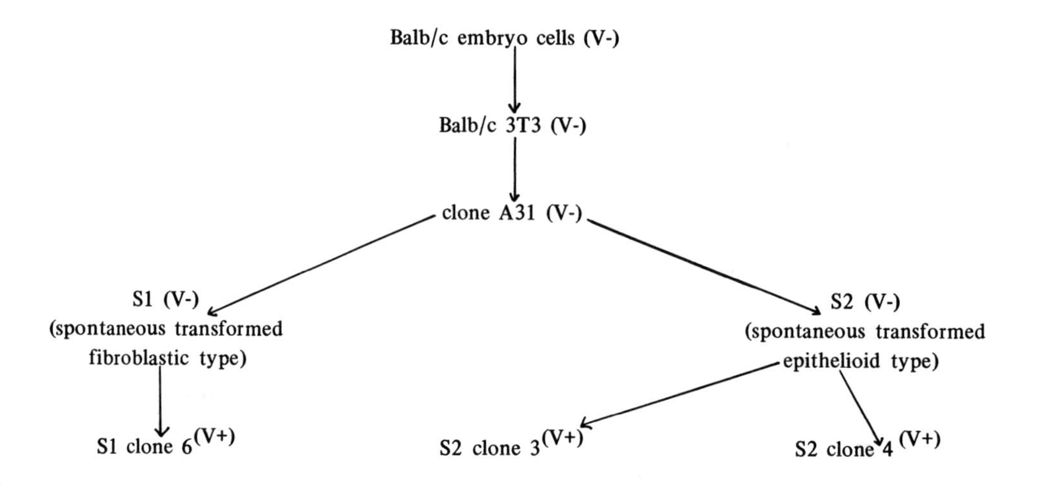

Fig 1. Geneology of various long-term cultured clones derived from Balb/c 3T3 mouse embryo cells. (V-)–virus negative. (V+)–spontaneously producing type C virus.

have appeared as rare variants in the clone A31 cultures (see Fig 1). They have been recloned several times, with selection of the most extreme cell type among the clones, based on altered morphology and the degree of loss of contact inhibition of cell division. Clone S1 is a transformed fibroblastic cell; clone S2 is an epithelioid variant. Figure 2 shows a colony of A31 on the left and, on the right, a colony of S2 cells. Each of these subclones of A31 (that is, R4, S1, S2) grows to high saturation densities, is able to grow in serum factor-free medium, and is tumorigenic in newborn Balb/c mice under conditions where the control A31 cells do not produce tumors. Table IV shows that S2 reaches saturation densities as great as have been seen with any of the cells transformed by exogenously added virus.

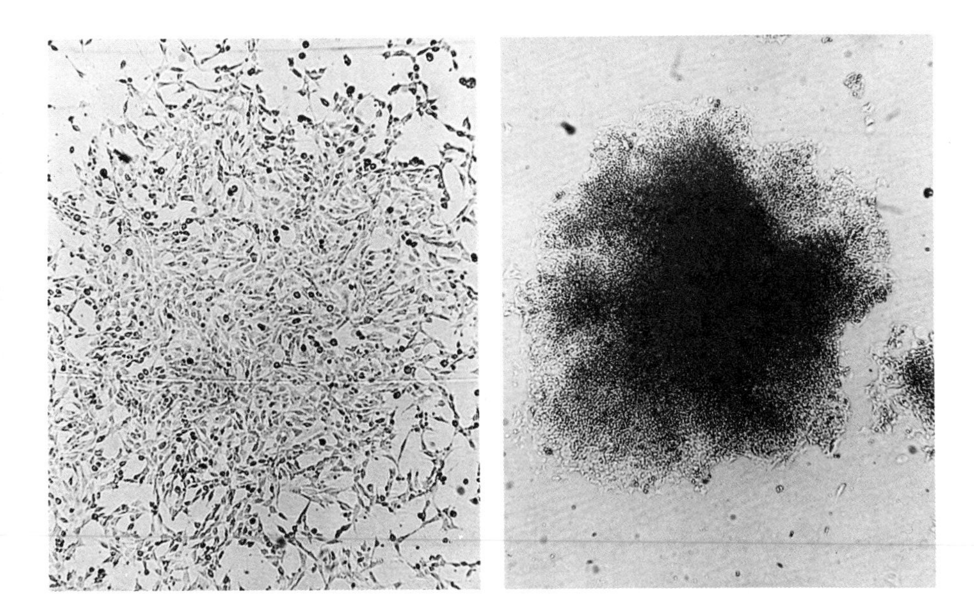

Fig 2. On the left is a colony of Balb/c 3T3 clone A31 fixed 10 days after inoculating a single cell onto a Petri dish. On the right is a colony of S2, a spontaneously transformed subclone derived from A31. Note the considerably denser, more compact colony morphology. The individual cells are epithelioid and tend to readily grow over one another. X 20. (Reduced 59%.)

The epithelioid variant, S2, has on two separate occasions, spontaneously released virus without any added inducers, as have six of 11 subclones of S2 and a subclone (Sl-6) from the transformed fibroblastic clone, S1. With both clones S1 and S2, virus-free cells give rise to virus producing cells anywhere from 40 to 200 cell generations after cloning. Recloning of the virus-producing mass cultures led to subclones that release virus with high probability and others that release virus with much lower probability.

Continued selection for those clones from S1 and S2 that are most transformed in their cell culture properties and most active in type C virus production has resulted in subclones with high virus titers being produced by highly transformed cells. Clone S2-3 produces large quantities of the endogenous virus allowing for the first time the possibility of testing the biologic and biochemical properties of an endogenous type C virus that has *never* been "horizontally" transmitted (26). By comparison, the endogenous virus from AKR cells spreads horizontally, infecting neighboring uninduced

TABLE IV

Biological Properties of Mouse Cell Lines Studied for the Induction of Endogenous Type C Viruses

Cell line	Colony morphology	Saturation density[a] (cells/plate ($x10^{-5}$))	Colony-forming ability On 3T3 monolayers (40)	In factor free medium (41)
Random Bred Swiss Series				
3T3 clone 42	normal	0.8-1.2	<0.1	<0.1
SV40 transformed				
Subclone 5	transformed	6-8	>30	>50
3T6	transformed	8-10	>20	>50
3T12-1	transformed	8-12	>50	>50
Balb/c Series				
Balb/c 3T3				
Clone A31	normal	1.0-1.2	<0.1	<0.1
"Spontaneously" transformed				
Clone S2	transformed	>20	>60	>50
"Radiation" transformed				
Clone R4	transformed	6-10	>60	>20
Balb/3T12-2	transformed	4-6	>20	>20

[a]Maximum number of cells per plate determined as previously described (39).

AKR cells, and the untransformed Balb/3T3 clones produce only low levels of virus and release it only transiently. If all mouse cells contain endogenous type C virus information, it follows that virus isolated from one cell and then propagated in another cell may represent a mixture of viruses and/or a mixture of viral information in a single particle.

While "spontaneous" transformation of mouse cells in tissue culture into neoplastic cells is associated with an increased probability of "spontaneous" release of endogenous virus, this association does not, in itself, demonstrate that the endogenous viral information plays an etiologic role in the cell transformation process. The association between type C virus and tumors may, in part, reflect the fact that the tumor cell is more likely to release its endogenous virus than is a more highly controlled normal cell. The control system that prevents expression of endogenous type C virus may have been permanently altered in transformed cells such as R4 and S2 that continue to produce type C virus; in contrast, Balb/3T3 cells or normal Balb/c embryo cells produce type C virus only transiently and only in response to an inducing agent (22). Type C virus can be induced in the original diploid Balb/c embryo cultures, but is at a lower level and lasts for an even shorter period of time than in the continuous aneuploid line, Balb/3T3; diploid cells may then have tighter control over expression of the endogenous type C virus than do aneuploid cells.

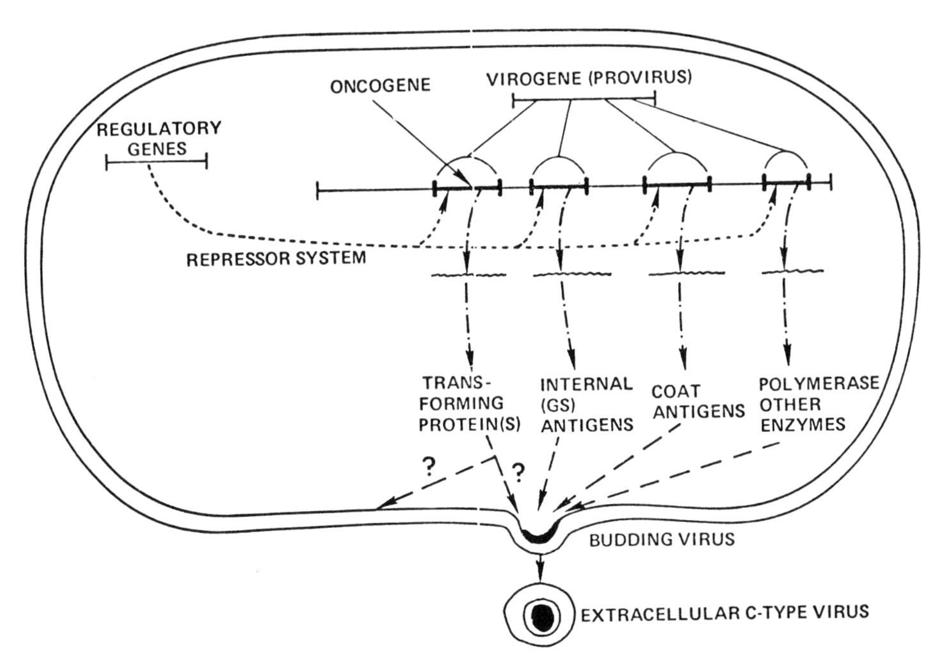

Fig 3. One possible model for control of virogenes and oncogenes.

THE ONCOGENE HYPOTHESIS: IMPLICATION OF ENDOGENOUS VIRUS IN CELLS

The oncogene hypothesis may be summarized in Figure 3. The virogene is that genetic information which is responsible for the production of a complete type C virus. One portion of the virogene is the oncogene. The oncogene is responsible for the production of the transforming protein which in turn converts the normal cell into a tumor cell. The whole system is under the control of repressors which are produced by regulatory genes. Other components of the virogene are those genes that code for the production of the several type C virus group-specific antigens (27), the virus-specific envelope antigens (28), the reverse transcriptase (2,3), and possibly other viral-specific enzymes (29). These proteins become incorporated into the extracellular type C virus particle and thus are more accessible to purification and analysis. The transforming protein, however, may or may not actually be a part of the extracellular virus particle. This, however, is the protein that is of most interest, since it is postulated to be directly involved in the conversion of a normal cell into a malignant cell. We would envision agents such as radiation, certain mutagens, and the natural aging process as acting to interfere with the normal repressor system and thus favor the partial and/or complete expression of virogene information. If the partial expression includes oncogene expression, the cell is "switched on" and would have a pronounced selective advantage relative to surrounding cells and therefore may be able to proliferate and give rise to a cancer. The various lines of evidence that have strengthened the original hypothesis have been recently summarized (30).

Recently, immunologic methods and molecular hybridization methods have become available that allow the detection of partial expression of virogene information in cells and will also allow the specific identification of virogene products. One

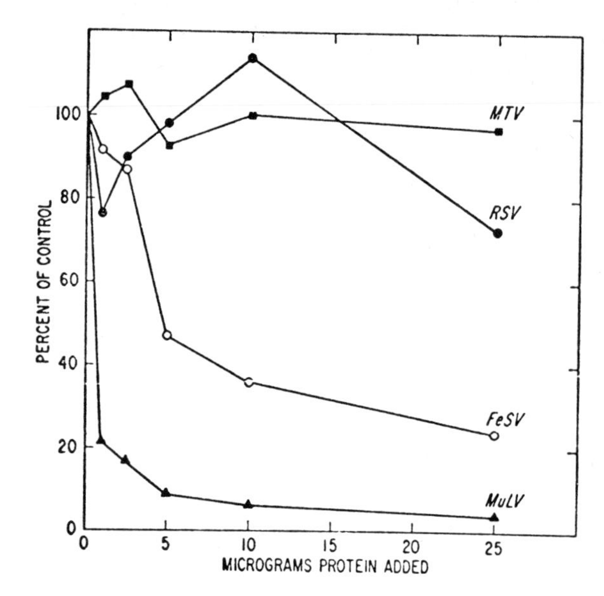

Fig 4. Inhibition of viral polymerase activity antiserum to the polymerase of mouse leukemia virus (Rauscher strain).

MuLV: Mouse leukemia virus polymerase (Rauscher strain)
FeSV: Feline sarcoma virus polymerase (Gardner strain)
MTV: Mouse mammary tumor virus polymerase
RSV: Avian sarcoma virus polymerase (Rous strain)
Incorporation in the absence of antiserum: 20 – 50,000 cpm ^{3}H-TTP (see [12]).

approach that our laboratory has taken has been to purify the viral reverse transcriptase and use them as antigens to produce antibodies in rabbits. These antibodies in turn are assayed by their ability to inhibit the reverse transcriptase reaction (11,12). Using these it has been possible to show, for example, that the reverse transcriptase in the virus particles is antigenically different from cellular DNA polymerases (7). However, the reverse transcriptase of type C viruses isolated from different mammalian species seem to be enough related to one another that an antibody produced to one will inhibit the other; for example, antibody to murine type C virus polymerase will also inhibit the feline type C virus polymerase (Fig 4).

Antibody made in rabbits to the mouse leukemia virus polymerase will inhibit the other well-studied mammalian type C virus polymerases, but will not inhibit either the avian type C virus polymerases or the reverse transcriptase associated with various viruses that may or may not have oncogenic potential. Using the reverse transcriptase and antibodies to them, it has been possible to identify viral-specific enzymes in particles and in tissues under conditions where it has not been possible to demonstrate infectivity (22,25), perhaps because a susceptible host cell has not yet been found. This seems to be the case with most of the induced viruses from cell clones derived from low leukemic strains; the viruses have very little capacity to reinfect the cells from which they were activated.

The oncogene hypothesis makes several testable predictions (Table V). The first is that all somatic cells should contain the genetic information to produce a type C virus that can be detected by using the DNA product made from type C viral RNA

TABLE V

Some Implications of the Oncogene Hypothesis

1. All somatic cells have DNA homologous to the Type C virus RNA of that species (virogene).

2. Transformed and tumor cells have new RNA homologous with the DNA produced from Type C viral RNA (oncogene).

3. Transformed and tumor cells have new viral specific protein(s), the product of the oncogene (transforming protein).

4. It should be possible to isolate *cell* mutants that release virus without added exogenous agents (virogene repressor mutants).

5. Type C viruses derived from closely related species should have closely related specific antigens, e.g., gs antigens, polymerase.

of that particular species. For example, normal cat cells should have in their DNA a complete copy of cat leukemia type C virus RNA, and the hamster cell should have in its DNA the genetic information for making a hamster type C virus. A second prediction is that transformed cells, whether transformed by exogenously added tumor viruses or by radiation, chemical carcinogens, or even "spontaneously," should contain new messenger RNA sequences that are not found in normal cells and that are common to all transformed and tumor cells of a particular species. The new RNA would be the product of the oncogene and in turn code for the production of the transforming protein(s). It should be possible to isolate cell mutants that at the nonpermissive temperature, because of a temperature-sensitive repressor, would be superinducible or would, possibly, spontaneously produce type C virus without an exogenous inducer. An additional prediction, if the general hypothesis is correct, and type C viral information has been a stable part of the evolution of vertebrates, would be that the type C viruses derived from closely related species would be closely related to one another in the antigenic properties of their characteristic proteins. Two such proteins are now available—the major group-specific antigen and the reverse transcriptase. The isolation of type C viruses from reptiles, birds, as well as mammals would suggest that they have evolved as the organism has evolved for many millions of years and that the species-specific proteins will have evolved in much the same way that serum albumins, globulins, and other proteins have evolved. The genetic relatedness of the group-specific antigens and the reverse transcriptases may well, then, be used as an index of the genetic relatedness of the species from which the type C virus was derived. Those viruses derived from higher mammals, especially primates, will be the most related to the viruses that come to be obtained from man. Our laboratory is concentrating at the moment on the primate type C viruses. These viruses should have enough genetic relatedness (31) so that an antiserum produced to the purified polymerase and the purified group-specific antigen should show some ability to recognize type C viruses isolated from human tissues.

With respect to human cancer, it is still not proven that viruses—any viruses—are natural etiologic agents in cancer. If and when type C viruses of human origin are definitely identified, the critical question will become how the process might be blocked. If they are transmitted horizontally, from person to person as an inapparent infection, or even from cell to cell within a single individual as the result of activation of virus from certain cells, then a conventional approach using viral vaccines may be feasible. If they are transmitted entirely, or predominantly, vertically then the hope would be to prevent expression of endogenous virus information. The steps of absorption, penetration, and integration of tumor virus information may not be critical events in the natural oncogenic process. The vertically transmitted virogene concept is not necessarily a pessimistic one. In fact, if there is specific oncogenic information present in cells on which a variety of carcinogens act, then rather than there being a hundred or a thousand different causes of cancer, there would be one, or relatively few, specific metabolic effects—the results of the activation of this normally repressed information.

The availability of the established cell lines that are well characterized and have been transformed by RNA tumor viruses, DNA tumor viruses, radiation, and chemical carcinogens, as well as "spontaneously," should allow the controlled study of many of the questions posed above. The "spontaneously" transformed cells, such as 3T12, would be expected to have partial expression of the endogenous viral information. Revertants of sarcoma virus-transformed clones (32,33,34) and "abortively" transformed clones (35,36) that show no expression of transformation but still have the exogenously added SV40 information in them (37) provide important additional classes of cells to be analyzed. The "nonproducer" cells transformed by mouse sarcoma virus (38) and the spontaneously transformed tumorigenic subclones (26) of Balb/3T3 are particularly valuable because they may serve as a model for naturally occurring tumors. With these various classes of well-characterized cells, it may be possible to determine what the critical difference is between a normal cell and a transformed cell. This is a formidable and perhaps hopeless task when one compares, for example, normal liver and hepatomas in terms of enzymes, antigens, nucleic acids, or membrane alterations. With cloned lines in defined systems with differing degrees of expression of exogenous and/or endogenous virus information, however, we may well be able to resolve what events are critical for the initiation and the maintenance of the transformed state.

REFERENCES

1. Vigier, P.: RNA oncogenic viruses: structure, replication, and oncogenicity. Prog. Med. Virol., 1:240, 1970.
2. Temin, H. M., and Mizutani, S.: RNA-dependent DNA polymerase in virions of Rous sarcoma virus. Nature, 226:1211, 1970.
3. Baltimore, D.: RNA-dependent DNA polymerase in virions of RNA tumour viruses. Nature, 226:1209, 1970.
4. Todaro, G. J., Aaronson, S. A., Scolnick, E. M., and Parks, W. P.: RNA-dependent DNA polymerase in viruses and in cells. In: the 2nd Lepetit Colloquium: The Biology of Oncogenic Viruses. North Holland Publishing Company, Amsterdam, p 207, 1971.
5. Parks, W. P., Scolnick, E. M., Todaro, G. J., and Aaronson, S. A.: RNA-dependent DNA polymerase in primate syncytium-forming ("foamy") viruses. Nature, 229:257, 1971.
6. Parks, W. P., and Todaro, G. J.: Biologic properties of syncytium-forming ("foamy") viruses. Virology, 47:673, 1972.
7. Ross, J., Scolnick, E. M., Todaro, G. J., and Aaronson, S. A.: Separation of murine cellular and murine leukemia virus DNA polymerase. Nature New Biol., 231:163, 1971.

8. Goodman, N. C., and Spiegelman, S.: Distinguishing reverse transcriptase of an RNA tumor virus from other known DNA polymerases. Proc. Natl. Acad. Sci., 68:2203, 1971.
9. Baltimore, D., and Smoler, D.: Primer requirement and template specificity of the DNA polymerase of RNA tumor viruses. Proc. Natl. Acad. Sci., 68:1507, 1971.
10. Gerwin, B. I., Todaro, G. J., Zeve, V., Scolnick, E. M., and Aaronson, S. A.: Separation of RNA-dependent DNA polymerase activity from the murine leukemia virion. Nature, 228:435, 1970.
11. Aaronson, S. A., Parks, W. P., Scolnick, E. M., and Todaro, G. J.: Antibody to the RNA-dependent DNA polymerase of mammalian C-type RNA tumor viruses. Proc. Natl. Acad. Sci., 68:920, 1971.
12. Parks, W. P., Scolnick, E. M., Ross, J., Todaro, G. J., and Aaronson, S. A.: Immunologic relationships of reverse transcriptases from RNA tumor viruses. J. Virol., 9:110, 1972.
13. Livingston, D. M., Scolnick, E. M., Parks, W. P., and Todaro, G. J.: Affinity chromatography of RNA tumor virus reverse transcriptase on a solid phase immunoadsorbent. Proc. Natl. Acad. Sci., 69:393, 1972.
14. Theilen, G. J., Gould, D., Fowler, M., and Dungworth, D. L.: C-type virus in tumor tissue of a woolly monkey (Lagothrix spp.) with fibrosarcoma. J. Natl. Cancer Inst., 47:881, 1971.
15. Kawakami, T., Huff, S. D., Buckley, D. M., Dungworth, D. L., Synder, S. P., and Gilden, R. V.: C-type virus associated with gibbon lymphosarcoma. Nature New Biol., 235:170, 1972.
16. Scolnick, E. M., Parks, W. P., Todaro, G. J., and Aaronson, S. A.: C-type virus reverse transcriptases: immunological properties. Nature New Biol., 235:35, 1972.
17. Aaronson, S. A., Hartley, J., and Todaro, G. J.: Mouse leukemia virus: "spontaneous" release by mouse embryo cells after long term in vitro cultivation. Proc. Natl. Acad. Sci., 64:87, 1969.
18. Huebner, R. J., and Todaro, G. J.: Oncogenes of RNA tumor viruses as determinants of cancer. Proc. Natl. Acad. Sci., 64:1087, 1969.
19. Pollock, E. J., Aaronson, S. A., and Todaro, G. J.: X-irradiation of Balb/3T3: sarcoma-forming ability and virus induction. Int. J. Rad. Biol., 17:97, 1970.
20. Freeman, A. E., Kelloff, G. J., Gilden, R. V., Lane, W. T., Swain, A. P., and Huebner R. J.: Activation and isolation of hamster-specific C-type RNA viruses from tumors induced by cell cultures transformed by chemical carcinogens. Proc. Natl. Acad. Sci., 68:2386, 1971.
21. Lowy, D. R., Rowe, W. P., Teich, N., and Hartley, J. W.: Murine leukemia virus: high frequency activation in vitro by 5-Iododeoxyuridine and 5-Bromodeoxyuridine. Science, 174:155, 1971.
22. Aaronson, S. A., Todaro, G. J., and Scolnick, E. M.: Induction of murine C-type viruses from clonal lines of virus-free Balb/3T3 cells. Science, 174:157, 1971.
23. Klement, V., Nicolson, M. D., and Huebner, R. J.: Rescue of the genome of focus-forming virus from rat non-productive lines by 5-Bromodeoxyuridine. Nature New Biol., 234:12, 1971.
24. Aaronson, S. A.: Chemical induction of focus-forming virus from non-producer cells transformed by murine sarcoma virus. Proc. Natl. Acad. Sci., 68:3069, 1971.
25. Todaro, G. J.: Detection and characterization of RNA tumor viruses in normal and transformed cells. In: Perspectives in Virology. Edited by M. Pollard. Academic Press, Inc., New York, New York, p 81, 1972.
26. Todaro, G. J.: "Spontaneous" release of type C viruses from clonal lines of "spontaneously" transformed Balb/3T3 cells. Nature, 240:157, 1972.
27. Gilden, R. V., and Oroszlan, S.: Group-specific antigens of RNA tumor viruses as markers for subinfectious expression of the RNA virus genome. Proc. Natl. Acad. Sci., 69:1021, 1972.
28. Aoki, T., Boyse, E. A., Old, L. J., de Harven, E., Hämmerling, U., and Wood, H. A.: G (Gross) and H-2 cell-surface antigens: location on Gross leukemia cells by electron microscopy with visually labeled antibody. Proc. Natl. Acad. Sci., 65:569, 1970.

29. Mölling, K., Bolognesi, D. P., Bauer, H., Büsen, W., Plassman, H. W., and Hausen, P.: Association of viral reverse transcriptase with an enzyme degrading the RNA moiety of RNA-DNA hybrids. Nature New Biol., 234:240, 1971.
30. Todaro, G. J., and Huebner, R. J.: The viral oncogene hypothesis: new evidence. Proc. Natl. Acad. Sci., 69:1009, 1972.
31. Scolnick, E. M., Parks, W. P., and Todaro, G. J.: Reverse transcriptases of primate viruses as immunological markers. Science, 177:1119, 1972.
32. MacPherson, I. A.: Reversion in hamster cells transformed by Rous sarcoma virus. Science, 148:1731, 1965.
33. Todaro, G. J., and Aaronson, S. A.: Properties of clonal lines of murine sarcoma virus transformed Balb/3T3 cells. Virology, 38:174, 1969.
34. Fischinger, P. J., Nomura, S., Peebles, P. T., Haapala, D. K., and Bassin, R. H.: Revision of murine sarcoma virus transformed mouse cells: variants without a rescuable sarcoma. Science, 176:1033, 1972.
35. Stoker, M.: Abortive transformation by polyoma virus. Nature, 218:234, 1968.
36. Smith, H., Scher, C., and Todaro, G. J.: Induction of cell division in medium lacking serum growth factor by SV40. Virology, 44:359, 1971.
37. Smith, H. S., Gelb, L. D., and Martin, M. A.: Detection and quantitation of Simian Virus 40 genetic material in abortively transformed Balb/3T3 clones. Proc. Natl. Acad. Sci., 69:152, 1972.
38. Aaronson, S. A., and Rowe, W. P.: Nonproducer clones of murine sarcoma virus transformed Balb/3T3 cells. Virology, 42:9, 1970.
39. Pollack, R., Green, H., and Todaro, G. J.: Growth control in cultured cells: selection of lines with increased sensitivity to contact inhibition and loss of tumor producing ability. Proc. Natl. Acad. Sci., 60:126, 1968.
40. Aaronson, S. A., and Todaro, G. J.: The basis for the acquisition of malignant potential by mouse cells cultivated in vitro. Science, 162:1024, 1968.
41. Jainchill, J., and Todaro, G. J.: Stimulation of cell growth in vitro by serum with and without growth factor. Relation to contact inhibition and viral transformation. Exp. Cell Res., 59:137, 1970.

19. HERPESVIRUS HOMINIS: FROM LATENCY TO CARCINOGENESIS?

Laure Aurelian

Departments of Animal Medicine and Microbiology
Johns Hopkins University School of Medicine
Baltimore, Maryland

Herpesviruses are formally defined as large enveloped virions with an icosahedral capsid consisting of 162 capsomeres and arranged around a DNA core (1). The viruses meeting the structural and architectural criteria for inclusion into the herpesvirus group share in common unique features of their replicative processes (2). Included in the group are herpes simplex, varicella zoster, the cytomegaloviruses and the herpesviruses isolated from various animals (3).

Man harbors at least two (4,5), and possibly more (6,7) types of herpes simplex virus, antigenically related. It has been demonstrated, however, that the virus associated with genital infections (herpesvirus type 2 or HSV-2) differs from that isolated from eruptions on other parts of the body (herpesvirus type 1 or HSV-1) with respect to physical, biologic, and immunologic properties (4 - 8). New lesions at different sites of the body may be established by natural or artificial inoculation of autologous or entirely different virus strains (8).

Infection with HSV-1 occurs between six months and five years of age and is usually inapparent (9). Primary infection with HSV-2 occurs later in life and is transmitted by sexual contact (10). The infection is generally inapparent and recovery occurs without serious consequences.

The observation that cervical carcinoma essentially behaves like a venereally transmitted disease (11,12), more prevalent in women who have multiple sex partners and who begin heterosexual activity early in life (13), has led to the suspicion that herpesvirus type 2 might be its causative agent. The problem of the association of this virus with squamous carcinoma of the human cervix constitutes the focus of this presentation.

CLUES POINTING TO THE ONCOGENICITY OF THE HERPES SIMPLEX VIRUS

The clues pointing to the oncogenicity of herpes simplex virus fall into four categories. First, it has been reported that patients with severe and frequently recurring labial herpetic infections develop squamous cell carcinoma of the lip (14 - 16). The association of type 2 herpesvirus with squamous cancer of the human cervix was reported on the grounds of epidemiologic studies (8,17,18) indicating a frequency of antibody to HSV-2 greater in cervical neoplasia than in control groups.

A second clue comes from work with cells in vitro. Thus, a high percentage of aneuploidy has been described for patients with preinvasive and invasive cervical cancer (19), although the question of specific structural aberrations of the chromosomes has not yet been settled (20). In this context it is of particular significance, that herpes simplex virus types 1 (21) and 2 (Katayama and Aurelian, unpublished data) have been shown to cause pulverization of the chromosomes of permissive cells and chromatid breaks in the chromosomes of nonpermissive ones (22). Furthermore, it has been reported that cells infected with herpes simplex virus acquire the capacity to synthesize an antigen (designated G) characteristic of some human tumors, as well as of cell lines of malignant origin (23). G antigen has not been found in cells of a variety of nonmalignant adult and embryonic human tissues (24). Third, there is guilt by association. Herpes-type particles have recently been shown to give rise to renal adenocarcinoma in frog embryos (25), to Marek's disease of chickens (26), and to lympho-proliferative disease in monkeys (27). Likewise, they were found in association with cells from human lymphoma (28) described by Burkitt (EB virus). Finally, HSV-2, attenuated by ultraviolet irradiation, has been shown to transform hamster cells in vitro (29).

PREDICTED ASSOCIATION OF HERPESVIRUS TYPE 2 WITH HUMAN CERVICAL CELLS

Recent studies have indicated that the mean age for infection with HSV-2 precedes by six years that for development of cervical atypia (30). This observation suggests that in order to consider HSV-2 as a possible etiologic candidate for cervical cancer, the virus must be able to persist for indefinite periods of time in the host it has infected. In this context, the ability of herpesviruses to persist, following primary infection (latency) is of particular significance. Thus, for their lifetime, individuals with primary herpetic infections may be afflicted with recurrent eruptions of the lips, genitals, or cornea. What makes these infections unique is that recurrences occur in the presence of circulating antibody and are triggered by specific physical or emotional stress. Prominent among these are fever, menstruation, and emotional stress (31).

The mechanisms of herpesvirus persistence and recurrence in patients with latent herpetic infections are still unknown. Two hypotheses have been proposed (32). The first invokes a slow but constant turnover, at the site of the recurrent lesion, of infected cells in which herpesvirus multiplies. Since infectivity of herpesvirus is associated with complete virions (2,33) this hypothesis predicts that infectious virus could be isolated from the site of the lesion in the interim between recurrences. To date, many such attempts at isolation have failed (31). The second hypothesis is based on work done with cells in culture. It envisions virus persistence as dependent on the arrest of virus multiplication in the infected cells at some very early point, before it expresses the functions leading to inhibition of host macromolecular synthesis (31), an essential prerequisite for the synthesis of structural components of the virus (34). A sufficient explanation for the induction of tumors in people with this latent infection would be that the neoplastic character of the cell depends on the presence of virus-specific protein(s) (Ca-Protein) coded by the latent viral genome. Secondary effects on the course of differentiation and evolution of the new cell type would therefore be expected.

The concept that evolves from these interpretations (Table I) is that of a viral genome, at different levels of transcription. A large proportion of women have antibody to type 2 herpesvirus (8,17,35, and Table V), but only some develop cancer

TABLE I
Predicted HSV-2 Components in Squamous Cells of the Human Cervix

Virus components	Normal squamous cells: Productive infection primary or recurrence	Normal squamous cells: Non productive infection "latency"	Tumor cells: Biopsied	Tumor cells: Exfoliated
1. Infectious virus	+	-	-	(?)[a]
2. Particles (complete or incomplete)	+	-	-	(?)[a]
3. Structural antigens	+	-	-	+
4. Ca-Protein(s)	(?)[b]	-	+	(?)[b]
5. Viral RNA	+	-	+	+
6. Viral DNA	+	+	+	+

[a]This prediction is based on the hypothesis that explains virus persistence in latently infected cells as dependent on the arrest of virus multiplication at some early point, before it expresses the functions responsible for inhibition of host macromolecular synthesis (31). Under this hypothesis, virus can be made to manifest itself in the latently infected cells by exposure of the cells to conditions of stress. Exfoliated tumor cells, having been exposed to alkaline medium (43) and complete or partial malnutrition are expected to contain virus particles and even infectious virus. Unsettling in terms of these predictions are two clinical observations. First, herpetic lesions are not observed together with cervical neoplasia and, second, herpetic inclusion bodies are never observed in exfoliated tumor cells but only in normal squamous ones (30). Indeed, electron microscopy and isolation studies failed to show the presence of virus particles or infectious virus in exfoliated tumor cells (45 and Fig 4).

[b]A sufficient explanation for the induction of tumors in people with latent HSV-2 infection would be that the neoplastic character of the cell depends on the presence of virus-specific protein(s) coded by the latent viral genome. These protein(s) designated "Ca-Protein(s)" would be virus- and tumor-specific, raising speculation as to the possibility of their being synthesized also during the productive infection. If Ca-Protein(s) result from the expression of information different from the one expressed in the productive infection, they should be associated only with tumor cells; on the other hand, if Ca-Protein(s) result from the unregulated expression of the same genetic information they should be present in the productive infection as well, provided this is studied at the time interval of maximum synthesis. Studies with oncogenic DNA viruses have indicated that virus- and tumor-associated antigens are also made during the productive infection. It should be pointed out, however, that despite their presence in transformed cells the role these antigens play in oncogenesis is still unknown; the crucial question of the nature of the viral gene function responsible for transformation has yet to be answered. The studies described in this chapter indicate that some complement fixing antigen(s) tumor- or virus-specific also appear to be made early in the productive infection.

of the cervix, suggesting that latently infected cells harbor the viral genome in a completely repressed state. This interpretation is consistent with the observation that virus or viral components cannot be detected at the healed site of the recurrent lesion, facial (31) or genital (Aurelian, unpublished data), in the interim between recrudescences. Ca-Protein(s) are not expected in these cells.

A partially transcribed genome is associated with virus-induced transformation. Besides viral DNA and virus-specific RNA, tumor cells are expected to contain Ca-

Protein(s). Lastly, recurrence is due to a totally depressed viral genome, resulting in the production of virus structural components (proteins and DNA) as well as infectious virus. In contrast to the situation observed in the interim between recrudescences, virus can be isolated with relative ease from lesions observed at the time of eruptions (4,6,31).

As already discussed, virus replication and recurrent herpetic eruptions are induced by conditions of stress. A priori, it would be expected that since stress can induce the expression of the genetic potentiality of a totally repressed genome it should also be able to induce that of a partially repressed one, such as that associated with the tumor cells. Thus, viral structural components should be present in stressed but not in unstressed tumor cells. These predictions can be tested experimentally.

HERPESVIRUS TYPE 2 AND SQUAMOUS CANCER OF THE HUMAN CERVIX

HSV-2 Isolated from Cervical Tumor Cells Grown in Tissue Culture

A line of cervical tumor cells designated S332G was established from a biopsy obtained from a case of intraepithelial carcinoma of the cervix. It was characterized morphologically by atypical cells with large lobulated nuclei and approximately 10% multinucleated cells (36), and biologically by a very slow rate of growth. Electron microscopy of thin sections revealed large epitheloid cells with prominent membranous organelles, elaborate microvilli and small condensations of chromatin at the nuclear periphery. Packed fibrils 5 - 8 nm* in diameter and consistent with tonofibrils of squamous cells were observed surrounding the membranous cytoplasmic elements. Three series of experiments were done with these cells. First, extracts of S332G cells were serially passed on permissive HEp-2 cells in order to determine whether they carry infectious HSV-2. Second, to inquire into the presence of structural viral antigens, S332G cells were stained by the indirect immunofluorescent procedure, using sera made in rabbits against 24-hour extracts of HSV-2-infected HEp-2 cells. The sera, designated Ra-2, were adsorbed with uninfected HEp-2 cells until they failed to stain uninfected HEp-2 cells. Their specificity for structural HSV-2 proteins was ascertained by the observation that Ra-2 sera do not stain HEp-2 cells infected with HSV-2 for less than 4 hours, the time shown to be that of onset of structural proteins synthesis (37). Last, electron microscopy was performed to determine the presence of complete or incomplete virions in S332G cells.

At three transfers (3rd, 5th, and 8th), S332G cells failed to show evidence of virus, virus antigens, and complete or incomplete virions. However, cellular degeneration occurred spontaneously in one out of replicate culture flasks, at the 10th, 12th, 15th, and 18th transfers. The degenerated cells stained preferentially with Ra-2 serum (Table II) and virus was isolated each time on HEp-2 cells. Electron microscopy revealed intranuclear particles with specific herpesvirus morphology, characteristic nuclear alterations (38), and enveloped cytoplasmic particles 140 - 160 nm* in diameter (Fig 1). Of particular significance are the relatively high number of unenveloped cytoplasmic particles and the linear intranuclear structures 20 - 22 nm* in diameter,

*1nm = 10^{-6} mm.

TABLE II

Fluorescence of Cervical Tumor Cell Lines Stained with Ra-2,[a] Hu-2,[b] and Ra-1[c] Sera by the Indirect Immunofluorescent Procedure

Sera	Dilution of sera	Degenerated S332G cells	Viable S332G cells	Viable JaCha cells
Ra-2	1:5	3-4+[d]	0	0
	1:10	3-4+	0	0
	1:20	3-4+	0	0
	1:40	0	0	0
Ra-1	1:5	2-3+	0	0
	1:10	0	0	0
	1:20	0	0	0
	1:40	0	0	0
PBS[e]	–	0	0	0
Hu-2	1:10	NT[f]	0	4+
	1:40	NT	0	2-3+
Normal human serum	1:5	NT	1+	2-3+
	1:10	NT	0	1+
	1:40	NT	0	0

[a]Rabbit anti-HSV-2 serum.
[b]Human serum containing neutralizing antibody to HSV-2.
[c]Rabbit anti-HSV-1 serum.
[d]Intensity of fluorescence.
[e]Phosphate buffered saline pH 7.1.
[f]Not tested.

both of which have been described as characteristic of HSV-2-infected cells (39,40). At the same transfers, S332G cells in duplicate cultures that had not degenerated (viable) did not react with Ra-2 sera (Table II), failed to induce cytopathogenic effects to and yield virus when passed on HEp-2 cells, and virus particles were not observed on electron microscopic examination. The immunologic specificity of the four isolates from the degenerated S332G cells (S-1, S-2, S-3, and S-4) as type 2 herpesvirus was further ascertained by the observation that Ra-2 sera were more effective in neutralizing the isolates than anti-HSV-1 sera (6,36).

Spontaneous degeneration was consistently associated with an increase in the pH of the medium. In order to study the effect of pH on virus induction, S332G cells were grown for 3 weeks in the presence or absence of a 5% CO_2 atmosphere, and the pH of the medium determined at 2-hour intervals for the first 10 hours after transfer and once a day during the next 5 days. Normal growth and an even pH of 7.2 - 7.3 were characteristic of the cultures grown under CO_2, whereas extreme pH ranges (7.7 - 8.7) and poor growth were observed in cultures grown without CO_2. At the end of 3 weeks, typical cytopathogenic changes were observed in the latter and the cells

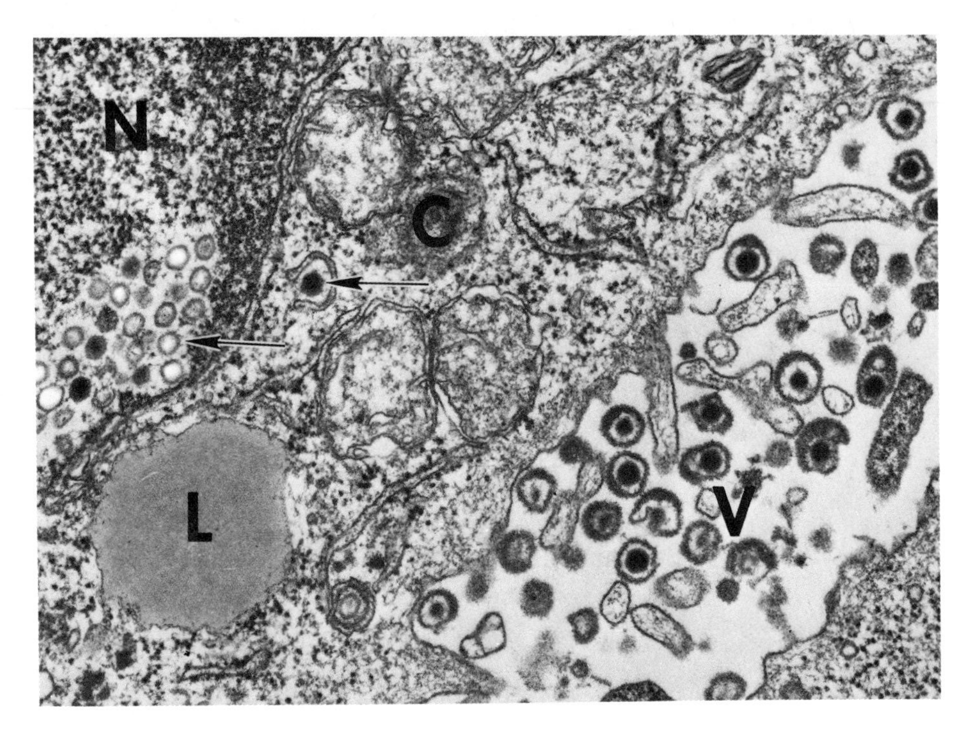

Fig 1. HEp-2 cell infected with S-2 isolate containing nucleocapsids (*arrows*) in nucleus (N) and enveloped virions (V) in intercellular space. (C) cytoplasm, (L) large cytoplasmic droplet X 58,000). (Reduced 70%.)

stained preferentially with Ra-2 serum. Virus isolated from these cells on HEp-2 cultures was identical to HSV-2 in terms of plaque morphology and immunologic specificity, as determined by plaque reduction neutralization tests (36).

Critical to the interpretation of our data is the failure to detect the presence of herpesvirus in viable cells, either prior to the onset of cell degeneration or in replicate cultures which continued to grow beyond the time of the appearance of cell degeneration. In none of the viable cultures tested from the 3rd until the 50th passage were herpesvirus antigens detectable by fluorescent antibody staining; nor was there any cytopathogenic effect on HEp-2 cells or evidence of virus by electron microscopy after five serial passages of S332G extracts in such cells. Therefore, there was no evidence of contaminating herpesvirus in the original specimen and no indication of a "chronic" infection (41,42). The most likely interpretation of our findings is that cervical carcinoma cells are "latently" infected with type 2 herpesvirus and that under certain conditions (such as high pH) virus replication is induced and infectious virus made. This interpretation implies that some or all of the S332G cells harbor the viral genome in a repressed state. Whether pH is the only type of stress required for virus induction is not completely clear at present; the possibility that any stress leading to cell death could induce virus replication cannot be excluded.

A second line of cervical tumor cells designated IaCha was established from a case of atypia of the cervix. Cellular degeneration was not observed when the cells were grown in a medium of high pH or containing 50 μg/ml of 5′-iododeoxyuridine or 5′-bromodeoxyuridine. Passage of the cells on HEp-2 cells failed to yield infectious virus at the 3rd, 8th, and 15th transfers, and the cells did not react with Ra-2 sera. Fluorescence was obtained, however, with a human serum designated Hu-2 and containing neutralizing antibody to HSV-2. A human serum without antibody to either herpesvirus type did not react with JaCha cells (Table II). Of particular interest is the observation that whereas the Ra-2 sera used in these studies stain HSV-2-infected HEp-2 cells only starting at 6 hours after infection, the Hu-2 serum also reacts with HEp-2 cells infected with HSV-2 for four hours suggesting that one or more antigens not detected by Ra-2 sera are identified by Hu-2 serum. Essentially similar results were reported by Nahmias (8), who described the presence of HSV-2 immunofluorescence in 5% of the cells from a third culture of human cervical tumor cells grown in vitro. In this culture, as in JaCha cells, immunofluorescence was observed without previous exposure of the cells to conditions of stress, and only some anti-HSV-2 sera, and not others, were capable of detecting the viral antigens.

In conclusion, whereas, evidence for the association of HSV-2 with cervical cancer cells grown in culture was obtained in all the lines studied so far, the mechanisms of viral persistence and induction appear to differ; since the different cultures were established by various methods from biopsies obtained from different patients, these may be a function of individually determined regulatory mechanisms.

HSV-2 Antigens and Cervical Tumor Cells in Vivo

The difference in the function of animal cells in the artificial in vitro environment of cell culture and in the whole animal (31) raises the question of the nature of the association of HSV-2 with cervical neoplastic cells in vivo. This study is made possible by the observation that prior to exfoliation, tumor cells on the surface of the neoplastic lesions are exposed to suboptimal conditions consisting of partial malnutrition and glandular secretions of relatively high pH (43). Thus, tumor cells in biopsy specimens, frozen immediately upon collection, constitute unstressed tumor cells, whereas, exfoliated tumor cells are their stressed counterparts. Frozen sections and impressions of biopsy specimens (biopsied tumor cells) were obtained from 29 patients with preinvasive and invasive cervical carcinoma. Exfoliated cells were collected from the same patients at the same time by the impression of a small wet sponge directly to the surface of the neoplastic lesion or by the irrigation method of Davis (44), using 0.1M phosphate buffered saline pH 7.1 (PBS). Slides were fixed in cold methanol (-40°C) for immunofluorescent staining and in 4% glutaraldehyde for electron microscopy. To inquire into the presence of HSV-2 antigens in the cervical tumor cells, slides were stained in duplicate with Ra-2 sera and PBS instead of antibody. Fluorescence was not observed (Fig 2) in biopsied neoplastic cells from all patients studied in this series (45). Furthermore, electron microscopy of cells from patients 553 and 565 (Table III) did not reveal virions and cytoplasmic changes previously shown to be associated with the synthesis of herpesvirus antigens (46). On the other hand, tumor cells on the surface of the neoplastic lesion as well as exfoliated tumor cells from 25 of the 29 (86%) patients studied in this series (Table III) reacted with Ra-2 sera. The proportion of reactive cells was independent of the stage of the disease, but varied with the patient and ranged between 0 and 40%. Fluorescence appeared as a diffuse cytoplasmic mass, sometimes with granules; nuclear fluorescence was not observed (Fig 3). Normal squamous differentiated cells did not react with Ra-2 serum.

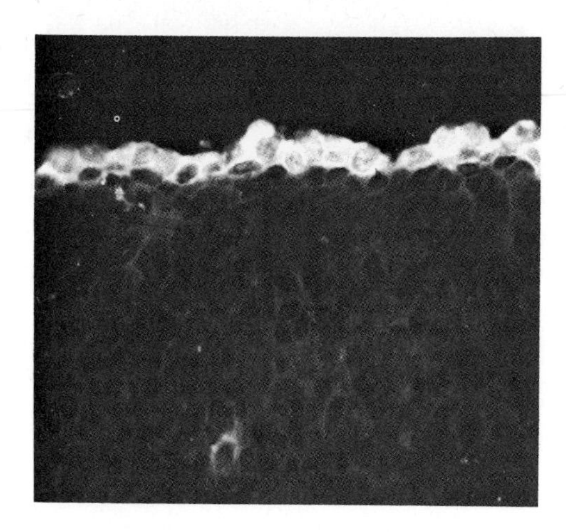

Fig 2. Frozen section of biopsy from a case of carcinoma in situ (No. 462) stained with Ra-2 serum. Fluorescence is observed only in tumor cells on the surface of the neoplasic lesion X 320). (Reduced 75%.)

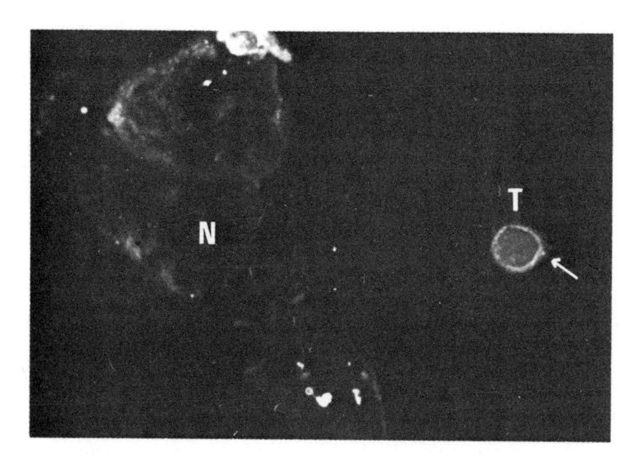

Fig 3. Exfoliated tumor cell (T) from a patient (No. 506) with invasive carcinoma of the cervix stained with Ra-2 serum displaying diffuse and granular (*arrow*) cytoplasmic fluorescence. Note absence of fluorescence of normal squamous (N) cells (X 480). (Reduced 75%.)

The specificity of the reaction for HSV-2 antigens was ascertained by the following control experiments: (a) Ra-2 sera reacted with HEp-2 cells infected with HSV-2 for 6 - 24 hours, but not with uninfected cells or cells infected with HSV-2 for 1 - 4 hours; (b) staining was not observed with phosphate buffered saline, preimmunized rabbit sera adsorbed with uninfected HEp-2 cells and antisera to adenovirus 18 and mycoplasma orale, whereas it was present in cells stained with antisera to HSV-1 displaying the expected cross-reactivity (4,5,6); (c) anaplastic cells reacted with two human sera from subjects without cancer but with neutralizing antibody to either herpesvirus type (Hu-2 and Hu-1 sera); and (d) adsorption of Ra-2 or Hu-2 sera with HSV-2-infected HEp-2 cells, until they failed to stain HSV-2-infected cells, removed their reactivity for atypic cells. Nonspecific staining was absent in 90% of patients studied in this series (Table III).

TABLE III

Proportion of Neoplastic Cells Reacting with Ra-2 Serum and Goat Anti-Rabbit Gamma Globulin[a]

Case	Percent exfoliated cells[b] staining with			Percent biopsied cells staining with
	Ra-2	PBS	Specific[c]	Ra-2[d]
Atypia				
533	46	4	42	0
539	36	0	36	0
555	11	11	0	0
556	42	0	42	0
549	6	0	6	0
576	3	0	3	0
586	0	0	0	0
585	18	0	18	0
575	2	0	2	0
583	20	0	20	0
587	18	0	18	0
Mean	18	1	17	0
In situ				
524	46	2	44	0
547	22	0	22	0
554	25	7	18	0
552	18	0	18	0
565	16	0	16	0
578	5	4	1	0
580	7	0	7	0
543	0	0	0	0
544	2	1	1	0
553	0	0	0	0
589	22	0	22	0
Mean	15	1	14	0
Invasive				
506	20	0	0	0
542	29	2	27	0
591A	32	7	25	1
593A	32	0	32	0
1147	29	0	29	0
1149	14	8	6	0
541	14	2	12	0
Mean	24	3	21	0

[a]Whereas biopsied tumor cells were obtained only from 29 patients summarized in this table, exfoliated tumor cells were obtained from a total of 61 patients, with results closely approximating those described here.

[b]Average number of cells counted for each patient is at least 25.

[c]Calculated by the subtraction of the proportion of cells staining with PBS from that staining with Ra-2 serum.

[d]Calculated as above; nonspecific staining observed for cases 555, 524, 554, and 591A.

To date, exfoliated cells were obtained from a total of sixty-one patients diagnosed as preinvasive and invasive cervical cancer. Fluorescence with Ra-2 sera was observed in atypic cells from 88% of those with preinvasive and 94% of those with invasive cancer (45,47).

The association of HSV-2 with cervical neoplasia appears to be specific. Exfoliated cervical cells from control subjects did not show evidence of herpesvirus antigens unless they were diagnosed as herpetic cervicitis (Table IV). Negative cases also included four with squamous metaplasia of the cervix, one case of mesodermal tumor of the endometrium, cells from pleural effusions from two cases of lymphosarcoma, one case of ovarian embryonal cell carcinoma and one case of embryonal rhabdomyosarcoma, and, finally, impressions of four cases of breast carcinoma. Consid-

TABLE IV

Proportion of Control Cells Reacting with Ra-2 Serum and Goat Anti-Rabbit Gamma Globulin[a]

Case	Average no. of cells counted	Percent staining		
		Ra-2	PBS	Specific[b]
Squamous metaplasia				
502	100	0	0	0
503	80	3	3	0
514	25	0	0	0
530	75	4	4	0
Carcinoma of endometrium				
519	50	0	0	0
Carcinoma of breast				
1	100	2	2	0
2	50	0	2	0
3	100	0	0	0
4	100	0	0	0
Lymphosarcoma				
560	200	10	10	0
561	123	0	0	0
Ov. emb. cell carcinoma				
559	75	5	9	0
Emb. rhabdomyosarcoma				
562	150	0	0	0
Herpetic cervicitis				
493	25	80	0	80
494	100	85	8	77

[a]From Aurelian et al. (45).
[b]Calculated by the subtraction of the proportion of cells staining with PBS from that staining with Ra-2 serum.

ering the likely hypothesis that the only difference between exfoliated tumor cells and those in biopsy specimens obtained from the same patients at the same time is the exposure of the tumor cells on the surface of the neoplastic lesions to glandular secretions of high pH (43), the presence of HSV-2 antigens in exfoliated but not biopsied tumor cells suggests that the cells harbor the viral genome in a repressed state. This interpretation is consistent with the observation that cells removed from the tumor biopsy and grown in culture do not show the presence of structural HSV-2 antigens unless exposed to conditions of stress such as high pH (Table II).

Twenty anaplastic cells (patients 565, 593A, and 1149) containing cytoplasmic fluorescent antigens and examined by electron microscopy did not reveal the presence of complete or incomplete virus particles (45). However, changes associated with the synthesis of herpesvirus antigens (46) and absent from biopsied cells were seen. These consisted of cytoplasmic vesicles, fragmentary membrane-like structures, and irregular electron-dense granules (Fig 4), as well as masses of closely packed osmiophilic bodies probably representing the fluorescent cytoplasmic granules (Fig 3). Proteins specified by herpesviruses are made in the cytoplasm and migrate to the nucleus for virus assembly (37). The absence of nuclear staining and virus particles from exfoliated cervical tumor cells suggests that in anaplastic cells migration does not occur. It is noteworthy in this context that arginine deprivation of herpesvirus-infected cells causes the absence of nuclear fluorescence and inhibits the formation of virus particles (48); although the major viral proteins are synthesized and accumulate in the cytoplasm, their migration to the nucleus does not occur (49). This interpretation is

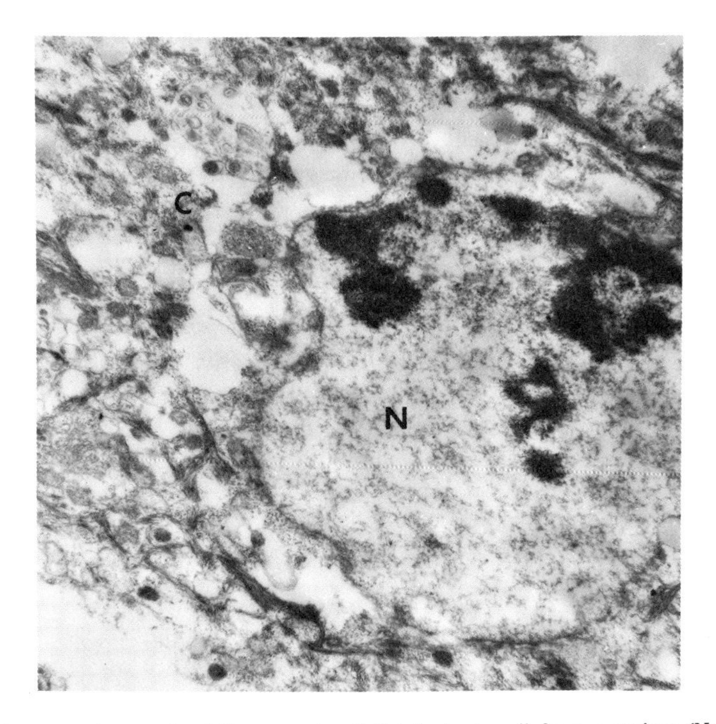

Fig 4. Electron micrograph of fluorescent exfoliated atypic cell from a patient (No. 565) with carcinoma in situ. Note cytoplasmic vesicles, disorganized membranous structures, and dense granules. Cytoplasm (C). Chromatin clumping at periphery of nucleus (N). (X 25,000.) (Reduced 70%.)

consistent with the observation that all patients studied in this series were free of active herpetic lesions at the time of collection of their exfoliated cells.

Basic Hypotheses

The presence of herpesvirus antigens in exfoliated tumor cells from 90% of patients studied in this series and the isolation of HSV-2 from cervical tumor cells grown in vitro strengthen the previously shown association between HSV-2 and cervical cancer. However, the role played by the virus in the causation of the tumor is still unknown. Three hypotheses may be proposed to account for this association.

The preferential hypothesis, based on the observation that herpesviruses grow better in actively replicating cells (50), suggests that infection with the virus follows the development of the neoplastic lesion. The roles of atypia and carcinoma in situ in the pathogenesis of cervical carcinoma are still unsolved. However, studies of incidence and prevalence rates have suggested that all or a large proportion of invasive carcinoma are preceded by carcinoma in situ. More recent data have also shown progression of epithelial changes from atypia and carcinoma in situ to invasive carcinoma (19,51). Accordingly, the preferential hypothesis predicts that the proportion of patients with early neoplastic lesions whose tumor cells contain HSV-2 antigens is lower than that of patients with invasive cervical cancer. Likewise, according to this hypothesis, the prevalence of antibody to HSV-2 in patients with preinvasive lesions should be lower than that in invasive cases. The promiscuity hypothesis envisions the association of HSV-2 with cervical cancer as due to the promiscuous nature of the cervical cancer population, which by definition is more apt to have all venereal diseases (11 - 13). Accordingly, cervical cancer groups should have an incidence of venereal diseases, other than HSV-2, significantly higher than that in control populations. Lastly, the etiology hypothesis interprets the association of HSV-2 with cervical cancer as resulting from the causative role, direct or co-carcinogenic, played by HSV-2. According to this hypothesis, one or more specific proteins (Ca-Protein) responsible for the neoplastic character of the cells must be associated with cervical neoplasia, since it is predicted that a partially transcribed HSV-2 genome is associated with virus-induced transformation. These proteins should not be found in normal tissue. In principle, it should be possible to differentiate between the three hypotheses described above by testing the predictions they make.

IS THE ASSOCIATION OF HSV-2 WITH CERVICAL NEOPLASIA DUE TO THE PREFERENTIAL GROWTH OF THE VIRUS IN TUMOR CELLS?

Two observations argue against the preferential hypothesis. First, as shown in Table III and discussed in the paragraph on HSV-2 antigens in cervical tumor cells in vivo, the proportion of patients with atypia, carcinoma in situ, and invasive cancer whose exfoliated tumor cells contain HSV-2 antigens is essentially identical and approximates 90%. Second, the results of a sero-epidemiologic study, designed to inquire into the prevalence of antibody to HSV-2 in patients with preinvasive and invasive cervical cancer, indicate that the occurrence of antibody to HSV-2 is similar in patients with invasive cancer (100%) and those with precancerous lesions (Table V). Of thirty-six women with cervical atypia, thirty-four (95%) had antibody to HSV-2, and

TABLE V

The Occurrence of Neutralizing Antibody to HSV-2 in Patients with Carcinoma of the Cervix as Compared with Other Groups[a]

Group	No. tested	Mean age	Mean economic decile[b]	Positive for HSV-2		Positive for HSV-1	
				No.	%	No.	%
Atypia	36	30	3.4	34	95	32	89
Matched controls	36	30	4.1	16	44	31	84
Carcinoma in situ	32	35	3.8	32	100	24	75
Matched controls	32	32	3.9	17	53	26	81
Invasive carcinoma	42	53	4.4	42	100	35	83
Matched controls	42	50	4.7	28	67	38	90
Total cases	110	40	4.0	108	98	90	82
Total controls	110	38	4.3	61	55	94	85
Carcinoma of other sites							
Vagina	1			0		1	
Uterus	1			0		1	
Other	8			5		8	
Total	10			5		10	

[a] From Aurelian et al. (35).
[b] Socioeconomic deciles of resident census tracts.

all thirty-two patients with carcinoma in situ (100%) were also positive for anti-HSV-2 antibody. The prevalence of antibody to HSV-2 is significantly greater in the carcinoma patients than in the matched control groups ($P<0.0005$ as determined by χ^2) and in patients with solid tumors at sites other than the cervix. Against the preferential hypothesis are also two observations made by Naib et al, namely, that typical herpesvirus-induced changes are not detected in neoplastic but in normal cervical cells and that a high incidence of neoplastic conversion is observed upon follow-up of patients with cytologic evidence of previous herpetic infection (30). Herpetic lesions were not observed in subjects studied in this series of experiments, suggesting the presence of type 2 antibody in their sera is due to a previously acquired infection, latent at the time of our studies.

THE PROMISCUITY HYPOTHESIS AND ITS PREDICTIONS

According to the promiscuity hypothesis, venereal diseases other than HSV-2 should be associated with cancer patients. The prevalence of trichomoniasis demonstrated cytologically on single as well as cumulative specimens and the frequency of positive serum tests for syphilis were studied in the same case-control groups used for

TABLE VI

Prevalence of Three Venereal Diseases in Carcinoma Groups and Controls of the Lower Socioeconomic Class[a]

Group	No. of cases	Mean age	Percent positive for HSV-2	Percent positive for trichomonas		Percent positive for STS
				Cumulative smear reports	Single smear reports	
Atypia	24	30	92	88	48	0
Matched controls	24	30	50	68	41	0
Carcinoma in situ	23	34	100	74	23	8
Matched controls	23	30	61	35	20	27
Invasive carcinoma	23	54	100	70	Not done	17
Matched controls	23	50	74	63	Not done	17
Total cases	70	39	97	77	36	8
Total controls	70	37	61	56	31	17

[a]Prevalences calculated for subjects in lower socioeconomic deciles. From I. Royston and L. Aurelian (52).

the study of HSV-2 antibody in order to inquire into this prediction (52). Two series of experiments were done. In the first series, a subject was considered positive for trichomoniasis if at any time she had had a Papanicolau smear revealing trichomonas organisms. Because a patient with suspected cervical neoplasia is apt to have several smears taken at various time intervals, the probability of detecting the organisms exceeds the probability of detecting them in a control patient who averages fewer smears. An effort was made to eliminate this bias in the second series of analyses, performed only for the preinvasive cases and their matched controls, by examining only one Papanicolau smear for each person. For the carcinoma group, this smear was the one obtained on the first visit to the cervical clinic; for the control, the one obtained closest to the date of blood collection. As shown in Table VI the results of single smear reports indicate that the prevalence of trichomonas does not differ significantly in the carcinoma group as compared to the control population; when cumulative reports are tabulated, the prevalence is somewhat higher for patients than controls, though not significantly. A person was considered positive for syphilis if at any time she had a positive STS test. Significant differences between carcinoma patients and controls were not observed (Table VI). The failure to show an association between cervical cancer and two venereal diseases other than HSV-2 argues against the interpretation that the association of HSV-2 with cervical cancer results solely from the independent association of the two diseases with promiscuity.

TUMOR-SPECIFIC HSV-2 ANTIGENS AND THE ETIOLOGY HYPOTHESIS

The humoral antibody detected in neutralization studies performed by various laboratories (17,18,35) is directed toward antigens that are HSV-2 but not tumor specific. Indeed, as shown in Table V, as many as 67% of control women without cancer or with cancer at sites other than the cervix show evidence of this antibody. Likewise, the antigen present in exfoliated tumor cells does not appear to be tumor specific as it reacts well with sera from women without cancer but with previous HSV-2 infection (Hu-2 sera). The etiology hypothesis predicts that the neoplastic character of the cells is determined by the synthesis of one or more proteins (Ca-Proteins) coded by the latent HSV-2 genome; such proteins are tumor as well as HSV-2 specific.

One method of inquiry into the validity of this prediction is the study for antibody to Ca-Protein(s) in sera from cancer of the cervix and control groups. Such a study is a priori based on two assumptions. First, on the basis of the nature of neoplastic disease, it must be assumed that antibody to Ca-Protein(s) is non-avid. Accordingly, the test used for its detection must be particularly sensitive. The Marcus modification (53) of the micro-quantitative complement fixation test of Wasserman and Levine fulfills these demands. In this test the OD* units fixed are determined by subtracting the reaction mixture OD* from the antibody control OD and the percentage of complement fixation is computed. Sera titers are expressed as the reciprocal of the highest serum dilution fixing at least 10% of the complement. The second assumption is that Ca-Protein(s) are synthesized in the permissive infection as well as in neoplasia. Herpesvirus DNA is double stranded (54), of a molecular weight of 95 X 10^6 to 99 X 10^6 daltons (55). According to current accounting practices, it has sufficient information to specify the sequence of 133,000 amino acids. This amount of information seems astronomical, even by comparison to viruses that, like the herpesviruses, replicate in the nucleus (papova or adenovirus) and that have an envelope (myxovirus). A point worth considering is that the genetic information of the virus, expressed in the nonpermissive infection (neoplasia), differs from that expressed in the permissive one. If this interpretation is accurate, Ca-Protein(s) should not be found in permissive cells infected in vitro (Table I). Another interpretation is that Ca-Protein(s) are synthesized early in the permissive infection, however, under these conditions virus replication proceeds normally and the synthesis of Ca-Protein(s) is inhibited. Unregulated synthesis of Ca-Protein(s) and inhibition of virus replication beyond this stage, would be characteristic of neoplasia (Table I). This particular interpretation was tested in this study. The antigen designated as AG-4 was made 4 hours after infection of permissive (HEp-2) cells with HSV-2. The sera studied were obtained from patients diagnosed as preinvasive and invasive cancer, treated or untreated before the time of blood collection, and controls, without cancer or with tumors at sites other than the cervix. These included cases of carcinoma of the lung, vagina, and ovaries. These sera had previously been used to search for neutralizing antibody to HSV-2 (35 and Table V). As shown in Tables VII and VIII the prevalence of complement-fixing antibody is significantly higher in patients with cervical cancer (68%) than in matched control groups (7%), or in patients with tumors at sites other than the cervix (0). The marked gradation in incidence of antibody (invasive cancer 89%; carcinoma in situ 69%;

*Optical density.

TABLE VII

Antibody to Herpes Virus Antigens in Sera of Patients with Cervical Cancer

	Antibody				Antibody		
Case	CF[a]	NT-HSV-2[b]	NT-HSV-1[c]	Case	CF	NT-HSV-2	NT-HSV-1
		Atypia				Invasive	
143	- (0)[d]	+	+	1	+ (16)	+	+
144	- (0)	+	+	9	+ (16)	+	+
146	- (0)	+	+	23	+ (16)	+	-
173	+ (16)	+	+	4	+ (32)	+	+
54	- (0)	+	+	6	+ (32)	+	+
7	+ (8)	-	+	10	+ (8)	+	-
43	- (0)	+	+	11	+ (32)	+	+
21	+ (8)	+	+	13	+ (16)	+	-
31	+ (32)	+	+	17	+ (8)	+	+
89	- (0)	+	-	20	+ (8)	+	+
175	- (0)	ND[e]	ND	30	+ (8)	+	-
2	- (0)	+	+	71	+ (16)	ND	ND
247	+ (2)	+	+	19	+ (8)	+	+
				114	+ (32)	+	+
		In situ		145	- (0)	ND	ND
				170	+ (4)	ND	ND
44	+ (8)	+	+	183	- (0)	ND	ND
48	+ (4)	+	+	268	+ (64)	+	-
64	+ (64)	+	+				
72	- (0)	+	+				
74	- (0)	+	+				
102	+ (32)	+	-				
97	+ (16)	+	+				
118	+ (64)	+	+				
113	+ (32)	+	+				
154	- (0)	+	+				
156	+ (4)	+	+				
199	+ (8)	+	-				
48	+ (4)	+	+				
98	+ (8)	+	+				
220	- (0)	+	+				
219	- (0)	+	+				

[a]Complement fixing antibody.
[b]Neutralizing antibody to HSV-2.
[c]Neutralizing antibody to HSV-1.
[d]Complement fixing titers in parentheses, are expressed as the reciprocal of the highest serum dilution fixing at least 10% of the complement.
[e]Not done.

atypia 38%) fits well the progression believed to occur in cervical neoplasia from atypia to invasive cancer (19,51). The observation that fixation is not obtained with sera from patients treated by irradiation or hysterectomy at least one month prior to blood

TABLE VIII

Antibody to Herpesvirus Antigens in Sera of Subjects without Cancer or with Cancer at Sites Other Than the Cervix

	Antibody				Antibody		
Case	CF[a]	NT-HSV-2[b]	NT-HSV-1[c]	Case	CF	NT-HSV-2	NT-HSV-1
	CaCx treated[d]				Subjects without cancer		
227	- (0)[e]	+	+	55	- (0)	+	+
22	- (0)	+	-	8	+ (8)	-	-
217	- (0)	+	-	91	- (0)	+	-
32	- (0)	+	+	59	- (0)	+	+
49	- (0)	+	-	61	- (0)	+	+
34	- (0)	+	+	62	- (0)	+	+
149	- (0)	+	+	84	- (0)	+	+
174	- (0)	+	+	78	- (0)	-	+
39	- (0)	+	+	90	- (0)	-	+
210	- (0)	+	+	122	- (0)	-	+
385	- (0)	+	+	119	- (0)	-	+
				66	- (0)	+	+
	Subjects without cancer			67	- (0)	+	+
				68	- (0)	ND	ND
28	- (0)	-	+	69	- (0)	+	-
27	- (0)	-	+	75	- (0)	+	-
29	+ (16)	+	+				
35	- (0)	+	+		Tumors at sites other than the cervix		
237	- (0)	+	-				
269	- (0)	-	+	163	- (0)	ND	ND
51	- (0)	-	-	165	- (0)	+	-
26	- (0)	+	+	195	- (0)	ND	ND
42	- (0)	-	+	133	- (0)	+	+
46	+ (64)	-	+	571	- (0)	ND	ND
50	- (0)	-	+	96	- (0)	ND	ND
52	- (0)	-	-	569	- (0)	ND	ND
53	- (0)	-	+	136	- (0)	+	+
47	- (0)	-	+	184	- (0)	+	+
5	- (0)	+	+				
37	- (0)	+	+		Spontaneous revertant[g]		
63	- (0)	ND[f]	ND				
65	- (0)	+	+	189	- (0)	+	+

[a] Complement-fixing antibody.
[b] Neutralizing antibody to HSV-2.
[c] Neutralizing antibody to HSV-1.
[d] Sera obtained from patients with cervical carcinoma 1 month or longer after therapy (irradiation or/and radical surgery).
[e] Complement-fixing titers (in parentheses) are expressed as the reciprocal of the highest serum dilution fixing at least 10% of the complement.
[f] Not done.
[g] Serum obtained from a case of carcinoma in situ that had spontaneously reverted was collected at the first follow-up visit at which the tumor was no longer observed.

collection further ascertains the tumor specificity of the antibody detected in this study (Table VIII).

Critical to the interpretation of our results is the observation that extracts from uninfected permissive (HEp-2) cells or HEp-2 cells infected with HSV-2 for less than 4 hours, do not react with the sera positive for AG-4, suggesting that the complement-fixing antibody associated with cervical cancer (Table VII) is virus specific. The exact nature, structural or functional, of AG-4 antigen, is unknown at present. It appears to be an antigen that like the postulated Ca-Protein(s) is tumor as well as virus specific. It is not clear, however, whether it indeed is Ca-Protein(s); the role it may play in oncogenesis must await further investigation.

CONCLUSIONS AND PERSPECTIVES

The persistence of herpes simplex virus in man as a model system for the interaction of herpesviruses with their host cells has been discussed in detail by Roizman (31). Considerably less information exists as to the mechanisms of association of herpesvirus with tumor cells. The observations cited here suggest that, like herpes simplex virus in control subjects with recurrent disease, the virus, generally unidentifiable, can be made to manifest itself. Biopsied cervical tumor cells in vivo and, at least in one case, grown in tissue culture (S332G cells), do not show evidence of virus presence unless previously exposed to certain conditions of stress, such as high pH. This unity in behavior is not tarnished by apparent differences. Thus, whereas in vitro, virus induction consists in the synthesis of virus structural components as well as the formation of infectious virus and is accompanied by the destruction of the host cells, in vivo virus induction does not proceed beyond the stage of synthesis of structural viral proteins. This conclusion is consistent with two well-authenticated observations. First, active herpetic lesions of the cervix are rarely, if ever, observed in association with cervical neoplasia (Davis, personal communication) and have not been seen at the time of cell collection from sixty-one patients studied in this series and, second, herpes-induced changes are not seen in tumor but rather in normal cervical cells (30). The abortion of the virus replicative cycle beyond the synthesis of structural viral antigens might be a function of the short life-span of the exfoliated cells.

Of particular interest is the observation that lines of cervical tumor cells established from different patients differ with respect to their association to HSV-2. Thus, unlike S332G cells, JaCha cells show evidence of HSV-2 antigens without being previously exposed to conditions of stress, however, they do not contain infectious virus, and exposure to alkaline medium does not induce the synthesis of structural HSV-2 antigens and virus formation. Furthermore, the antigens observed in JaCha cells, although apparently herpetic in nature, differ from the structural antigens observed in stressed S332G cells. This conclusion is based on the observation that, unlike Ra-2 serum, the Hu-2 serum which stains JaCha cells also reacts with HEp-2 cells infected with HSV-2 for three to four hours, suggesting that it may contain antibody to nonstructural or "early" structural HSV-2 antigens (37).

At this time it is not clear whether stress conditions other than high pH can induce virus replication. A survey of the extensive literature on recurrent herpetic infections shows that these vary from physical stress, such as strong wind and sun, to emotional stress and psychotherapy, to drug or hormone treatment, and menstruation.

It is possible that virus induction is a function of cell death; any factor causing cell damage would induce virus replication. A prerequisite for virus induction in cervical tumor cells is the survival in these cells of the complete viral genome and a regulatory mechanism that can be derepressed. Thus, unless the patient suffers from a well authenticated case of recurrent herpetic disease, there is no a priori reason why it should be assumed that the complete viral genome survives in the cervical cells or that the regulatory mechanisms of persistence can be derepressed. To date, not enough clinical information has been accumulated actually to establish the frequency rate of recurrent cervical herpetic disease in the general population. The establishment of new tissue culture lines from biopsies of cervical cancer will allow further studies on the mechanisms of association of HSV-2 with cervical tumor cells.

Our interpretation of the association of HSV-2 with cervical tumor cells implies that some or all the tumor cells harbor the viral genome in a partially repressed state, resulting in the presence in these cells of cancer-associated protein(s). The results of complement fixation studies designed to inquire into this prediction have indicated that, indeed, patients with cervical cancer have antibody to an antigen (AG-4) made early in the reproductive cycle of HSV-2 in permissive cells. This antibody is not present in sera from control subjects or from cervical cancer cases given therapy prior to blood collection. Of particular interest is that therapy had no effect on the neutralizing antibody to HSV-2 (Table VIII). The nature and function of the AG-4 antigen and the question of its possible qualification for the role of Ca-Protein(s) are not known at present. Studies now in progress in our laboratory are directed toward the detection of viral DNA and Ca-Protein(s) in cervical neoplastic cells as well as to a better understanding of the nature of AG-4 antigen and its relationship to the postulated Ca-Protein(s).

The findings presented in this paper strengthen the association of HSV-2 with cervical carcinoma; however, is HSV-2 responsible for the tumor in which it is found? The answer is not clear and the evidence, other than that which has been cited, is very meager. As pointed out earlier (Table V), HSV-2 is widespread and readily isolated throughout the world. Yet cervical cancer is not quite that common and varies in frequency rates in different populations. The possibility that the specific cell harboring the virus and the expression of those viral genes needed to transform the cells are singly or both determined by the genotype of the host cannot be excluded.

SUMMARY

Two types of herpesvirus immunologically and biologically distinct have been isolated from man; herpesvirus type 1 (HSV-1), associated with facial lesions and herpesvirus type 2 (HSV-2), isolated from smegma samples and cervical lesions and shown to be venereally transmitted. On the basis of sero-epidemiologic data, it has been suggested that HSV-2 may play an etiologic role in squamous carcinoma of the human cervix, a disease more prevalent in women who have multiple sex partners and who begin heterosexual activity early in life.

Findings providing evidence for the association of HSV-2 with cervical cancer are presented. Exfoliated but not biopsied tumor cells from patients with cervical carcinoma contain structural HSV-2 antigens as revealed by immunofluorescence. Virus particles are not observed. Normal squamous cells from the same subjects or from controls without the disease and cells from a number of tumors at sites other than the cervix do not react with anti-HSV-2 serum. A herpesvirus was isolated from spontaneously degenerating cultures of cervical tumor cells grown in vitro. The virus was identified as HSV-2 on the basis of biologic and immunologic properties. Herpesvirus antigens and virus particles were not seen in duplicate cultures of viable tumor cells. These results are discussed in terms of the virus-host cells interaction; it appears that cervical tumor cells harbor the viral genome in a repressed state (latency). Under conditions of stress, such as high pH, the viral genome may express its genetic potentialities resulting in the formation of viral components and/or infectious virus.

Studies designed to differentiate between three possible interpretations of the association of HSV-2 with cervical tumor cells include (1) a sero-epidemiologic study for the prevalence of neutralizing antibody to HSV-2 in patients with cervical cancer at various stages of the disease (atypia, in situ, invasive) and in matched control groups and (2) an inquiry into the presence of a tumor-specific viral antigen (AG-4) in the same patients and control groups. The results of this study are interpreted to suggest that partial expression of the viral genome is associated with this neoplasia.

ACKNOWLEDGMENTS

Deepest gratitude is due to Dr. J. D. Strandberg for excellent electron microscopical support, Drs. H. J. Davis and C. Julian for collection of sera and cervical tumor samples, Dr. R. Marcus for his patience in teaching us his complement-fixation test and his continuous interest while we were struggling with it, and to Dr. L. V. Melendez for his help and collaboration. The expert technical support of Mrs. B. Schumann and Mr. E. Chow is gratefully acknowledged.

This study was supported in part by Grant VC-16 from the American Cancer Society and by contract NIH-NCI-71-2121 from the Special Virus Cancer Program, NCI.

REFERENCES

1. Lwoff, A., and Tournier, P.: The classification of viruses. Am. Rev. Microbiol., 20:45, 1966.
2. Roizman, B.: The herpesviruses. A biochemical definition of the group. Current Topics in Microbiology and Immunology, 49:1, 1969.
3. Wilner, I. B.: A classification of the Major Groups of Human and Other Animal Viruses. Burgess Publishing Co., Minneapolis, Minnesota, 3rd ed., p 78, 1966.
4. Dowdle, W. R., Nahmias, A. J., Maxwell, R. W., and Pauls, F. P.: Association of antigenic type of herpesvirus hominis with site of viral recovery. J. Immunol., 99:974, 1967.
5. Plummer, G., Waner, J. L., and Bowling, C. P.: Comparative studies of type 1 and type 2 "herpes simplex" viruses. Brit. J. Expt. Path., 49:202, 1968.
6. Ejercito, P. M., Kieff, E. D., and Roizman, B.: Characterization of herpes simplex virus strains differing in their effect on social behaviour of infected cells. J. Gen. Virol., 3:357, 1968.
7. Terni, M., and Roizman, B.: Variability of herpes simplex virus–isolation of two variants from simultaneous eruptions at different sites. J. Infect. Dis., 121:212, 1970.
8. Nahmias, A. J., Naib, Z. M., and Josey, W. E.: Herpesvirus hominis Type 2 infection–association with cervical cancer and prenatal disease. Perspectives in Virology, VII:73, 1971.

9. Burnet, F. M., and Lush, D.: Herpes simplex–studies on the antibody content of human sera. Lancet, I:629, 1939.
10. Nahmias, A. J., Dowdle, W. R., Naib, Z. M., Josey, W., McClone, A., and Domsecik, G.: Genital infection with type 2 herpesvirus hominis–a commonly occurring venereal disease. Brit. J. Vener. Dis., 45:294, 1969.
11. Rotkin, I. D.: Sexual characteristics of a cervical cancer population. Am. J. Pub. Health, 57:815, 1967.
12. Terris, M., and Oalman, M. C.: Carcinoma of the cervix–an epidemiologic study. JAMA 174:1847, 1960.
13. Rotkin, I. D.: Adolescent coitus and cervical cancer–association of related events with increased risk. Cancer Res., 27:603, 1967.
14. Wyburn-Mason, R.: Malignant change following herpes simplex. Brit. Med. J., 2:615, 1957.
15. Kvasnicka, A.: Relationship between herpes simplex and lip carcinoma III. Neoplasma (Bratislava), 10:199, 1964.
16. Kvasnicka, A.: Relationship between herpes simplex and lip carcinoma. IV–selected cases. Neoplasma (Bratislava), 12:61, 1965.
17. Nahmias, A. J., Naib, Z. M., and Josey, W. E.: Association of genital herpesvirus with cervical cancer. Int. Virol., 1:187, 1969.
18. Rawls, W. E., Tomkins, W. A. F., and Melnick, J. L.: The association of herpesvirus type 2 and carcinoma of the cervix. Am. J. Epid., 89:547, 1969.
19. Jones, H. W., Jr., Katayama, K. P., Stafl, A., and Davis, H. J.: Chromosomes of cervical atypia, carcinoma in situ and epidermoid carcinoma of the cervix. Obst. and Gynec., 30:790, 1967.
20. Auesperg, N., and Wakonig-Vaartaja, T.: Chromosome changes in invasive carcinoma of the uterine cervix. Acta Cytol., 14:495, 1970.
21. Stitch, H. F., Hsu, T. C., and Rapp, F.: Viruses and mammalian chromosomes. I. Localization of chromosome aberrations after infection with herpes simplex virus. Virology, 22:439, 1964.
22. Hampar, B., and Ellison, S. A.: Chromosomal aberrations induced by an animal virus. Nature, 192:145, 1961.
23. McKenna, J. M., Davis, F. E., Prier, J. E., and Kleger, B.: Induction of neoantigen (G) in human amnion ("WISH") cells by herpesvirus A. Nature, 212:1602, 1966.
24. McKenna, J. M., Sandelson, P., and Blakemore, W. S.: Studies of the antigens of human tumors. I. Demonstration of a soluble specific antigen in HeLa cells and some human tumors. Cancer Res., 24:754, 1964.
25. Mizell, M., Toplin, I., and Isaacs, J. J.: Tumor induction in developing frog kidneys by a zonal centrifuge purified fraction of the frog herpes-type virus. Science, 165:1134, 1969.
26. Purchase, H. G., Witter, R. L., Okozaki, W., and Burmester, B. R.: Vaccination against Marek's disease. Perspectives in Virology, VII:91, 1971.
27. Melendez, L. V., Hunt, R. D., Daniel, M. D., Garcia, F. G., and Fraser, C. E. O.: Herpesvirus saimiri. II. Experimentally induced malignant lymphoma in primates. Lab. An. Care, 19:378, 1969.
28. Epstein, M. A., Achong, B. G., and Barr, J. M.: Virus particles in cultured lymphoblasts from Burkitt's lymphoma. Lancet, 2:702, 1964.
29. Duff, R., and Rapp, F.: Properties of hamster embryo fibroblasts transformed in vitro after exposure to ultra-violet irradiated herpes simplex virus type 2. J. Virol., 8:469, 1971.
30. Naib, Z. M., Nahmias, A. J., Josey, W. E., and Kramer, J. H.: Genital herpetic infection–association with cervical dysplasia and carcinoma. Cancer, 23:940, 1969.
31. Roizman, B.: In: Of Microbes and Life. Edited by J. Monod and E. Borek. Columbia University Press, New York, p 189, 1971.
32. Roizman, B.: An inquiry into the mechanisms of recurrent herpes infections of man. Perspectives in Virology, IV:283, 1966.

33. Smith, K. O.: Relationship between the envelope and the infectivity of herpes simplex virus. Proc. Soc. Expt. Biol. and Med., 115:814, 1964.
34. Aurelian, L., and Roizman, B.: Abortive infection of canine cells by herpes simplex virus II. Alternative suppression of synthesis of interferon and viral constituents. J. Mol. Biol., 11:539, 1965.
35. Aurelian, L., Royston, I., and Davis, H. J.: Antibody to genital herpes simplex virus: association with cervical atypia and carcinoma in situ. J. Nat. Cancer Inst., 45:455, 1970.
36. Aurelian, L., Strandberg, J. D., Melendez, L. V., and Johnson, L. A.: Herpesvirus type 2 isolated from cervical tumor cells grown in tissue culture. Science, 174:704, 1971.
37. Spear, P. G., and Roizman, B.: The proteins specified by herpes simplex virus. I. Time of synthesis, transfer into nuclei and properties of proteins made in productively infected cells. Virology, 36:545, 1968.
38. Strandberg, J. D., and Aurelian, L.: Replication of canine herpesvirus. II. Virus development and release in infected dog kidney cells. J. Virol., 4:480, 1969.
39. Couch, E. F., and Nahmias, A. J.: Filamentous structures of type 2 herpes virus hominis infection of the chorioallantoic membrane. J. Virol., 3:228, 1969.
40. Schwartz, J., and Roizman, B.: Similarities and differences in the development of laboratory strains and freshly isolated strains of herpes simplex virus in HEp-2 cells–electron microscopy. J. Virol., 4:879, 1969.
41. Hampar, B., and Copeland, M. L.: Persistent herpes simplex virus infection in vitro with cycles of cell destruction and regrowth. J. Bacteriol., 90:205, 1965.
42. Hampar, B.: Persistent cyclic herpes simplex virus infection in vitro. II Localization of virus, degree of cell destruction and mechanisms of virus transmission. J. Bacteriol., 91:1959, 1966.
43. Moghissi, K. S.: Cyclic changes of cervical mucous in normal and progestin treated women. Fertil. Sterie., 17:663, 1966.
44. Davis, H. J.: The irrigation smear. A cytologic method for mass population screening by mail. Amer. J. Obst. Gynec., 84:1017, 1962.
45. Aurelian, L., Strandberg, J. D., and Davis, H. J.: HSV-2 antigens absent from biopsied cervical tumor cells–a model consistent with latency. Proc. Soc. Expt. Biol. Med., 140:404, 1972.
46. Nazerian, K., and Purchase, H. G.: Combined fluorescent-antibody and electron microscopy study of Marek's disease virus-infected cell cultures. J. Virol., 5:79, 1970.
47. Royston, I., and Aurelian, L.: Immunofluorescent detection of herpesvirus antigens in exfoliated cells from human cervical carcinoma. Proc. Nat. Acad. Sci., 67:204, 1970.
48. Courtney, R. J., McCombs, R. M., and Benyesch-Melnick, M.: Antigens specified by herpes viruses. I. Effect of arginine deprivation on antigen synthesis. Virology, 40:379, 1970.
49. Mark, G. E., and Kaplan, A. S.: Synthesis of proteins in cells infected with herpesvirus VII. Lack of migration of structural viral proteins to the nucleus of arginine-deprived cells. Virology, 45:53, 1971.
50. Aurelian, L.: Factors affecting the growth of canine herpesvirus in dog kidney cells. Applied Microbiology, 17:179, 1969.
51. Koss, L. G.: Concept of genesis and development of carcinoma of the cervix. Obst. Gynec. Survey, 24:850, 1969.
52. Royston, I., and Aurelian, L.: The association of genital herpesvirus with cervical atypia and carcinoma in situ. Am. J. Epid., 91:531, 1970.
53. Marcus, R. L., and Townes, A. S.: The occurrence of cryoproteins in synovial fluid; the association of a complement-fixing activity in rheumatoid synovial fluid with cold-precipitable protein. J. Clinical Investigation, 50:282, 1971.
54. Russell, W. C.: Herpes virus nuclei acid. Virology, 16:355, 1962.
55. Frenkel, N., and Roizman, B.: Herpes simplex virus–genome size and redundancy studied by renaturation kinetics. J. Virology, 8:591, 1971.

20. AN ATTEMPT TO CORRELATE RNA-TEMPLATED DNA POLYMERASE ACTIVITY WITH VIRUS-LIKE PARTICLES IN HUMAN MILK: MURINE MAMMARY TUMOR VIRUS (MuMTV) AS A MODEL SYSTEM[1]

Arnold S. Dion, Nurul H. Sarkar and Dan H. Moore

Institute for Medical Research
Copewood Street
Camden, New Jersey 08103

INTRODUCTION

Model systems have served as an invaluable aid in our understanding of molecular and developmental mechanisms. The possible viral etiology of human breast cancer has a number of precedents in laboratory animals, which include an association of viral particles with mammary tumors of the mouse (1), rat (2), and monkey (3); however, only in the mouse has the causative viral agent been purified from milk and shown to induce mammary tumors (4). Therefore, our investigations have involved the mouse as a model system for the investigation of the relationship between human milk-borne particles and RNA-templated DNA polymerase activity. The latter study was prompted by the present uncertainty regarding a correlation between human milk-borne particles and RNA-dependent DNA polymerase activity (5).

Justification for comparisons between MuMTV and particles found in some human milks is dependent upon the following observations:

1. morphological similarities between human and mouse milk-borne particles (5-7);
2. the possibility of cross antigenicity as suggested by the neutralization of MuMTV by sera of some breast cancer patients (8);
3. occurrence of particle bound 60 - 70S RNA and a resident RNA-templated DNA polymerase (9,10); and
4. nucleic acid hybridization studies (11).

A preliminary characterization of the reverse transcriptase activity of MuMTV, as analyzed by polyacrylamide gel electrophoresis, has been reported (12).

RELATIONSHIP BETWEEN PARTICLE NUMBER AND DNA SYNTHESIS

That normal mouse milk constituents do not possess DNA polymerase activity resulting in a labeled 70S RNA complex, is demonstrated in Figure 1. Equivalent milk

[1] This investigation was supported by Contract PH 43-68-1000 from the National Cancer Institute, Grant CA-08740 from the National Cancer Institute, General Research Support Grant FR-5582 from National Cancer Institute, and Grant-in-Aid M-43 from the State of New Jersey.

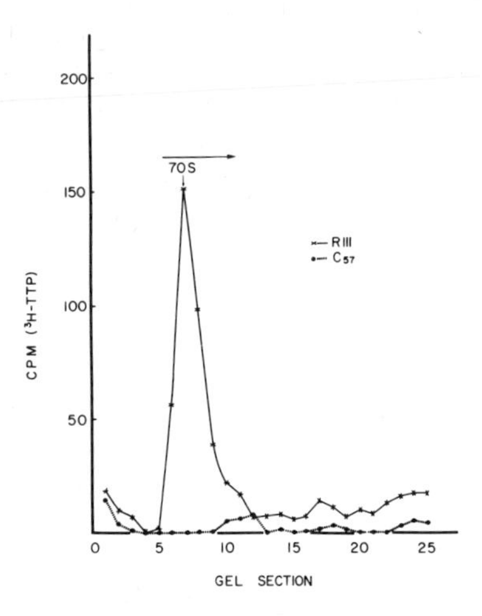

Fig 1. Comparison of RNA-dependent DNA polymerase activity of particulate fractions derived from RIII and C57BL mouse milk. 1 ml aliquots of each milk were skimmed, the particulate fractions pelleted and analyzed for RNA-directed DNA polymerase activity by PAGE (12). In all figures, electrophoretic mobilities are from left to right. Unless otherwise indicated, 1.5 mm sections were cut from each gel.

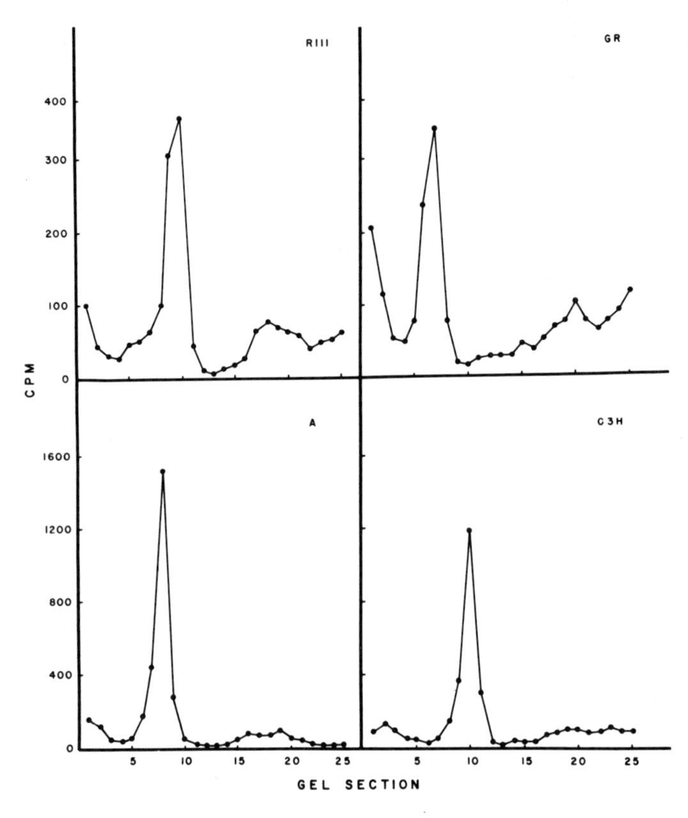

Fig 2. Gel electrophoresis patterns obtained after phenol/cresol extraction and ethanol precipitation of reaction products (30 minutes at 37°C) from each of 4 strains of mouse milk, each known to contain B particles of MuMTV. Reaction conditions and method of gel analysis are described in reference 12. It should be noted that these analyses by PAGE were not performed simultaneously.

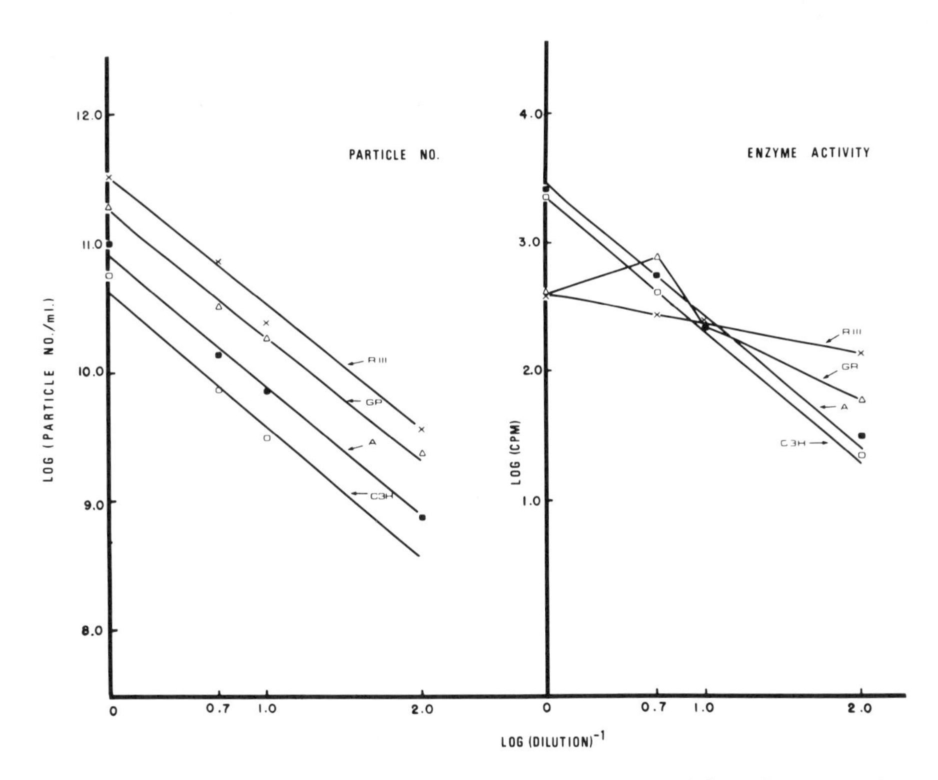

Fig 3. Relationship between number of milk-borne particles derived from 4 mouse strains and enzyme activity, viz., counts per minute (cpm) complexed to 70S RNA after analysis by PAGE. Aliquots of the designated mouse milk were skimmed and diluted with T.15NE (0.01 M Tris, 0.15 M NaCl, 0.002 M EDTA, pH 8.3). Undiluted samples represent number of particles and enzyme activity contained in 1 ml of whole milk after skimming and pelleting of virions, and have not been corrected for losses incurred during purification. Reaction conditions and analysis by PAGE described in reference 12. For particle counting a carefully calibrated suspension of latex beads was added to each sample after dilution, pelleting, and resuspension of virus. Viral particle number was determined from the ratio of virus to latex beads by electron microscopy.

samples from RIII and C57BL, the latter shown to be free of MuMTV and murine leukemia virus by immunochemical techniques, were assayed for RNA-dependent DNA synthesis by analysis of the reaction products by polyacrylamide gel electrophoresis (PAGE). Only the RIII milk sample, known to contain B particles of MuMTV resulted in DNA labeling of the 70S RNA peak. Similar results were reported by Schlom and Spiegelman (13), employing glycerol velocity gradient centrifugation.

Figure 2 illustrates the labeling patterns observed after PAGE of reaction products obtained from disrupted milk-borne virions derived from various mouse strains. Most of the labeled DNA is associated with a main RNA peak of activity of approximately 70S. To obtain a relationship between particle number and polymerase activity, parallel dilutions were prepared for electron microscopy and reverse transcriptase activity; results are shown in Figure 3. A linear diminution of ^{3}H-TTP label complexed to 70S RNA was observed over a two $\log_{10}$ dilution of virions derived from A and C3H mouse milk. In addition, constant activity on a per particle basis was found in these strains. In contrast, dilutions of RIII milk-borne virions appeared more active per particle with higher dilutions. Finally, a 1:5 dilution of GR milk-derived virions was more active than the undiluted sample or consequent higher dilutions. The increase in activity of diluted samples of RIII and GR virions may indicate the presence of inhibitor(s) in these milks, and may represent important phenomena in considering the reverse transcriptase activity of some human milks.

DNA SYNTHESIS BY DISRUPTED, RIBONUCLEASE-TREATED PREPARATIONS OF RIII MuMTV

Inhibition of DNA synthesis by pretreatment of detergent-disrupted virions with ribonuclease is an important criterion for determining whether this synthesis is RNA templated. However, Schlom and Spiegelman (13) and Dion and Moore (12) have demonstrated that this criterion is not met by ribonuclease-digested preparations of RIII MuMTV, when only incorporation of label into TCA-precipitable counts is monitored. The latter observation emphasizes the need for the simultaneous detection method of Schlom and Spiegelman (13), since in this method of assay little or no label was found associated with the 70S complex as shown in Figure 4; however, increased labeling was noted in a molecular weight range of approximately 1.6 times 10^6–7.8 times 10^4. Table I lists various conditions of pancreatic ribonuclease A digestion of disrupted MuMTV virions; in all instances, incorporation of ^{3}H-TTP was found to be greater in the treated preparations. Since these studies had been performed with virions directly pelleted from skim milk, extraviral nucleic acid contamination acting as template for DNA synthesis was possible; however, identical results were obtained when virions purified on sucrose gradients were assayed under conditions given in Figure 4. The latter, of course, does not eliminate the possibility of host-derived nucleic acid within the virion (14).

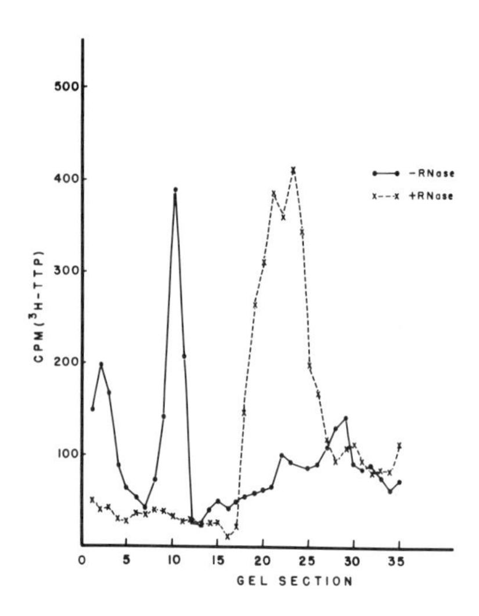

Fig 4. Gel patterns obtained from reaction products (60 minute incubation at 37°C) of RIII MuMTV after preincubation in the presence or absence of pancreatic ribonuclease A. Pelleted virions were resuspended in 0.01 M Tris (pH 8.3), disrupted with NP-40 in the presence of dithiothreitol, and preincubated with ribonuclease (24 μg/ml) for 60 minutes at 24°C. A control preparation was treated similarly in the absence of ribonuclease. Incubation conditions and sample preparation for PAGE as given in reference 12.

TABLE I

Enhancement of ^{3}H-TTP Incorporation by Pancreatic Ribonuclease Digesion of the Endogenous Template of RIII MuMTV

Conditions[a]			
Temperature	Time (min)	RNAse	pmoles incorporated[b]
37°C	0	-	0.45
37°C	0	24 μg/ml	0.84
37°C	5	-	0.74
37°C	5	24 μg/ml	1.19
24°C	60	-	1.89
24°C	60	48 μg/ml	2.91

[a]Conditions of preincubation.
[b]Determined as TCA-precipitable counts: 0.1 ml aliquot of a phenol extract was precipitated by the addition of 0.1 ml yeast RNA (300 μg/ml) as carrier and 0.1 ml TCA mix (equal volumes of 100% TCA, saturated sodium orthophosphate, and saturated sodium pyrophosphate). After chilling 10 minutes on ice, TCA-precipitable counts were collected on millipore filters, dried and counted in BBOT-Toluene (9).

TABLE II

Reverse Transcriptase Activity of Untreated and Ribonuclease-treated RNA Template of MuMTV: Deoxyribonucleoside Triphosphate Requirement

RNAse[a]	Conditions	pmoles ^{3}H-TTP incorporated[b]	%
-	complete	1.15	(100)
+ (24 μg/ml)	complete	1.52	132.4
-	- dCTP	0.20	17.7
+ (24 μg/ml)	- dCTP	0.21	18.4
-	- d(C,A,G)TP	0.06	5.3
+ (24 μg/ml)	- d(C,A,G)TP	0.07	5.8

[a]60 minute preincubation in the presence or absence of ribonuclease (RNAse) at 24°C.
[b]Incorporation of ^{3}H-TTP as determined by procedure given in Table I.

That both the untreated and ribonuclease-resistant templates are heterogeneous in base composition, is indicated by Table II. Both templates require the presence of all four deoxyribonucleoside triphosphates (dNTP's) for maximal incorporation of ^{3}H-TTP, obviating the possibility that the ribonuclease-resistant template consists mainly of polyadenylate sequences.

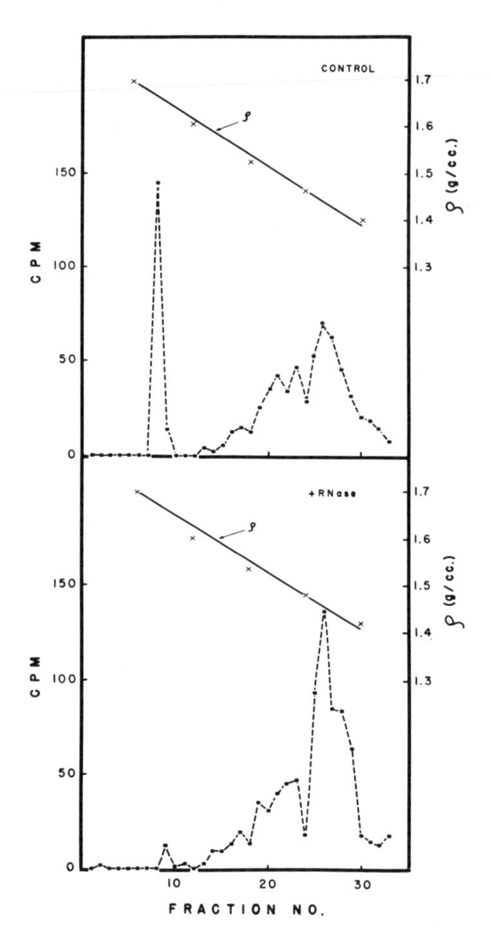

Fig 5. Cs_2SO_4 gradient centrifugation of reaction products obtained after preincubation in the presence (Fig 5, *bottom*) or absence (Fig 5, *top*) of pancreatic ribonuclease A, followed by a 30-minute incubation at 37°C. Preincubation conditions are given in the legend to Figure 4. Phenol/cresol extracts were ethanol precipitated, dried in vacuo, and resuspended in 2.5 ml T.15NE (pH 8.3). After adding an equivalent volume of saturated Cs_2SO_4, the samples were centrifuged at 20°C for 65 hours at 32,000 rpm (SW50.1). The cellulose nitrate tubes were punctured from the bottom and TCA-precipitable counts collected on millipore filters.

Isopycnic banding in Cs_2SO_4 gradients of reaction products obtained from untreated and ribonuclease-treated preparations is illustrated in Figure 5. In the control preparation, label is mainly associated with RNA, banding at a density of 1.66 g/cm^3; there is also evidence for heterodisperse label of intermediate density and label associated with DNA (1.42 - 1.45 g/cm^3). In contrast, ribonuclease treatment of disrupted virions prior to incubation results in little or no label associated with a density region corresponding to RNA; however, label was found in intermediate density regions and, predominantly, in the DNA region.

Preferential inhibition of DNA-dependent synthesis by actinomycin D, when compared to RNA-templated synthesis, has been reported (15,16). As shown in Table III, untreated and ribonucleased preparations of RIII MuMTV were assayed in the presence of actinomycin D (20 μg/ml) and a 27.5% inhibition of incorporation was observed for the untreated control, a level of inhibition in agreement with those reported by Manley et al (16) for RNA-dependent synthesis. In contrast, there was no inhibition of ^{3}H-TTP incorporation by actinomycin D in the preparations subjected to ribonuclease digestion. Analyses of the treated samples by gel electrophoresis yielded identical labeling patterns (not shown) in the presence or absence of actinomycin D. From the requirements for inhibition by this antibiotic (17), we tentatively conclude that a double-stranded DNA is not serving as template after ribonuclease digestion of the resident RNA template. We are currently investigating both the nature and the possible role of the ribonuclease-resistant template of RIII MuMTV.

TABLE III

Effect of Actinomycin D on ^{3}H-TTP Incorporation of Untreated and Ribonuclease-Digested MuMTV

Conditions[a]			pmoles incorporated[b]	
Temperature	Time (min)	RNAse	- Act. D.	+Act. D.
26°C	60	-	1.32(100)	0.95(72.5)
26°C	60	24 μg/ml	1.84(100)[c]	2.05(111.5)

[a]Conditions of preincubation.
[b]As determined by procedure given under Table I.
[c]Numbers in parenthesis refer to percent.

EFFECT OF OLIGO dT ON DNA SYNTHESIS BY RIII MuMTV VIRIONS

Recent reports, demonstrating the presence of extensive polyadenylate (poly A) sequences in oncogenic RNA viruses (18 – 20), led us to investigate the effect of exogenous oligo $(dT)_{10}$ (P-L Biochemicals, Inc.) in order to determine to what extent oligo dT enhances incorporation of ^{3}H-TTP by functioning as primer on a poly A template. These results are directly applicable to the problem of assaying human milks for RNA-dependent DNA polymerase for two reasons: 1. they provide an additional criterion for deciding whether virions found in human milks share the common property of other oncogenic RNA viruses of extensive poly A sequences, and 2. they provide a means of increasing assay sensitivity in monitoring human milks for RNA-templated DNA polymerase activity.

Table IV demonstrates the effect of various concentrations of oligo $(dT)_{10}$ on the incorporation of ^{3}H-TTP by disrupted RIII MuMTV virions. Rather high concentrations of oligo dT are needed for maximal enhancement; however, Duesberg et al (21) have reported a 1:1 correspondence by weight of oligo dT to template RNA for maximal priming, employing a purified template-enzyme assay.

The effect of oligo dT addition (13.2 μg/ml) on DNA synthesis by untreated and ribonuclease-digested preparations, as analyzed by PAGE, is illustrated in Figure 6 (*top*). In untreated preparations a stimulation of 70S RNA-associated label was observed; however, the majority of the enhanced labeling was associated with lower

TABLE IV

Enhancement of DNA Synthesis by the Addition of Oligo $(dT)_{10}$

μg oligo $(dT)_{10}$ / ml	pmoles ^{3}H-TTP incorp.[a] / 60 min.	pmoles (+oligo dT) / pmoles (- oligo dT)
0	1.1	-
2.64	4.3	3.9
5.28	5.7	5.2
13.20	9.8	8.9
26.40	11.0	10.0
39.60	10.9	9.9

[a]Determined as TCA-precipitable counts (Table I).

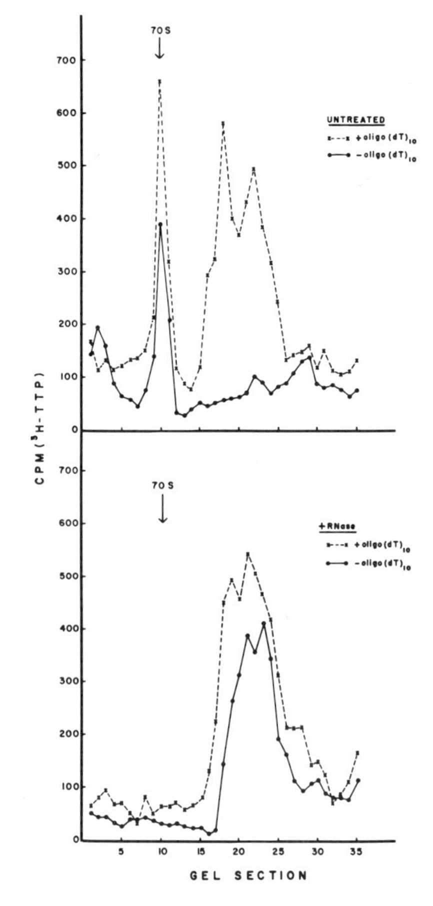

Fig 6. The effect of exogenous oligo $(dT)_{10}$ and pancreatic ribonuclease A on DNA synthesis by detergent-disrupted RIII MuMTV. Disrupted milk-borne virions were preincubated with (*bottom figure*) and without (*top figure*) ribonuclease, as indicated in the legend to Figure 4. Oligo $(dT)_{10}$ at a final concentration of 13.2 μg/ml was added where indicated just prior to incubation at 37°C for 60 minutes. Reaction conditions and method of analysis by PAGE are given in reference 12. These gels were sectioned at 1.2-mm intervals.

molecular weight complexes, which are coincident with the labeling pattern observed in preparations which had been subjected to ribonuclease digestion (Fig 6, *bottom*). These data suggest that similar low molecular weight labeling occurs under two types of conditions, i.e., either through digestion of the resident RNA or by addition of oligo $(dT)_{10}$.

As previously shown in Table II, omission of dNTP's greatly inhibited incorporation of ^{3}H-TTP. If oligo dT enhancement of incorporation is the result of priming complexed to poly A sequences, then a smaller diminution in incorporation should be observed when dNTP's are omitted, employing ^{3}H-TTP as label and oligo dT as exogenous primer. The expected results were observed and are given in Table V; the omission of all dNTP's, except for ^{3}H-TTP, still resulted in substantial incorporation. By comparing the gel patterns of labeled products derived from reactions in the presence or absence of oligo dT, with or without a complete complement of dNTP's and employing ^{3}H-TTP as label, the distribution of polyA sequences should be discernible. In Figure 7 (*top*) little or no labeling occurs when both oligo dT and dNTP's are omitted, indicating that no endogenous primer is complexed to a poly A sequence. In contrast, when ^{3}H-TTP labeling is performed in the presence of oligo dT (Figure 7, *bottom*) without a full complement of dNTP's, labeling is observed in the 70S region and in a low molecular weight region corresponding to approximately 250 - 400 nucleotides in length. It should also be noted that the major portion of oligo dT enhancement of incorporation in the low molecular weight region is eliminated. The latter suggests that this enhancement is not due to oligo dT priming from a poly A template.

The difficulty in obtaining labeled viral MuMTV RNA has prevented, for the present, obtaining direct evidence for poly A sequences in this virus; however, the above data constitute strong indirect evidence for such sequences in B particles, adding an additional common property between B- and C-type oncogenic RNA viruses.

TABLE V

Effect of Oligo (dT) Addition on the Deoxyribonucleoside Triphosphate Requirement of MuMTV

RNAse[a]	Conditions	pmoles incorp./60[b]	%
-	complete	10.65	(100)
+	complete	9.20	86.1
-	-dCTP	6.38	59.5
+	-dCTP	5.93	55.8
-	-d(C,A,G)TP	5.49	51.4
+	-d(C,A,G)TP	3.42	32.2

[a] Conditions of preincubation as given in Table II.
[b] Determined by procedure given under Table I.

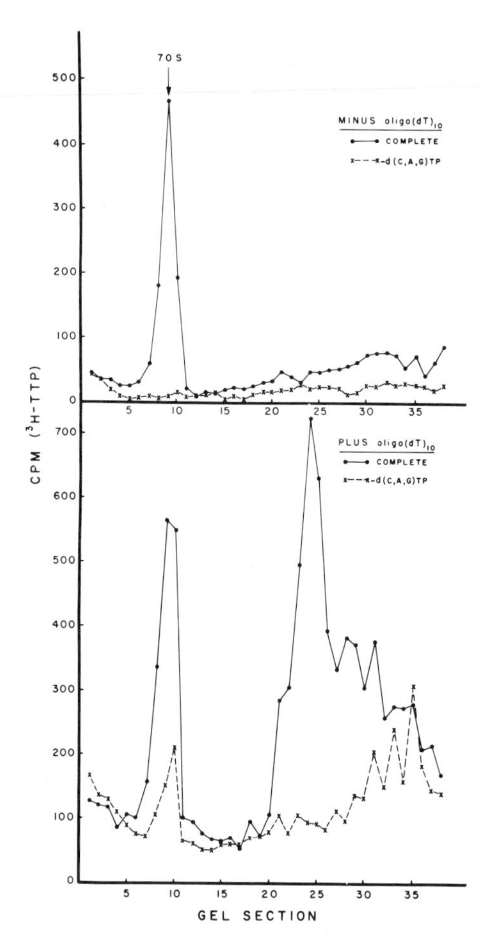

Fig 7. The effect of oligo $(dT)_{10}$ addition on DNA synthesis by disrupted RIII MuMTV in the presence and absence of nucleoside triphosphates. Oligo $(dT)_{10}$ was added at a final concentration of 13.2 μg/ml. The complete reaction mixture contained all four deoxyribonucleoside triphosphates at concentrations given in reference 12. -d(C,A,G) TP denotes the omission of deoxycytosine, deoxyadenosine, and deoxyguanosine triphosphates. All reaction mixtures contained ^{3}H-TTP (3,172 cpm/pmole). Gel sections were cut at 1.2-mm intervals.

PARTICLE MORPHOLOGY AND REVERSE TRANSCRIPTASE ACTIVITY FOUND IN HUMAN MILKS

As shown in Figure 8, three types of morphologically distinct particles have been reported to be present in human milks, and these have been designated as MS-1, MS-2, and MS-3 (5). MS-1 particles closely resemble B particles found in mouse milk relative to a number of parameters, e.g., average dimensions, and spike distribution and spacing. MS-2 particles resemble MS-1 particles in having spikes; however, the distribution of these spikes is uneven and these particles appear damaged. MS-3 particles appear similar to C-type particles with respect to both size and in having a smooth surface.

The relative frequency of particle types found in human milks is presented in Table VI. The majority of these milks contain particles, MS-3 being the most fre-

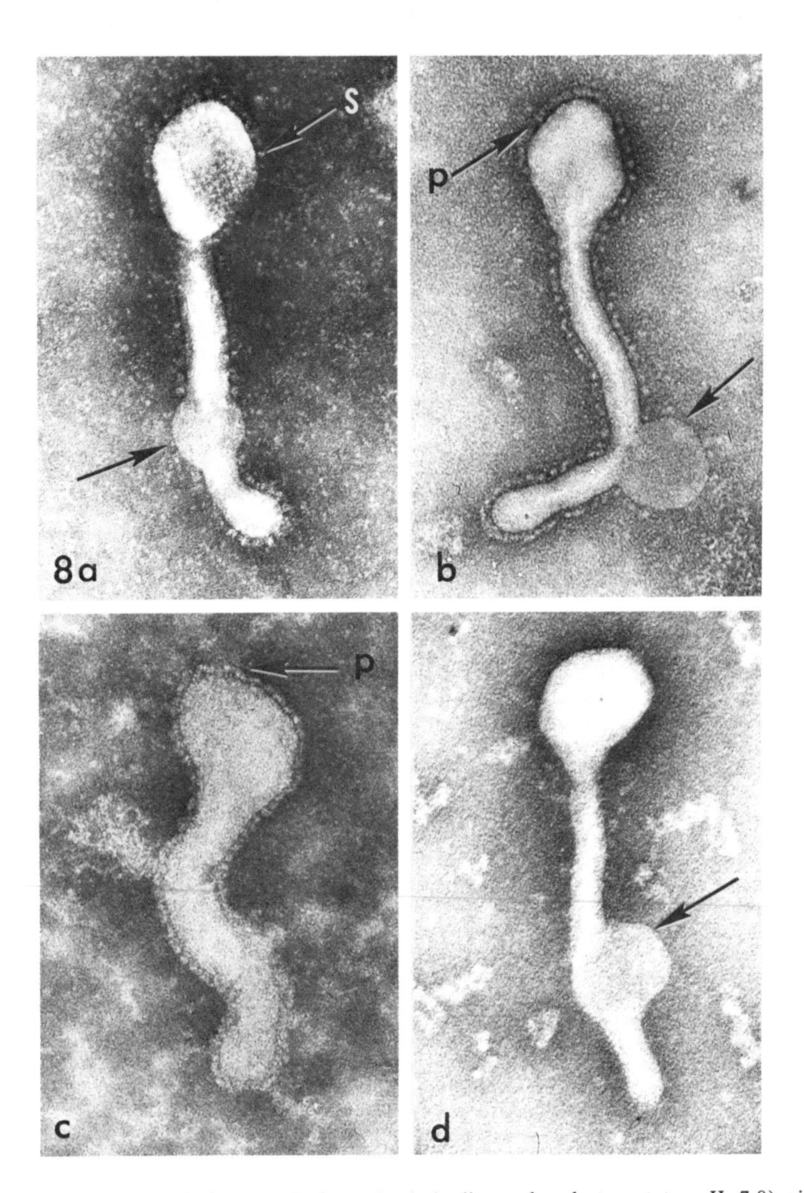

Fig 8. Morphology of the negatively stained (sodium phosphotungstate, pH 7.0) virus-like particles from human milk. Type MS-1 particle (Fig 8a) has a head and tail and is covered with spikes (S) about 100 Å long which resemble the spikes of the mouse mammary tumor virus. Type MS-2 particles (Figs 8b and c) have nonuniform projections (P) that are not identical with those of the MS-1 type. MS-3 particles (Fig 8d) also have a head and tail configuration, but the surface membrane is smooth, which is similar to the leukemia-sarcoma viruses. Blebs on tails (*arrow*) are observed on all of these particles, especially on type MS-3. Figures 8a and c are from Sarkar and Moore (5). Mag. 180,000 times. (Reduced 90%.)

quently observed. No human milk has ever been found to contain only MS-1 and MS-2 particles, in the absence of MS-3.

This distribution of particles and the damaged appearance of MS-2 particles suggests that perhaps some human milks contain a factor(s) destructive to the morphol-

TABLE VI

Relative Frequency of Particle-Types Found in Human Milks

Particle-type	Relative frequency
No particles	+
MS-1 only	–
MS-2 only	–
MS-3 only	+ + +
MS-1 and 2	–
MS-1 and 3	+
MS-2 and 3	+ +
MS-1, 2 and 3	+

ogy of MS-1 particles, resulting in the appearance of MS-2 and, possibly, some MS-3 particles (5). To test this, RIII whole milk containing MuMTV was diluted 1:10 with either whole or skimmed human milks and incubated for 18 hours at 37°C; the results are presented in Table VII. All of the human milks assayed resulted in some damage to RIII MuMTV with respect to both morphology and RNA-dependent DNA polymerase activity. In addition the cream fraction appeared to be most destructive to the parameters tested when whole or skimmed fractions from the same milk (H806) were assayed. Figure 9 illustrates the inhibitory activity of human milk on the RNA-

TABLE VII

Effect of Whole and Skimmed Human Milk on Reverse Transcriptase Activity of RIII MuMTV

Sample	Human milk	Intact/damaged virus	pmoles ^{3}H-TTP incorporated[a]	% Inh.
1	Control	2.30	1.05	(0)
2	H783 (skimmed[b])	0.76	0.32	69.5
3	H785 (skimmed[b])	0.72	0.31	70.4
4	H806 (skimmed[b])	1.08	0.66	37.0
5	H808 (skimmed[b])	0.97	0.39	62.8
6	F.D. (skimmed[b])	0.75	0.70	33.3
1	Control	1.41	1.00	(0)
2	H709 (whole)	0.64	0.73	27.0
3	H709 (skim[c])	0.75	0.65	35.0
4	H806 (whole)	0.00	0.13	87.0
5	H806 (skim)	0.35	0.59	41.0

[a] TCA-precipitable counts (Table I).
[b] Skimmed in the presence of EDTA.
[c] Skimmed in the absence of EDTA.

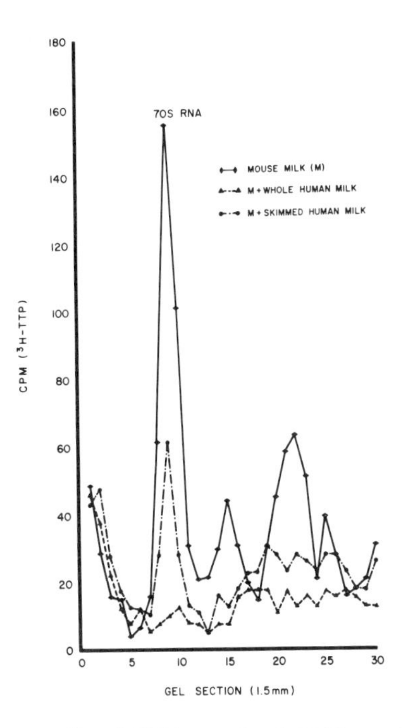

Fig 9. Inhibition by human milk of RNA-templated DNA synthesis by RIII MuMTV. RIII mouse milk was diluted 1:10 with either phosphate buffered saline (pH 7.4) or whole or skimmed human milk, and incubated 18 hours at 37°C. After pelleting the virions and detergent-disruption, RNA-directed DNA synthesis was determined according to the method of Dion and Moore (12).

templated DNA polymerase reaction of RIII MuMTV as analyzed by PAGE. Inhibition of RIII RNA-templated DNA synthesis by the addition of human milk has been reported (10).

With a knowledge of the complexity of human milks, an extensive study has been undertaken to determine whether a correlation exists between family history of breast cancer, particle morphology, and RNA-dependent DNA polymerase activity in a new series of human milk samples. At present we have examined 25 human milks, 11 were obtained from women with a history of breast cancer and 14 without.

For electron microscopic examination, 15 ml samples of human milks were prepared and fractionated as described (6). Six grids were prepared for each milk and these were examined by three electron microscopists, each scanning ten grid squares from two grids. For the assay of RNA-dependent DNA polymerase activity, 15 ml of milk were prepared for assay according to the method of Schlom, Spiegelman, and Moore (10), without trypsin treatment of the skim milk fraction. After a 30-minute incubation at 37°C in the presence of ^{3}H-TTP (6000 - 7000 cpm/pmole) the reaction was terminated, phenol/cresol extracted, and prepared for PAGE (12).

Only 1 sample, obtained from a woman without a prior history, contained an MS-1 particle which resembles MuMTV. From a total of 11 milks from women with histories, 8 (73%) contained MS-2 particles. If only milks from women with an immediate history, e.g., the mother, are considered, then a total of 5 out of 7, or 72%, contain MS-2 particles. Of the 14 milks obtained from women without a history, 9

(64%) contained MS-2 particles. Although the present sample size is relatively small, there is an indication of the lack of correlation between history of breast cancer and particles found in human milks. A similar conclusion had been reported earlier (5).

In general, when enzyme activity, i.e., RNA-dependent DNA polymerase is monitored by the simultaneous detection method by PAGE, a positive reaction is found for milks containing MS-1 particles; however, we have not accumulated sufficient numbers of these milks to establish a firm correlation. Of the 17 milks positive for MS-2 particles, 9 (53%) were also positive for enzyme activity; however, all of these milks also contained MS-3 particles. Therefore, it is not possible to definitely correlate enzyme activity with either MS-2 or MS-3 particles, since approximately the same percent (50%) of enzyme-positive milks contained MS-3 particles.

These preliminary studies point out the need for the development of techniques for the separation and characterization of the various particles present in human milks, especially in the case of MS-3 particles, which might indicate the presence of leukemia. Finally, our model studies with RIII MuMTV stress the need to establish the nature of inhibitors and factors responsible for particle damage and loss of RNA-templated DNA polymerase activity in human milks.

ACKNOWLEDGMENTS

The authors wish to express their appreciation for expert technical assistance rendered by the following: reverse transcriptase assays: Garland S. Fout and Mary Lou Orcutt; human milk collection and processing: Marilyn Mason; electron microscopic screening of human milk fractions: Anne Flacco and William Manthey.

REFERENCES

1. Moore, D. H.: Mouse mammary tumour agent and mouse tumours. Nature, 198:429, 1963.
2. Chopra, H. C., Bogden, A. E., and Zelljadt, I., et al: Virus particles in a transplantable rat mammary tumor of spontaneous origin. Europ. J. Cancer, 6:287, 1970.
3. Chopra, H. C., and Mason, M. M.: A new virus in a spontaneous mammary tumor of a rhesus monkey. Cancer Res., 30:2081, 1970.
4. Moore, D. H., Pillsbury, N., and Pullinger, B. D.: Titrations of bioactivity of fresh and treated RIII milk and milk fractions. J. Natl. Cancer Inst., 43:1263, 1969.
5. Sarkar, N. H., and Moore, D. H.: On the possibility of a human breast cancer virus. Nature, 236:103, 1972.
6. Moore, D. H., Sarkar, N. H., Kelly, C. E., Pillsbury, N., and Charney, J.: Type B particles in human milk. Tex. Rep. Biol. Med., 27:1027, 1969.
7. Moore, D. H., Charney, J., and Kramarsky, B., et al: Search for a human breast cancer virus. Nature, 229:611, 1971.
8. Charney, J., and Moore, D. H.: Neutralization of murine mammary tumour virus by sera of women with breast cancer. Nature, 229:627, 1971.
9. Schlom, J., Spiegelman, S., and Moore, D. H.: RNA-dependent DNA polymerase activity in virus-like particles isolated from human milk. Nature, 231:97, 1971.
10. Schlom, J., Spiegelman, S., and Moore, D. H.: Detection of high-molecular-weight RNA in particles from human milk. Science, 175:542, 1972.
11. Axel, R., Schlom, J., and Spiegelman, S.: Presence in human breast cancer of RNA homologous to mouse mammary tumour virus RNA. Nature, 235:32, 1972.

12. Dion, A. S., and Moore, D. H.: Gel electrophoresis of reverse transcriptase activity of murine mammary tumor virions. Nature New Biol. 240:17, 1972.
13. Schlom, J., and Spiegelman, S.: Simultaneous detection of reverse transcriptase and high molecular weight RNA unique to oncogenic RNA viruses. Science, 174:840, 1971.
14. Levinson, W. E., Varmus, H. E., Garapin, A.-C., and Bishop, J. M.: DNA of Rous sarcoma virus–its nature and significance. Science, 175:76, 1972.
15. McDonnel, J., Garapin, A. C., Levinson, W. E., et al: DNA polymerases of Rous sarcoma virus: delineation of two reactions with actinomycin. Nature, 228:433, 1970.
16. Manly, K. F., Smoler, D. F., Bromfeld, E., and Baltimore, D.: Forms of deoxyribonucleic acid produced by virions of the ribonucleic acid tumor viruses. J. Virol., 7:106, 1971.
17. Kahan, E., Kahan, F. M., and Hurwitz, J.: The role of deoxyribonucleic acid in ribonucleic acid synthesis. VI. Specificity of action of actinomycin D. J. Biol. Chem., 238:2491, 1963.
18. Lai, M. M. C., and Duesberg, P. H.: Adenylic acid-rich sequence in RNAs of Rous sarcoma virus and Rauscher mouse leukemia virus. Nature, 235:383, 1972.
19. Green, M., and Cartas, M.: The genome of RNA tumor viruses contains polyadenylic acid sequences. Proc. Natl. Acad. Sci. USA, 69:791, 1972.
20. Gillespie, D., Marshall, S., and Gallo, R. C.: RNA of RNA tumour viruses contains poly A. Nature New Biology, 236:227, 1972.
21. Duesberg, P., Helm, K. V. D., and Canaani, E.: Comparative properties of RNA and DNA templates for the DNA polymerase of Rous sarcoma virus. Proc. Natl. Acad. Sci. USA, 68:2505, 1971.

Note added in proof:

After submission of this manuscript for publication, two reports appeared confirming the presence of polyadenylate sequences in MuMTV (Schlom, J., et al.: Science 179:696, 1973), and the lack of endogenous primers complexed to polyadenylate sequences in C-type oncogenic RNA viruses (Reitz, M., et al.: Biochem. Biophys. Res. Comm. 49:1216, 1972).

21. THE RELEVANCE OF RNA TUMOR VIRUSES TO HUMAN CANCER[1]

S. Spiegelman, R. Axel, W. Baxt, S. C. Gulati, R. Hehlmann, D. Kufe and J. Schlom

Institute of Cancer Research and Department of Human Genetics and Development
College of Physicians and Surgeons
Columbia University, New York City 10032

INTRODUCTION

Over the past fifty years viral oncology has amassed a considerable fund of information on the B-type and C-type RNA tumor viruses. These have been implicated in a variety of cancers in a wide spectrum of animals. In many instances, definitive proof has been attained by inoculation of purified viruses in experiments that satisfy Koch's criteria for the identification of the causative agent of a particular disease. The issue of the present discussion centers on the following question: How much of the experience accumulated in animal viral oncology is transferable to human cancer?

It is obvious that until a human viral agent is isolated and a susceptible animal identified, the definitive experiment cannot be performed. Consequently, other experimental methodologies must be developed to provide information relevant to involvement of a virus in human cancer.

It was evident (1) that the discovery (2, 3) of a viral RNA-instructed DNA polymerase (reverse transcriptase) provided potentially powerful new molecular aids in the search for possible viral agents in human cancer. The presence of reverse transcriptase was soon shown to be a universal attribute of the RNA tumor viruses (4,5). Specific hybridizability of the synthetic DNA product to homologous viral RNA established (6) that the viral RNA was indeed functioning as the instructive template. It became evident that detection of the enzyme and use of its DNA product as a molecular probe could provide evidence of oncogenic viral information by procedures that were not previously available. From the results we describe below, it would appear that many of these expectations are being realized.

VIRAL-SPECIFIC RNA IN MOUSE MAMMARY CARCINOMA CELLS–A MODEL SYSTEM

If a suitable DNA probe were available, it was obvious that RNA-DNA hybridization (7) could be used to detect viral-specific RNA in human tumors. Analogous experiments have been performed with tissue culture cells infected with oncogenic DNA viruses (8,9). Similarly, viral-specific RNA has been found (10) in both

[1] The original research by the authors was supported by the National Institutes of Health, National Cancer Institute, Special Virus Cancer Program Contract 70-2049 and Research Grant CA-02332.

the nuclear and cytoplasmic fractions of mouse cells infected with the Moloney sarcoma-leukemia RNA virus complex.

A necessary prelude to any attempts at similar examinations of human tumors required a demonstration that these methods were applicable to tumors using a suitable animal system. In addition to the presence of similar particles in the milk (11), other parallels can be drawn between murine mammary tumor and human adenocarcinoma of the breast. The murine system was therefore adopted as an experimental model to develop procedures permitting the detection of viral-specific RNA in tumor tissue.

We have shown (6) that the murine mammary tumor virus (MMTV) contains a reverse transcriptase activity that can be used to generate radioactively labeled DNA complementary to MMTV-RNA. Before this DNA can be used as a probe for viral-specific RNA in tumor cells, it is important that it be adequately monitored for suitability in a hybridization test. In particular, it must be shown that it bands solely at the density of DNA in a Cs_2SO_4 gradient, and that it hybridizes to the homologous viral RNA and not to normal cellular RNA.

DNA was synthesized with a detergent-disrupted preparation of the murine mammary tumor virus (MMTV). The DNA product, freed of protein and RNA, was examined by equilibrium centrifugation in Cs_2SO_4 with the results described in Figure 1. It is evident that the ^{3}H-DNA synthesized bands in the expected region of this density gradient. It is important to emphasize that the alkali treatment must be rigorous enough to remove all the RNA present in the original reaction. Any residue of DNA-RNA hybrids formed during the reverse transcriptase reaction will make the product unusable as a probe for complementary RNA.

Figure 2 shows the outcome of annealing the ^{3}H-DNA product to the 70S RNA of the murine mammary tumor virus. As is usual in such hybridizations (6), approximately 50% of the DNA product is shifted to the RNA and hybrid regions of

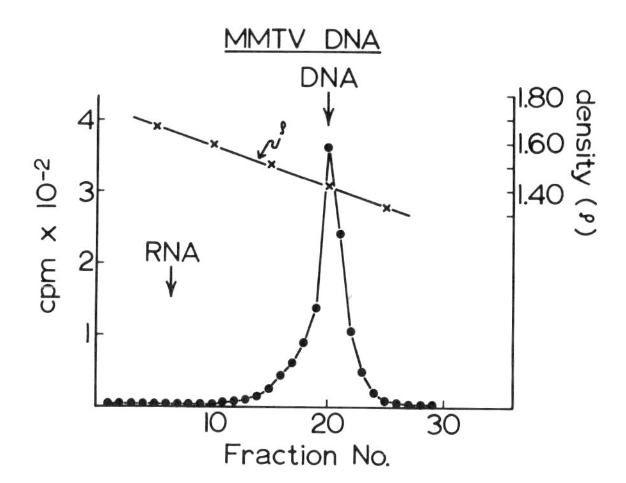

Fig 1. Cs_2SO_4 equilibrium density gradient centrifugation of the product of RNA-dependent DNA polymerase reaction of the mouse mammary tumor virus. The DNA product was synthesized and purified. It was then dissolved in 5.4 ml of 5mM EDTA, mixed with an equal volume of saturated Cs_2SO_4, and centrifuged at 44,000 rpm in a 50 Ti rotor (Beckman) for 60 hr at 20°C. 0.4-ml fractions were collected and processed for acid-precipitable radioactivity (12).

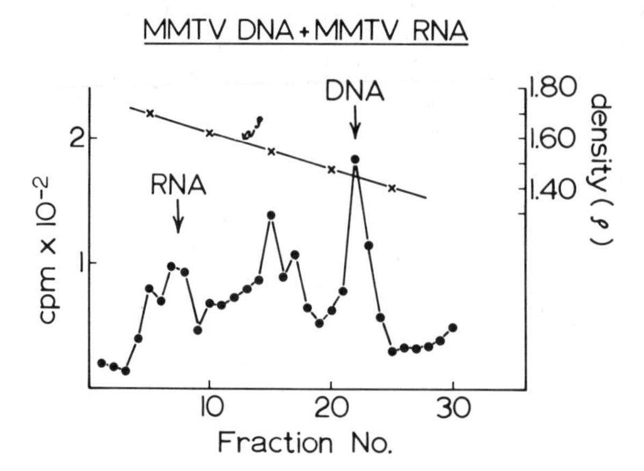

Fig 2. Cs_2SO_4 equilibrium centrifugation of viral ^{3}H-DNA after annealing to purified mammary tumor virus 70S RNA. The purified ^{3}H-DNA product (2000 cpm) was annealed to 2 μg of viral 70S RNA. After annealing, the reaction was subjected to Cs_2SO_4 gradient centrifugation as described in the legend to Figure 1 (12).

the density gradient. Except in very brief reactions, not all of the DNA formed is hybridizable to the RNA, since, in the absence of such agents as actinomycin D, DNA is synthesized, which is identical rather than complementary to the template.

Comparison of Figures 1 and 2 shows clearly that RNA molecules complementary to the ^{3}H-DNA product are readily detected by movement of the radioactive DNA toward the RNA region of the density gradient, due to the DNA:RNA hybrid complexes that form during the annealing reaction.

As an initial step, it was decided to see whether viral-specific RNA could be detected in the nuclear and cytoplasmic fractions of tumor tissue. RNA was prepared from nuclei. After the removal of nuclei, the supernatant was subjected to 180,000 times g centrifugation through a 25% sucrose column, as described (12). The pellet, which would contain both monosomes and polysomes, was used as a source of what is designated as polysomal RNA.

Annealing reactions with the cytoplasmic pellet fraction and the nuclear fraction gave results depicted in Figures 3A and B. It is evident that both fractions contain RNA molecules complementary to ^{3}H-DNA from mammary tumor virus. As a consequence of the annealing reaction, 30 - 35% of the ^{3}H-DNA has shifted to the regions of the gradient corresponding to RNA and hybrid density.

Several experiments were performed to examine the specificity of hybrid formation between pRNA from tumor cells and ^{3}H-DNA from mammary tumor virus. Figure 4A demonstrates that pRNA, prepared from normal mouse liver, exhibits no evidence of hybridization with mammary tumor viral ^{3}H-DNA. Further, breast tumor pRNA is unable to complex with ^{3}H-DNA homologous to the RNA of Rauscher leukemia virus, an oncogenic agent unrelated to the mouse mammary tumor virus (Fig 4B).

Finally, the hybridizability of mammary tumor viral ^{3}H-DNA to breast tumor pRNA was compared with its ability to complex to pRNA derived from normal breast tissue. The normal pRNA was prepared from lactating C57 female mice, a strain in

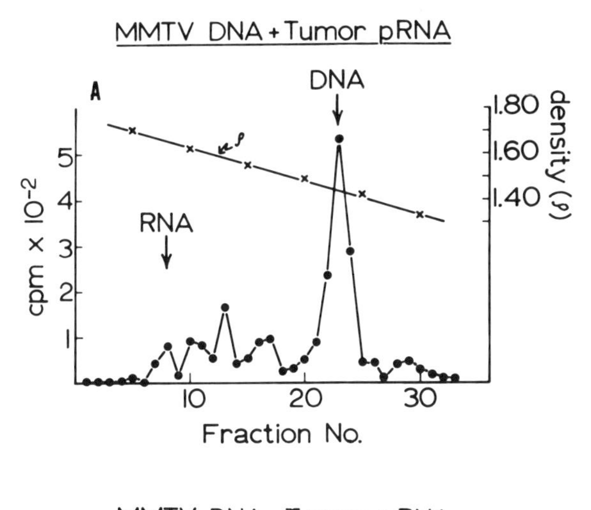

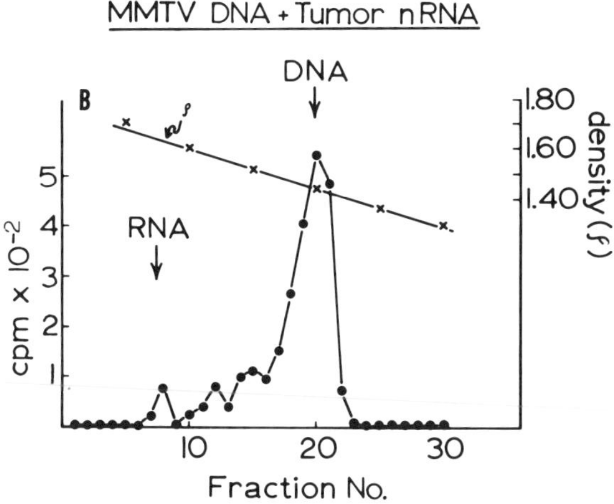

Fig 3. Cs_2SO_4 equilibrium centrifugation of viral ^{3}H-DNA after annealing to nuclear and polysomal RNA from mouse mammary tumors. (A) Polysomal RNA and (B) nuclear RNA were isolated from mouse mammary tumors, and 250 μg was annealed to viral ^{3}H-DNA at 37°C for 18 hr. The reactions were then subjected to Cs_2SO_4 equilibrium centrifugation (12).

which mammary cancer is extremely rare. The results are shown in Figure 5 in the form of a saturation curve. The pRNA from the tumor tissue exhibits its usual ability to hybridize to tumor virus DNA, with evidence of entering a saturation phase at about 100 μg. On the other hand, the pRNA from normal breast shows no evidence of hybridizability to the same DNA within the concentration range examined.

The experiments just described, detecting RNA-DNA hybrid formation between MMTV-DNA and mouse mammary tumor RNA, established the presence of viral-specific RNA in the nuclear and cytoplasmic fractions of tumor tissue. The negative response of breast tumor RNA to the unrelated DNA homologous to RLV-RNA (Fig 4) and the positive reaction with MMTV-DNA (Fig 3C) support the conclusion that the annealing reaction is specific. This is further strengthened by the failure to find RNA complementary to the MMTV-DNA in similar fractions derived from liver and normal breast tissues.

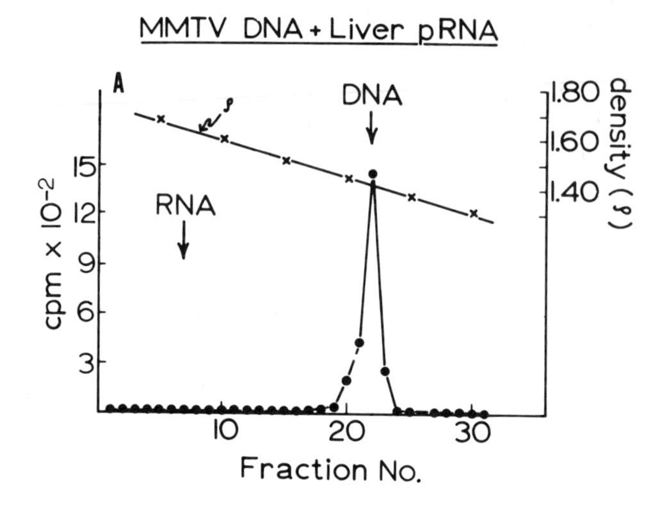

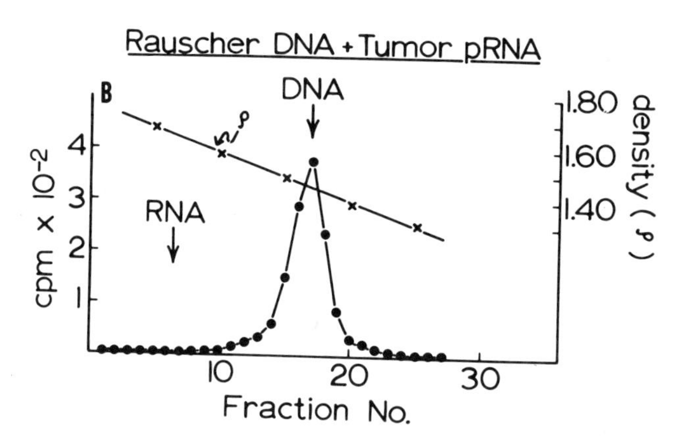

Fig 4A. Cs_2SO_4 equilibrium centrifugation of viral ^{3}H-DNA after annealing it with mouse liver polysomal RNA. Polysomal RNA was extracted from tumor-free female mice and 250 μg was annealed to purified viral ^{3}H-DNA (4000 cpm). After annealing, the reaction was subjected to Cs_2SO_4 equilibrium centrifugation, and fractions were collected and analyzed.

Fig 4B. Cs_2SO_4 equilibrium centrifugation of Rauscher leukemia virus ^{3}H-DNA after annealing to mouse breast tumor polysomal RNA. The virus was purified, and ^{3}H-DNA product was synthesized. 250 μg of mouse mammary tumor polysomal RNA was then annealed to Rauscher leukemia viral ^{3}H-DNA (2000 cpm), and the reaction was subjected to Cs_2SO_4 equilibrium centrifugation (12).

The techniques of molecular biology have been successfully used with purified viruses and cells in tissue culture. The experiments carried out with the mouse mammary cancer model demonstrate their applicability to actual tumor tissue. Because they provide us with the necessary technology to handle tumor specimens, the results have obvious implications for new experimental approaches to detect possible viral agents in human cancer.

MOUSE MAMMARY TUMOR VIRUS-SPECIFIC RNA IN HUMAN BREAST CANCER

It was evidently now technically feasible to obtain an experimental answer to the following question: "Can one find RNA molecules in human tumors that are

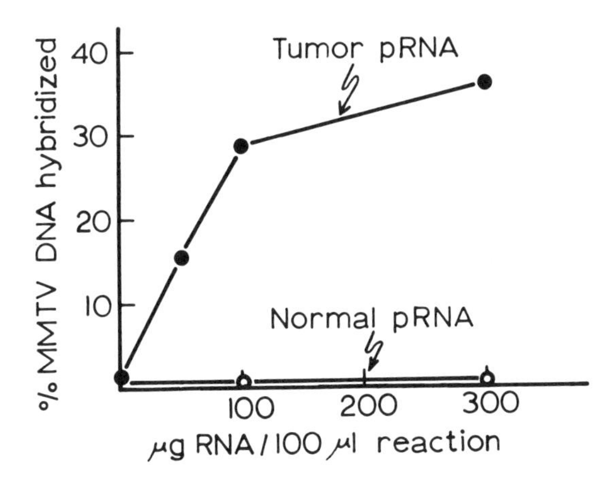

Fig 5. Comparison of annealing reaction between mammary tumor viral ^{3}H-DNA and mouse mammary tumor polysomal RNA and normal breast polysomal RNA. RIII mouse breast tumor polysomal RNA and polysomal RNA from lactating breasts of tumor-free C57 mice were extracted and annealed to ^{3}H-DNA from mammary tumor virus at various RNA concentrations. The individual annealing reactions were then analyzed by Cs_2SO_4 equilibrium centrifugation and the percent DNA hybridized was determined by the cpm of DNA sedimenting in the RNA and hybrid regions of the gradients. o——o normal polysomal RNA; •——• tumor polysomal RNA (12).

homologous to those of a putative human viral agent (for example, the particle found in human milk) or to an analogous animal virus of proven oncogenic potential?

A number of practical and theoretical considerations led to the choice of using the mouse mammary tumor virus (MMTV) for producing the radioactively labeled DNA to search for homologous RNA in human breast tumors. The human milk particles would have been a natural candidate had they been available in adequate quantities. Hopefully, this deficiency will eventually be rectified. Successful cultivation in tissue culture will lead to extraordinarily interesting experiments with human tumor material and permit a decision on the degree of homology between the human agent and the known animal mammary tumor viruses. At the time, the mouse virus was the only generally available agent of proven ability to induce breast cancer. Positive outcomes with it therefore provided information more directly relevant to the question of viral involvement in human breast cancer.

The synthesis of the MMTV ^{3}H-DNA and the rigorous proof of its specific complimentarity to MMTV 70S RNA and mouse tumors induced by MMTV were accomplished as described in the previous section. Extraction of RNA from human breast tumors and normal tissues was also as previously described (12,13). Figure 6A shows the sort of positive response one can observe when MMTV ^{3}H-DNA is annealed with pRNA preparations from human malignant breast tumors. This profile is in contrast with the negative outcomes observed when similar reactions were carried out with pRNA from normal breast tissue or from fibrocystic disease tissue (Fig 6B), where no MMTV ^{3}H-DNA is observed hybridized in the RNA region.

Since it is impractical to present the individual Cs_2SO_4 gradient profile of every tissue examined, a convention was adopted to permit a more convenient recording of our findings. To achieve the accuracy desired, 10-minute counts (cp10m) were taken. After correction for background counts, the sum of the ^{3}H counts in the RNA density region (between densities of 1.63 and 1.67 g/ml) was used to measure

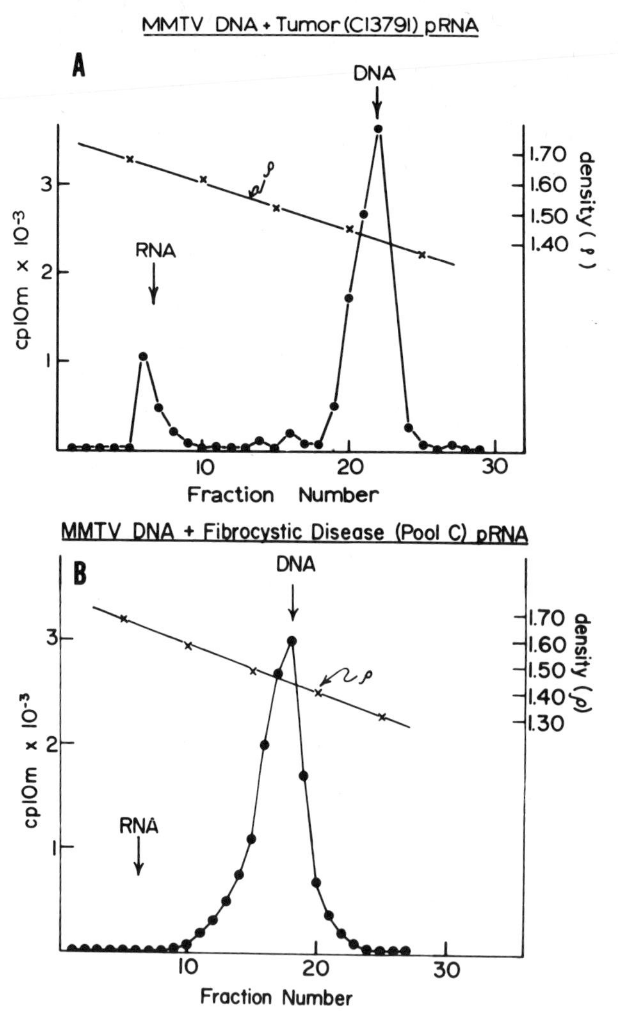

Fig 6. Cs_2SO_4 equilibrium density gradient centrifugation of MMTV ^{3}H-DNA after annealing to (A) 400 μg pRNA from human breast tumor (C13791) and (B) 400 μg pRNA from biopsy specimens of fibrocystic disease of the breast (13).

the amount of DNA complexed to RNA. An operational mean background and its standard deviation (S) was empirically determined by the total cp10m counts of 3 tubes in the negative regions (tubes 2,3,4) of each of 50 gradients. All specimens showing cp10m of less than 3 standard deviations in the RNA density region are considered as negative. Although this procedure may eliminate some positives, it provides a 99.9% confidence and better that those retained are significant. Figure 7 gives a convenient pictorial summary of the results obtained.

Of the 29 malignant tumors examined, 67% gave positive responses with values that can be assigned a better than one out of a thousand chance of being significant. None of the pRNA samples derived from normal breasts or the nonmalignant fibrocystic and gynecomastia tissues and fibroadenomas yielded positive reactions with the MMTV DNA. It should be noted that actually 21 different samples of fibrocystic disease tissues were examined in all. The small size of the specimens usually

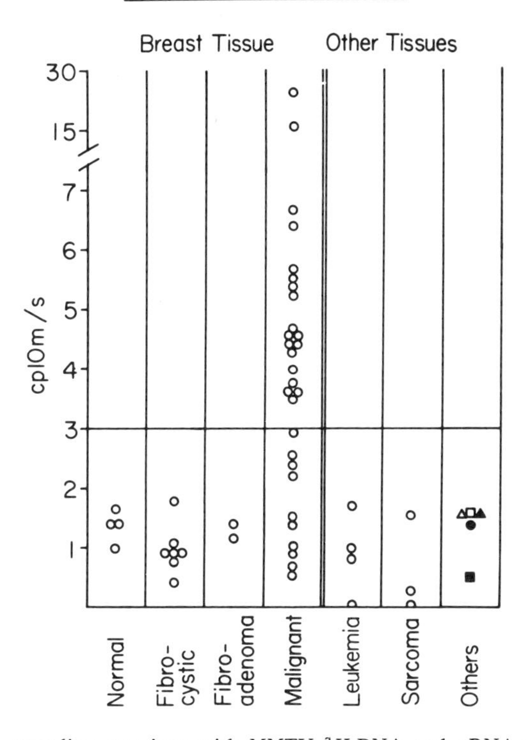

Fig 7. Results of annealing reactions with MMTV ^{3}H-DNA and pRNA from breast and other tissues. Annealing reactions were performed between MMTV ^{3}H-DNA and pRNA isolated from specimens of normal breast, fibrocystic disease of the breast, breast fibroadenomas, gynecomastia carcinoma of the breast, leukemic white blood cells, sarcomas, human placenta (□), human liver (○), and human intestine (▲). The reactions were then subjected to Cs_2SO_4 equilibrium density centrifugation. The amount of ^{3}H-DNA, expressed as cp10m corrected for background, banding in the density region of RNA (between densities 1.63 and 1.67) was determined for each reaction. An operational mean background and its standard deviation (S) was then determined from the cp10m of consistently negative portions (fractions 2,3,4) of fifty different gradient profiles. The number of DNA cp10m (corrected for background) banding in the RNA region of the gradient was then divided by the operational standard deviation (80 cp10m). Any reaction with cp10m in the RNA region less than three standard deviations is considered negative, thus providing 99.9% confidence that those reactions retained as positive (greater than 3 S) are significant (13).

available from patients with this disease made it necessary, except in one instance, to pool two or more in order to obtain sufficient pRNA for an adequate examination. Negative responses were also observed with RNA from normal human placenta, liver, and intestine.

Of great interest is the fact that positive responses were not observed with any of the human leukemias and sarcomas tested. Specificity can also be tested conversely. Breast tumor pRNA positive for a reaction with MMTV DNA can be challenged with a DNA homologous to the RNA of an unrelated oncogenic virus. We have shown that the synthetic DNAs homologous to the RNA of either the murine Rauscher leukemia virus (RLV) or of the avian myeloblastosis virus (AMV) do not cross hybridize with the corresponding RNA of the mouse mammary tumor agent. Consequently, RLV DNA and AMV DNA can be used to examine the specificity of the positive reaction observed between human breast cancer pRNA and MMTV DNA. The pRNA of three breast tumors showed positive reactions with MMTV ^{3}H-DNA and no detectable annealing reactions with RLV ^{3}H-DNA or avian myeloblastosis virus (AMV) ^{3}H-DNA.

What the experiments do imply is that there is information in the RNA of human breast tumors that is specifically homologous to the RNA of an animal breast tumor virus.

MOUSE LEUKEMIA VIRUS-SPECIFIC RNA IN HUMAN LEUKEMIC CELLS

The specific response between MMTV DNA and human breast tumors led us to undertake parallel investigations with human leukemic cells. An agent of established leukemogenic potential in mice, the Rauscher leukemia virus (RLV), was chosen to produce the radioactively labeled DNA required. This RLV ^{3}H-DNA probe was synthesized and purified as previously described and tested for annealing specificity to RLV 70S RNA and pRNA from spleen tumor cells infected with RLV.

The results of these experiments have been described in detail (14) and are summarized in Figure 8. Of 27 human leukemic white blood cell samples tested, 89% contained pRNA that specifically annealed to RLV DNA, but not to the DNA probes synthesized from MMTV or the unrelated AMV. Further, no control white blood cells or other normal tissues examined showed significant levels of annealing to the specific RLV RNA. It should be noted that white blood cells in the active phase of the disease of both acute and chronic leukemias showed positive reactions.

RLV-SPECIFIC RNA IN HUMAN SARCOMAS

The intimate association of leukemia viruses with sarcoma viruses of chickens, mice, cats, and monkeys led us to consider the possibility that a parallel situation may exist in the human disease. If correct, human sarcomas would also contain RNA that can specifically hybridize to DNA homologous to the RNA of the Rauscher murine leukemia virus. Our results (15) confirm this expectation. As shown in Figure 9, 68% of human sarcomas contain RNA that can hybridize to RLV DNA. These include fibrosarcomas, osteogenic sarcomas, and liposarcomas. Again, as with the leukemic cells, the sarcoma RNA does not hybridize with the DNA probes synthesized from MMTV or AMV. Again, RNA from normal human tissues, including fetal limb, did not show significant levels of hybridization with RLV DNA.

VIRAL-RELATED RNA IN HODGKINS' DISEASE AND OTHER HUMAN LYMPHOMAS

From the viewpoint of etiology and cellular pathology, lymphomas in mice are linked to the leukemias and sarcomas. In addition, it may be noted that some human lymphomas are accompanied by the clinical appearance of a peripheral leukemia. In any event, if human neoplasias continue to parallel those observed in mice, it might be expected that human lymphomas, like the leukemias and sarcomas, would contain RNA uniquely homologous to that of the mouse leukemia agent.

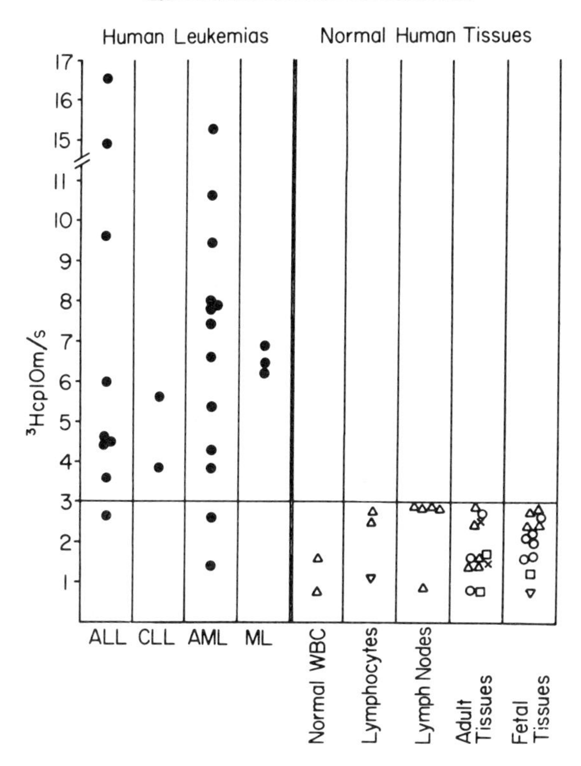

Fig 8. Results of hybridization reactions with RLV ^{3}H-DNA and pRNA from human leukemic white blood cells and normal human cells. The normal human tissues tested were: normal white blood cells, PHA-stimulated lymphocytes (△), the human lymphocyte cell line NC37 (▽), lymph nodes, other adult tissues: liver (△), spleen (X), intestine (○), striated muscle (□), and fetal tissues: liver (△), lung (▽), limbs (○), placenta (□). The reactions were subjected to Cs_2SO_4 equilibrium density centrifugation as described under Figure 1. The amount of RLV ^{3}H-DNA hybridized to cell RNA is expressed as described in the legend of Figure 7 (14).

We have found (16) a confirmation of this expectation. Human lymphomas, including Hodgkins' disease, lymphosarcomas, and reticulum cell sarcomas, have been found to contain RNA exhibiting homology to the RNA of the Rauscher leukemia virus, but not to the unrelated RNAs of either the mouse mammary tumor virus or of the avian myeloblastosis virus. As summarized in Figure 10, 69% of the RNAs derived from lymphomas yielded positive hybridizations with RLV DNA, whereas none of the 48 control tissues was positive. Furthermore, Hodgkins' lymphoma RNAs showed no significant reactions for either AMV DNA or MMTV DNA.

POSSIBLE COMPLICATIONS IN MOLECULAR HYBRIDIZATIONS

The experiments described in our hybridization studies on breast cancer, leukemia, sarcomas, and lymphomas were designed to detect in the neighborhood of 0.001 μg or less of complementary RNA in the pRNA preparations being tested. The sensitivity of the method is therefore being pushed toward its limit. Under the circumstances, it is necessary to exercise extreme care in the preparation of both the

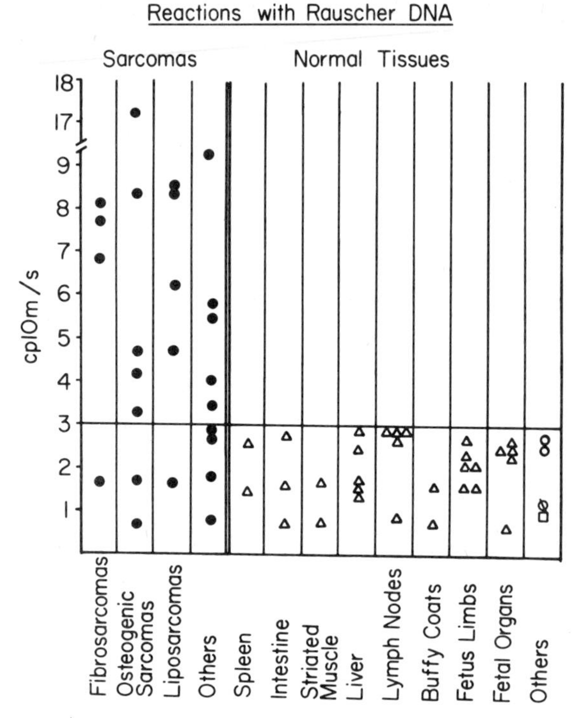

Fig 9. Results of hybridization reactions with RLV ^{3}H-DNA and pRNAs from human sarcomas and normal human cells. The pRNAs were isolated from sarcoma specimens (fibrosarcomas, osteogenic sarcomas, liposarcomas, leiomyosarcomas, neurofibrosarcomas, and rhabdomyosarcomas), normal adult tissues (spleen, intestine, striated muscle, liver, lymph nodes, and white blood cells), fetal tissues (limbs, liver, and lungs), placenta (ϕ), NC37 cells (□), and PHA-stimulated lymphocytes (○). The reactions were then subjected to Cs_2SO_4 equilibrium density centrifugation as described in Figure 4. The amount of RLV ^{3}H-DNA hybridized to cell RNA is expressed as described in the legend to Figure 7 (15).

pRNA and the ^{3}H-DNA used as a probe. Complete removal of the contaminating protein from the pRNA is necessary to avoid nonspecific trapping of ^{3}H-DNA in nonrelevant portions of the density gradient.

To insure the interpretability of the results, it is imperative that every purified ^{3}H-DNA preparation used be monitored for adequacy by being subjected to the following tests: 1. It must be shown to band cleanly in the appropriate DNA region of the density gradient to insure that it is free of contaminating viral RNA; 2. It must be extensively hybridizable to the homologous viral RNA used as the template in the original synthesis. This can exceed 90% in the presence of excess viral RNA if actinomycin D, at a level of 100 μg per ml, is included to prevent DNA-instructed DNA synthesis, a possibility particularly prevalent in crude enzyme preparations; and 3. The ^{3}H-DNA product should not hybridize to unrelated viral RNA or to RNA from normal tissues. We customarily use the RNAs from AMV, RLV, and MMTV as test objects for cross hybridizability checks. A suitable synthetic DNA complementary to any one of them will not hybridize to either of the other two or to pRNA from normal tissue. If a synthetic DNA passes this test, then specific hybridization to one of the challenging viral RNAs cannot be explained in terms of complexing with stretches of A, since all tumor virus RNAs contain such stretches (17 – 19).

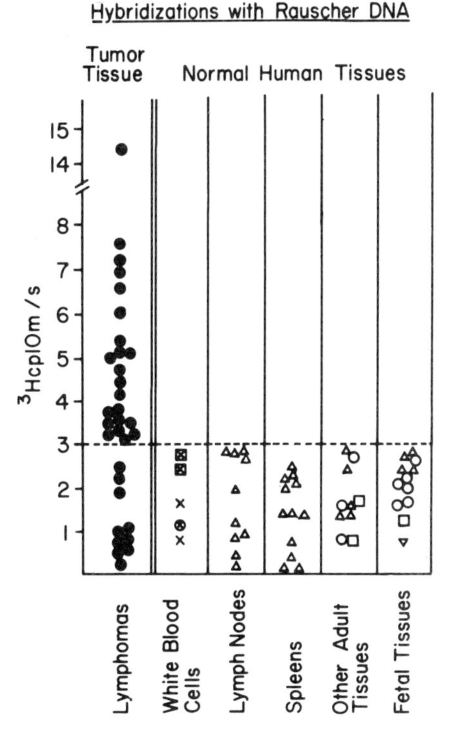

Fig 10. Results of hybridization reactions with RLV ^{3}H-DNA and pRNA from human lymphomas and normal human cells. The normal tissues tested are: normal white blood cells, PHA-stimulated lymphocytes (△), a human lymphocyte cell line, NC37 (▽), lymph nodes, spleens, other adult tissues: (liver (△), intestine (○), striated muscle (□), and fetal tissues: liver (△), lung (▽), limbs (○), placenta (□). The reactions were then subjected to Cs_2SO_4 equilibrium density centrifugation as described in Figure 4. The amount of RLV-^{3}H-DNA hybridized to cell RNA is expressed as described in the legend of Figure 7 (16).

It is not a trivial matter to obtain nucleic acid preparations that will satisfy all of these criteria. It is unfortunately not uncommon to obtain DNA preparations that fail the third requisite of specific hybridizability to its homologous RNA. This failure may stem from slight contamination with cellular DNA in the synthesis mixture, in which case the product will contain labeled DNA that will hybridize to both normal and tumor RNA. Another source of confusion is the possible complementary copying of long stretches of A, which are present in oncogenic RNAs and in normal cellular messages. ^{3}H-DNA with the resulting blocks of T will hybridize to any RNA containing corresponding blocks of A. One can readily test for the occurrence of this type of pairing by examining the effect of prehybridizing with unlabeled oligo rU.

Every ^{3}H-DNA used as a probe in the present and previous studies successfully passed the three tests outlined above. We cannot overemphasize the need for exercising the precautions noted in investigations along these lines. Unless the materials used satisfy the criteria stipulated, the resulting experiments are likely to generate more confusion than information.

Finally, it may be worth explicitly noting one limitation inherent in the kind of hybridization we can presently perform. The DNA probes generated by the reverse transcriptase do not constitute complete copies of the viral RNAs used as templates. Further, the degrees of homology to the RNAs of the animal viruses and to that of the putative human agents are unknown. Both of these features limit the effectiveness of these probes as detecting devices. Under these circumstances too much weight cannot be given to negative outcomes. When more efficient probes are synthesized, many of the issues considered in the experiments will have to be reexamined.

IMPLICATIONS OF VIRAL-SPECIFIC RNA IN HUMAN TUMORS

In many ways the most noteworthy feature of our studies of human neoplasias emerge when they are compared with one another, and to this end a composite of the results we have thus far accumulated is presented in Table I. Human breast cancer contains RNA homologous only to that of the murine mammary tumor virus. Human leukemias, sarcomas, and lymphomas all contain RNA homologous to that of the murine Rauscher leukemia virus and not to that of MMTV RNA. Finally, none of the human tumors contain RNA detectably related to the RNA of the avian myeloblastosis virus.

TABLE I

Homologies among Human Neoplastic RNAs and Animal Tumor Viral RNAs

Viral RNA	Human neoplastic RNAs			
	Breast cancer	Leukemia	Sarcoma	Lymphoma
MMTV	+	–	–	–
RLV	–	+	+	+
AMV	–	–	–	–

Note: The results of molecular hybridization between ^{3}H-DNA complementary to the various viral RNAs and pRNA preparation from the indicated neoplastic tissues. The plus sign indicates that hybridizations were positive and the negative sign, that none could be detected.

It is clear that with all of the human neoplasias examined, the specificity pattern of the unique RNA they contain is in complete accord with what has been observed in the corresponding viral-induced malignancies in the mouse.

The existence of MMTV-related RNA in human breast cancer and RLV-related RNA in human leukemia, sarcoma, and lymphomas does not, of course, establish a viral etiology for these diseases. One must now perform experiments designed to answer the following questions: 1. How large is the RNA being detected? 2. Is the viral-related RNA associated with a reverse transcriptase and is it located in structures characteristic of incomplete or complete virus particles? The requisite techniques have been developed to answer these questions with respect to the human neoplasias we have studied. We now turn our attention to a description of these techniques and their application to the analysis of malignant tissues.

THE SIMULTANEOUS DETECTION TEST FOR REVERSE TRANSCRIPTASE AND HIGH MOLECULAR WEIGHT RNA

Ideally, a method for detecting the presence of an RNA tumor virus in biological fluids should be simple, sensitive, and sufficiently discriminating so that a positive outcome can be immediately taken as an acceptable signal for the presence of the viral agent being sought. It is obvious that certainty can be increased by devising a test that simultaneously identifies two diagnostic features.

The oncogenic RNA viruses exhibit two biochemical properties unique to them as a group. They possess a large single-stranded RNA molecule with a sedimentation coefficient of 60 to 70S (20), often referred to as high molecular weight (HMW) RNA. They also contain "reverse transcriptase," an enzyme capable of using the viral RNA as a template to generate a DNA complementary copy (2,3,6).

The possibility of a concomitant test for 70S RNA and reverse transcriptase was suggested by previous (6,21,22) studies of the early reaction intermediates. The initial DNA product was found complexed to the 70S-RNA template. The structure could be detected by the unusual position of the newly synthesized small ^{3}H-DNA products in Cs_2SO_4 equilibrium, glycerol velocity gradients, and acrylamide gel electrophoresis. That these complexes were held together by hydrogen bonds was demonstrated by mild denaturation (heating for 10 min at 68°C in 50% formamide), following which the ^{3}H-DNA was found in its expected positions in both the Cs_2SO_4 and glycerol gradients (6,21).

It seemed probable that these features of the early stages of the reaction could be used to design the assay required by subjecting isotopically labeled products to sedimentation analysis. If the labeled DNA product is found complexed via hydrogen bonds to a 70S RNA molecule, evidence is provided for the presence of a reverse transcriptase using a 70S-RNA template and, hence, for an RNA tumor virus in the material being examined. It was on this basis that Schlom and Spiegelman (23) developed the simultaneous detection test which we now describe.

Pellets containing suspected virus are resuspended in 0.33% of the detergent NP-40 and 0.1 M dithiothreitol (DTT). After incubation for 10 min at 4°C to achieve virus disruption, a standard 15-min reverse transcriptase reaction is carried out with an aliquot. After the product is freed of protein by treatment with sodium dodecyl sulfate and phenol, it is subjected to a sedimentation analysis in a 10 to 30% glycerol gradient with suitable size markers to determine the apparent size distribution of the DNA synthesized.

Assays of this nature were carried out on milks from RIII mice, which carry the mouse mammary tumor virus, and from C-57 mice that are free of this agent. The results obtained are shown in Figures 11A and 11B, the legend of which describes all the pertinent experimental details. It is clear that only the milk known to contain the mammary tumor virus (Fig 11A) shows evidence of tritiated DNA in the 70S position of the gradient. Note, however, that both gradients contain tritiated DNA sedimenting at 6S or less, ascribable to the considerable content of cellular fragments, a constant feature of all milks. These fragments are invariably associated with cellular DNA polymerases as well as template DNA, a combination that can, and does, lead to the synthesis of radioactively labeled DNA during the assay. However, this introduces no

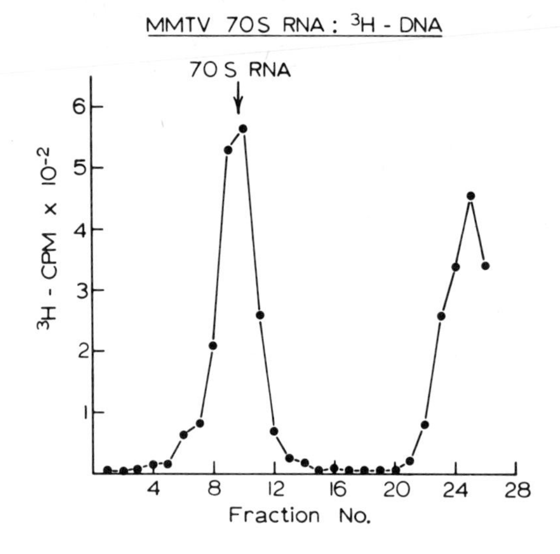

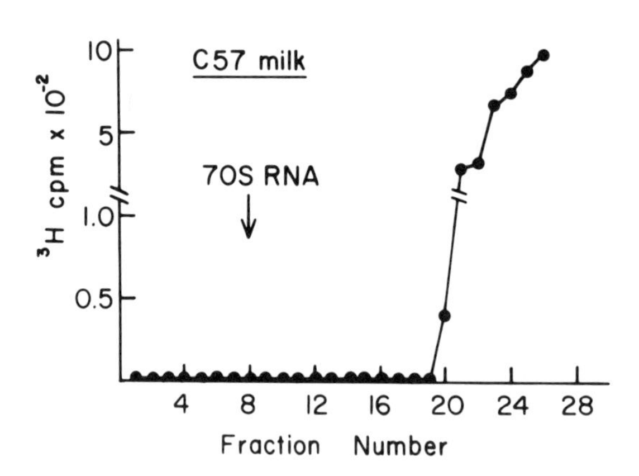

Fig 11. Detection of the HMW-RNA:^{3}H-DNA complex of MMTV in mouse milk. One ml of milk and one ml of 0.15 M EDTA (pH 7.5) were mixed and centrifuged at 3,000 X g for 10 min at 4°C. The clear "milk-plasma" zone between the lipid and precipitated casein layers was removed X g for 1 hr at 4°C. The resulting pellet was resuspended in 45 μl of 0.01 M Tris (pH 8.3) times g for 1 hr at 4°C. The resulting pellet was resuspended in 45 μl of 0.01 M Tris (pH 8.3) containing 0.33% Nonidet P-40 (Shell) and 0.1 M dithiothreitol (DTT) and kept at 4°C for 10 min. This was then added to a standard reverse transcriptase reaction mixture (125 μl final volume) containing in μmoles: 6.25 Tris HCl (pH 8.3), 1 of $MgCl_2$, 1.25 NaCl (instead of KCl to avoid precipitation when SDS is added in a subsequent step), 0.2 each of the unlabeled deoxyribonucleoside triphosphates (dATP, dGTP, dCTP) and ^{3}H-dTTP to a final specific activity of 8,900 cpm/pmole. All reactions were carried out at 37°C for 15 min and terminated by the addition of NaCl and sodium dodecyl sulfate (SDS) to final concentrations of 0.2 M and 1% respectively. After addition of an equal volume of phenol-cresol mix, the mixture was shaken at 25°C for 5 min and centrifuged at 5,000 X g for 5 min at 25°C. The aqueous phase was then carefully layered over a preformed 10 to 30% linear glycerol gradient and centrifuged for 40,000 rpm for 3 hr at 4°C in a SW-41 (Spinco) rotor. External markers employed were ^{3}H-70S avian myeloblastosis virus RNA and tritium-labeled 28S and 18S RNA from NC-37 cells. Fractions were collected from below and aliquots assayed for acid-precipitable radioactivity. (A) RIII milk, (B) C57 milk (23).

ambiguity in this assay, since the DNA synthesized by a contaminating cellular enzyme is readily distinguished by its gradient position from the DNA synthesized on the large viral RNA template.

A number of simple experiments can be performed to establish that the labeled DNA found in the 70S region is in that part of the gradient because it is complexed to a large RNA molecule and that it is formed as a result of the reverse transcriptase reaction.

The 70S region of the gradient in Figure 11A is pooled, precipitated, and redissolved. Half the sample serves as a control and the other is exposed to mild denaturation to rupture hydrogen bonds (incubation at 68°C for 10 min in 50% formamide). The control and the treated samples are then subjected to cesium sulfate equilibrium centrifugation. It is evident that the ^{3}H-DNA of the untreated aliquot (Fig 12) resides in the RNA region (1.650 – 1.680) of the density gradient, as would be expected if small DNA pieces are complexed to relatively large RNA molecules. On the other hand, a profile of the denatured sample shows that all the ^{3}H-DNA has shifted to the DNA region (1.420 – 1.450) of the equilibrium gradient.

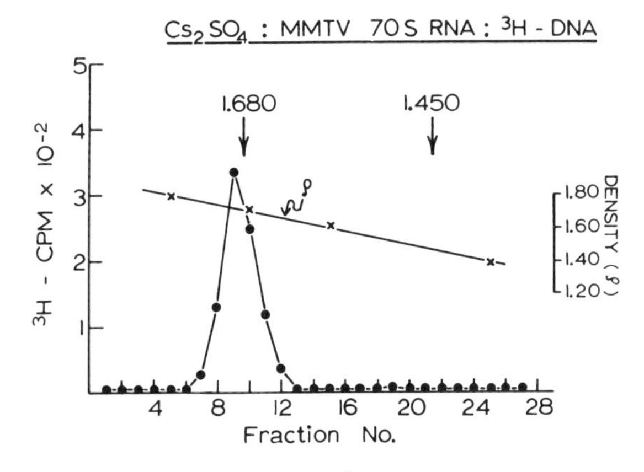

Fig 12. Cs_2SO_4 analysis of the HMW-RNA: ^{3}H-DNA complex. The 60 – 70S region of the Figure 11A was pooled, precipitated by the addition of 2 volumes of ethanol and yeast RNA carrier (15 μg/ml final concentration), stored at -20°C overnight, and centrifuged at 16,000 X g for 20 min at 0°C. The resulting pellet, resuspended in 0.002 M EDTA and saturated Cs_2SO_4 to give a final density of 1.550 g/cm^3 was centrifuged at 44,000 rpm in a 50 Ti rotor (Spinco) for 60 hr at 20°C; fractions were collected and processed for acid-precipitable radioactivity (23).

Such features as sensitivity to ribonuclease and the requirement for all four deoxyribonucleoside triphosphates can also be employed to demonstrate that the appearance of the 70S RNA:^{3}H-DNA complex is in fact the result of a reverse transcriptase reaction. If the reaction described in Figure 11 is carried out after preincubation of the enzyme preparation with ribonuclease, the glycerol velocity gradient shows no evidence of any tritiated DNA in the high molecular weight region of the gradient. Furthermore, we have shown (23) that if one of the deoxyribonucleoside triphosphates is omitted, no DNA is found in the 70S position in a velocity gradient.

It is of some interest to emphasize the increased sensitivity that derives from using the gradient analysis of the product. For example, the total reaction of Figure 11 was examined for sensitivity to ribonuclease by incorporation of tritiated TTP into acid-insoluble counts, and very little difference was found. Had we depended only on this observation, we would have concluded that there was little or no reverse transcriptase activity in the material. This is not an uncommon situation when examinations are made with crude material, because of large amounts of cellular DNA polymerases and DNA templates usually present. Sedimentation analysis of the reaction product, however, quickly reveals the presence of the reverse transcriptase reaction and its sensitivity to ribonuclease.

It is well known (24) that the 70S RNA of oncornaviruses is easily converted into components that sediment at 35S. It is therefore not surprising to find that occasionally, when a sedimentation analysis of a reaction is run, tritiated DNA is found complexed to molecules sedimenting at either 70S or 35S, or both.

APPLICATION OF THE SIMULTANEOUS DETECTION TEST TO HUMAN MILK PARTICLES

Particles morphologically similar to the type-B mouse mammary tumor virus (MMTV), Mason-Pfizer monkey virus, and type-C murine leukemia-sarcoma viruses have been observed in samples of human milk (25,26). We have recently shown (11) that human milks contain particles with a density of 1.16 to 1.19 g/ml and that these particles contain the RNA-instructed DNA polymerase. The demonstration of the viral origin of this reverse transcriptase activity in human milk is of obvious importance, since it has been shown that the causative agents of both murine mammary tumors and murine leukemia appear in, and can be transmitted via, mother's milk. We undertook to see whether particles isolated from human milk contained 60 – 70S RNA, and whether this RNA is used as a template in the reverse transcriptase reaction (27).

The simultaneous detection of HMW RNA and reverse transcriptase in the human material is identical to that used for mouse milk. A distinct peak of acid-precipitable ^{3}H-TTP, observed in the 60 – 70S or 35S region of the gradient, constitutes a positive. Numerous human milk preparations have been examined with this assay and Figure 13 shows a representative positive outcome. A ^{3}H-DNA:RNA complex is observed sedimenting at 67S in a milk from a woman with a strong familial history of breast cancer.

The results we obtained provided evidence that human milks contain particles that exhibit two biochemical characteristics unique to the known RNA tumor viruses, a 70S RNA template associated with a reverse transcriptase.

APPLICATION OF THE SIMULTANEOUS DETECTION TEST TO MALIGNANT TISSUES

Attempts to detect reverse transcriptase and 70S RNA in tumor extracts are inevitably complicated by the presence of cellular nucleases, polymerases, and RNA. The fractionation procedure developed with the aid of the mouse mammary tumor as a model was designed (28) to minimize these sources of confusion by disruption of the cells in the presence of EDTA to destroy ribosomal structures and the prior removal of

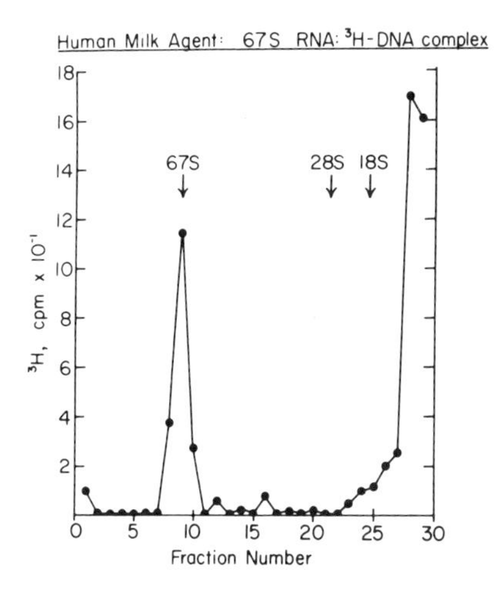

Fig 13. Detection of HMW-RNA ^{3}H-DNA complex in human milk. (a) 75 ml of human milk and 75 ml of 0.15 M EDTA (pH 7.5) were mixed and centrifuged at 3000 g for 10 min. The clear zone between the lipid and precipitated casein layers was removed and incubated at 37°C for 30 min in the presence of trypsin (Worthington Biochemicals, Freehold, New Jersey) at a final concentration of 1 mg/ml. Lima bean trypsin inhibitor was then added to a final concentration of 0.5 mg/ml and the sample was layered over an 8 ml 20% glycerol column in a SW27 (Spinco) centrifuge tube. After centrifugation at 98,000 g for 60 min at 4°C, the resulting pellet was resuspended in 45 μl of 0.01 M tris(hydroxymethyl)aminomethane (pH 8.3) containing 0.33% NP-40 detergent and 0.1 M (DTT) and kept at 4°C for 10 min. This suspension was then added to a standard reverse transcriptase reaction mixture and processed as described in Figure 11 (27).

nuclei and mitochondria. The resulting supernatant fluid is then layered on a 20% glycerol column and centrifuged at 100,000 times g for one hr, to yield a pellet (P-100) that should contain principally membrane structures and viral particles. Further, when an equivalent quantity of normal lactating breast tissue from tumor-free NIH Swiss mice is processed to yield an identical P-100 fraction, it exhibits no ability to incorporate ^{3}H-TTP into rapidly sedimenting structures (Fig 15).

A representative outcome of a simultaneous detection assay on the P-100 pellet from a mouse mammary tumor extract is shown in Figure 14A. Here we see a well-defined peak of ^{3}H-labeled DNA in the 70S position of the glycerol gradient. If the P-100 fraction (or the nucleic acid derived from it) is pretreated with ribonuclease, no evidence of DNA product is seen in the 70S region of the gradient (Fig 14B).

We have further shown (28) that when the rapidly sedimenting (70S) ^{3}H-DNA complex synthesized by mouse mammary tumor extract is subjected to Cs_2SO_4 equilibrium density gradient centrifugation, all of the ^{3}H-DNA synthesized bands at the density of RNA (1.64 - 1.68 g/ml). When this 70S complex is denatured by heating to 98°C for 10 min to eliminate hydrogen bonding, all of the ^{3}H-DNA bands at the density of DNA (1.44 g/ml).

The response to ribonuclease and the behaviors in glycerol and Cs_2SO_4 gradients before and after heat denaturation all serve to demonstrate a ribonuclease-sensitive synthesis of an RNA:^{3}H-DNA complex with a sedimentation coefficient of 70S. These findings, however, do not yet establish that a reverse transcriptase has been

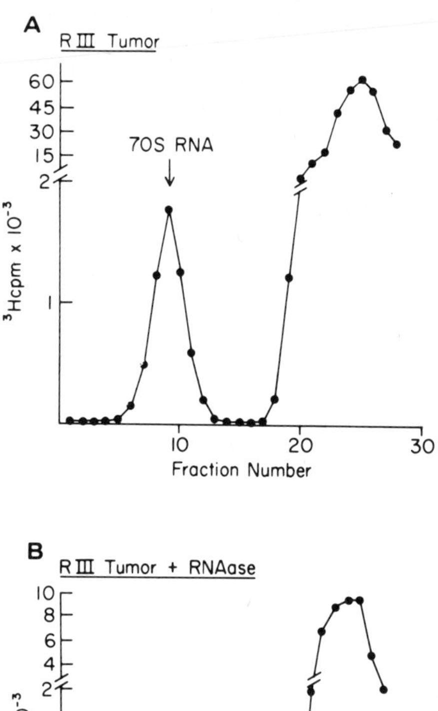

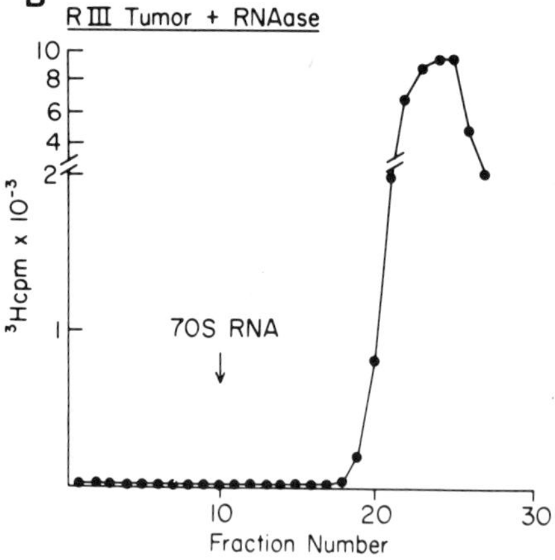

Fig 14. Detection of 70S RNA-^{3}H-DNA complex in mouse mammary carcinoma tissue. (A) A standard 125 μl endogenous reverse transcriptase reaction was performed on the P-100 fraction derived from a mouse mammary tumor (Methods). After incubation for 15 min at 37°C, the nucleic acid complex was extracted with phenol-cresol, pH 8.6, and the aqueous phase was layered over a linear glycerol gradient (10 to 30% glycerol in TNE) and centrifuged for 210 min, 4°C, at 40,000 rpm in a Spinco SW-41 rotor. 0.4 ml fractions were collected and assayed for acid-precipitable radioactivity. (B) Following detergent-disruption, the P-100 fraction derived from mouse mammary tumor was preincubated in the presence of RNase A (50 μg/ml) and RNase 1 (50 μg/ml for 15 min at 25°C. A standard endogenous reverse transcriptase reaction was then performed as described above (28).

identified in the tumor cell extracts. Rigorous proof requires a demonstration that the DNA synthesized can back hybridize to the putative template, which in the present case is the 70S RNA of the mouse mammary tumor virus. To this end, a standard endogenous reverse transcriptase reaction was performed with the P-100 fraction from mouse mammary tumors. Following velocity centrifugation analysis of the reaction products, the ^{3}H-DNA sedimenting in the 70S region of the glycerol gradient was

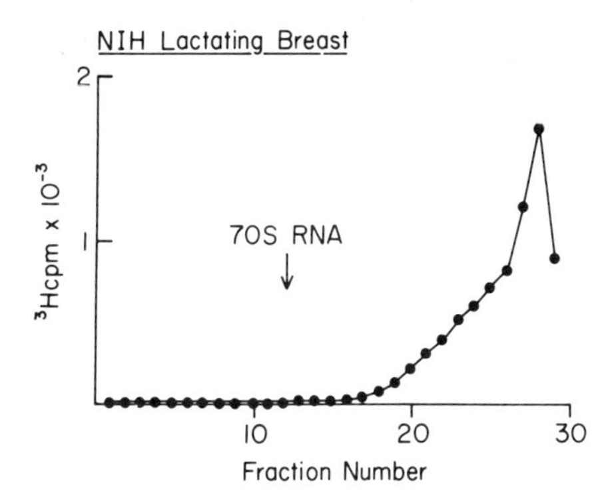

Fig 15. Assay for 70S RNA-^{3}H-DNA complex in lactating breast tissue. Five grams of lactating breast tissue from tumor-free lactating NIH mice were processed for 70S RNA-^{3}H-DNA as described in the legend to Figure 14 (28).

pooled and precipitated with ethanol. The resulting nucleic acid pellet was then subjected to extensive alkali digestion to remove all RNA present.

Figure 3A shows a Cs_2SO_4 equilibrium gradient analysis of the outcome of an annealing reaction between this ^{3}H-DNA product (R-III-^{3}H-DNA) and 70S RNA of the murine mammary tumor virus isolated from the milk of Paris R-III mice. It is evident that approximately 85% of the DNA product is shifted from the DNA region to the RNA and hybrid region of the gradient due to DNA:RNA hybrid complexes that form during the annealing reaction. When this R-III-^{3}H-DNA is annealed with an equivalent amount of 70S RNA isolated from Rauscher leukemia virus, no evidence of hybrid formation is detected. All of the RIII-^{3}H-DNA remained in the DNA region of the gradient (Fig 16).

USE OF ^{3}H-DNA SYNTHESIZED WITH TUMOR FRACTION AS A PROBE FOR HOMOLOGOUS RNA IN MAMMARY TUMOR

We have previously shown (12) that MMTV ^{3}H-DNA synthesized by MMTV isolated from the milk of Paris R-III mice can be used as a molecular probe to detect the presence of viral-specific RNA in mouse mammary tumor. It was therefore of interest to determine whether the R-III-^{3}H-DNA synthesized by mammary tumor extracts could serve as a similar probe for the detection of MMTV RNA in tumor tissue. Figure 17A shows the outcome of an annealing reaction with R-III-^{3}H-DNA and pRNA (see Methods) isolated from mouse breast tumors. As a consequence of the annealing reaction, 70% of the ^{3}H-DNA has shifted to the RNA and hybrid region of the gradient. When a similar pRNA fraction obtained from the livers of tumor-free NIH mice is annealed with R-III-^{3}H-DNA, no evidence of hybrid formation is observed (Fig 17B).

LOCALIZATION OF REVERSE TRANSCRIPTASE AND VIRAL RNA

The very fact that the P-100 fraction yields a 70S RNA:^{3}H-DNA complex in an endogenous reaction already implies that the reverse transcriptase must be asso-

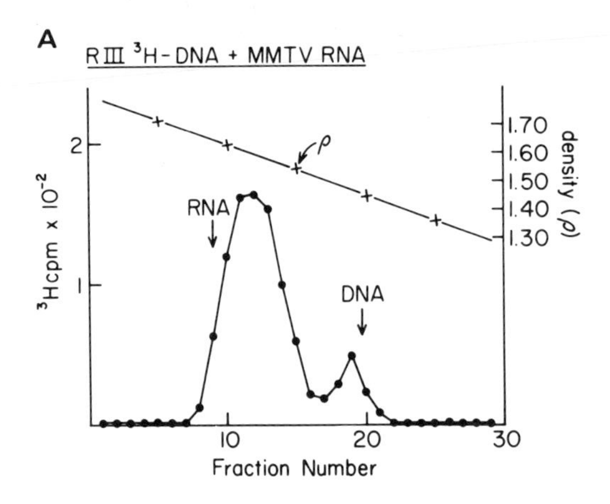

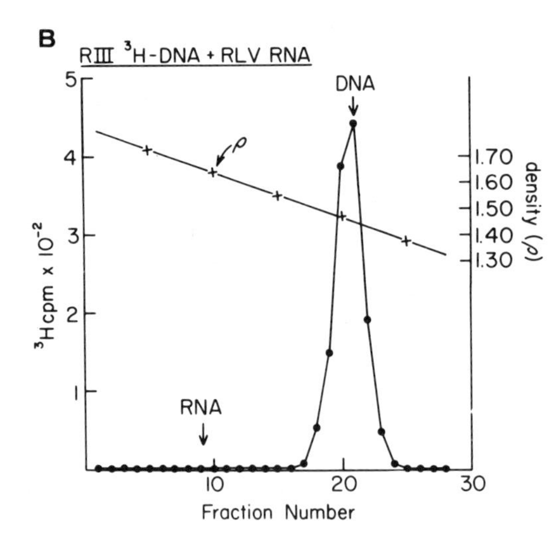

Fig 16. Cs_2SO_4 equilibrium density gradient centrifugation of RIII-^{3}H-DNA after annealing to purified MMTV and RLV 70S RNA.

A standard endogenous reverse transcriptase reaction was performed with the P-100 fraction from mouse mammary tumors. 70S RNA-^{3}H-DNA was obtained following velocity centrifugation and digested with 0.4 M NaOH, 37°C, for 18 hrs to remove all RNA present. The ^{3}H-DNA product (RIII-^{3}H-DNA) was then annealed to (A) 0.2 μg MMTV 70S RNA and to (B) 0.2 μg RLV 70S RNA. After annealing, the reaction was subjected to Cs_2SO_4 gradient centrifugation as described in Methods (28).

ciated with the template it uses. It was of interest to see whether this association could be identified with a particulate element possessing a density characteristic of an RNA tumor virus. At the same time one could determine whether all the viral-related RNA was confined to one particular density fraction.

An experiment was therefore designed to permit the simultaneous detection of 70S RNA and reverse transcriptase and the independent identification of viral-specific RNA by molecular hybridization in a tumor extract subjected to density fractionation. A P-180 fraction was prepared from mouse breast tumor tissue now using Mg^{++} containing buffers to preserve the integrity of the polyribosomes. The pellet fraction

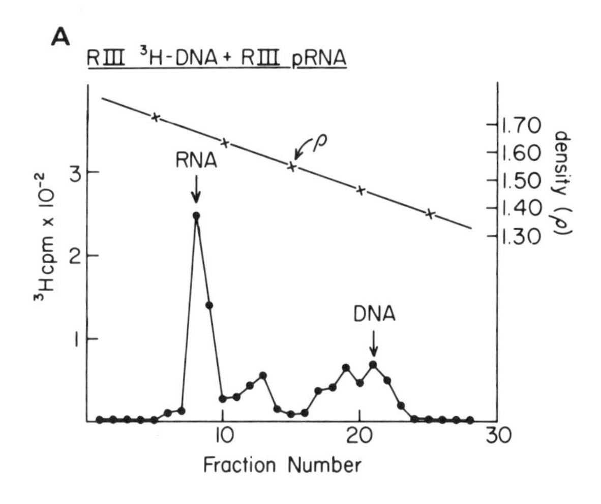

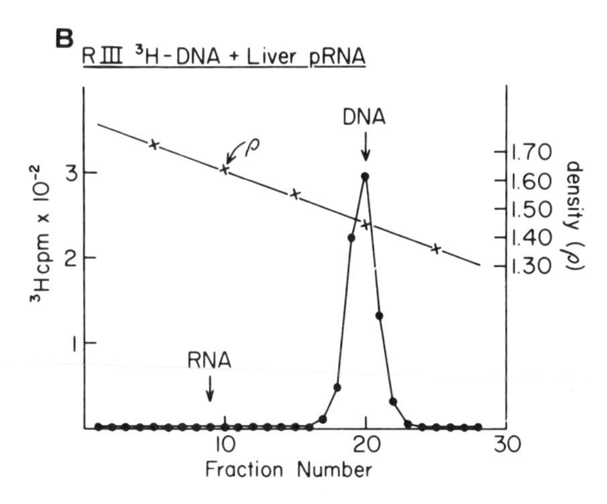

Fig 17. Cs_2SO_4 equilibrium centrifugation of RIII-^{3}H-DNA after annealing to RIII breast tumor and NIH mouse liver polysomal RNA. Polysomal RNA was isolated from mouse mammary carcinomas (A), and NIH mouse liver (B) and 100 μg of pRNA from each tissue was annealed to RIII-^{3}H-DNA. After annealing, the reactions were subjected to Cs_2SO_4 equilibrium centrifugation analysis (28).

was then subjected to sucrose density gradient centrifugation to permit equilibrium banding of viral particles and velocity sedimentation of polyribosomes. The absorbance profile (A_{260}) of this density gradient is shown as the broken curve in Figure 18A. The gradient was then divided into ten fractions. Half of each fraction was pelletted to assay for the presence of endogenous reverse transcriptase activity, whereas RNA was extracted from the other half for analysis by annealing reactions.

The bar graph of Figure 18A shows the amount of ^{3}H-DNA cpm found sedimenting in the 70S region of a glycerol velocity gradient following the performance of a reverse transcriptase reaction on each of the indicated ten sucrose fractions. These data show that the reverse transcriptase activity and the 70S RNA present in the P-180 fraction of the tumor extracts localize at a density between 1.16 and 1.19 g/ml, the density characteristic of the oncogenic RNA viruses. Figure 18B shows the outcome of

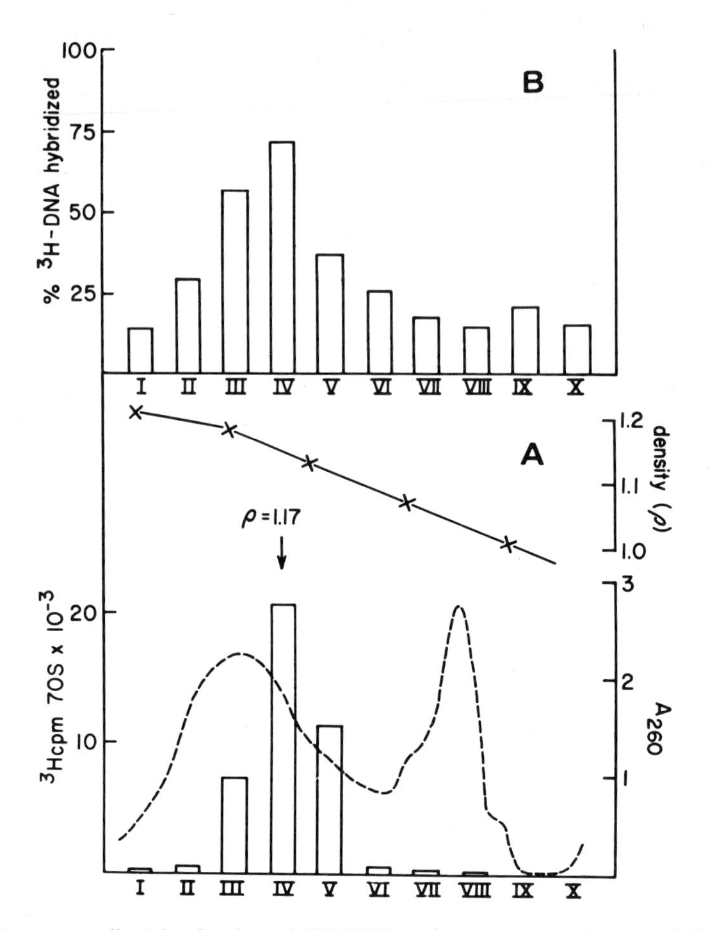

Fig 18. Sucrose gradient localization of 70S RNA and reverse transcriptase activity in extracts of mouse mammary tumors. A polysomal fraction (P-180) was prepared from mouse mammary tumors (Methods). The pellet was resuspended in TNM and layered on a linear gradient of 7 – 47% sucrose in TNM and spun at 25,000 rpm in a SW-27 rotor at 4°C for 195 min. The gradient was dripped from below and absorbance at 260 nm (- - -) was monitored in the flow through curvette of the Gilford 2400S recording spectrophotometer.

Ten equal fractions were collected and divided into two aliquots: one for reverse transcriptase assay and the other for RNA extraction. The amount of 70S-^{3}H-DNA synthesized by an endogenous reverse transcriptase reaction performed on each fraction was determined by glycerol velocity centrifugation (A). RNA obtained from an equal volume of each of the 10 fractions was then annealed to RIII-^{3}H-DNA. The annealing reactions were then analyzed by Cs_2SO_4 density centrifugation (B) (28).

annealing reactions between MMTV ^{3}H-DNA and RNA extracted from an equal volume of each gradient fraction. As expected, maximum viral-specific RNA is detected at a density of 1.17 g/ml coincident with the density of intact virions. However, it will be noted that significant viral-specific RNA is detected at densities where no viral activity is found. Whether this represents viral RNA associated with polyribosomes or some other structure is under further investigation.

The results described show that it is possible to identify 70S RNA-directed DNA synthesis in murine mammary tumor tissue. Further, it was demonstrated that the reverse transcriptase and its 70S RNA template are physically associated in a particle possessing the density (1.17 g/ml) characteristic of the oncornaviruses. Finally, it was independently demonstrated, by molecular hybridization, that a large percentage of the viral-specific RNA present in the mouse mammary tumor is found in a particle of the same density.

THE SEARCH FOR 70S RNA ASSOCIATED WITH A REVERSE TRANSCRIPTASE IN HUMAN LEUKEMIC CELLS

The successful use of the simultaneous detection test (23) to detect reverse transcriptase employing a 70S RNA template in murine mammary tumors encouraged the performance of similar experiments with human leukemic cells (29). To insure that a relevant reaction was being identified, finding a "70S"-DNA product of an endogenous reaction was supplemented with information that the reaction possessed the following features: 1. a requirement for all four deoxyriboside triphosphates to eliminate possible end addition reactions or physical trapping of isotope; 2. sensitivity to ribonuclease and banding at the proper density in Cs_2SO_4 gradient to establish that the DNA product is complexed to RNA; and, most important, 3. the product must be hybridizable to an "oncogenic" RNA.

It is not always possible to obtain enough labeled product to carry through all these tests on a given sample of tissue. However, as many as can be were carried out, with hybridizability of the DNA product to a relevant RNA providing the most diagnostically useful information.

Peripheral white blood cells were prepared from the buffy coats of whole blood samples as described previously (14). Cells from leukemic and nonleukemic patients were broken by the use of a Dounce homogenizer, and the nuclei were removed by low speed centrifugation. After trypsin digestion, the supernatant was centrifuged at 150,000 times g to yield a cytoplasmic pellet containing membranes and viral particles, if present. The pellet was then treated with NP-40 to disrupt possible viral particles and used in an endogenous reverse transcriptase reaction run in the presence of 50 to 100 μg/ml actinomycin D to inhibit DNA-instructed DNA synthesis. The product of the reaction with the putative nucleic acid template was freed of protein and analyzed on a glycerol sedimentation velocity gradient.

Figure 19 shows representative experiments examining the effects of ribonuclease treatment of the product and the omission of a deoxytriphosphate during the reaction. In Figure 19A, we see a typical 70S complex involving the DNA synthesized by the pellet fraction from the white blood cells of a patient with acute lymphoblastic leukemia. The fact that treatment of the product with ribonuclease eliminates the complex shows that the [^{3}H]-DNA is complexed to a 70S-RNA molecule. Figure 19B shows that omission of dATP from the reaction leads to a failure to form the 70S complex synthesized by the pellet fraction derived from the white blood cells of a patient with acute myelogenous leukemia. In similar experiments it was shown that omission of either dCTP or dGTP also resulted in the absence of the 70S-DNA complex. Such outcomes argue against nontemplated end addition reactions.

In some cases sufficient leukemic cells were obtained to synthesize enough of the 70S-DNA complex to permit a more complete characterization of the product and Figure 20 summarizes such an experiment. Figure 20A shows the profile of DNA

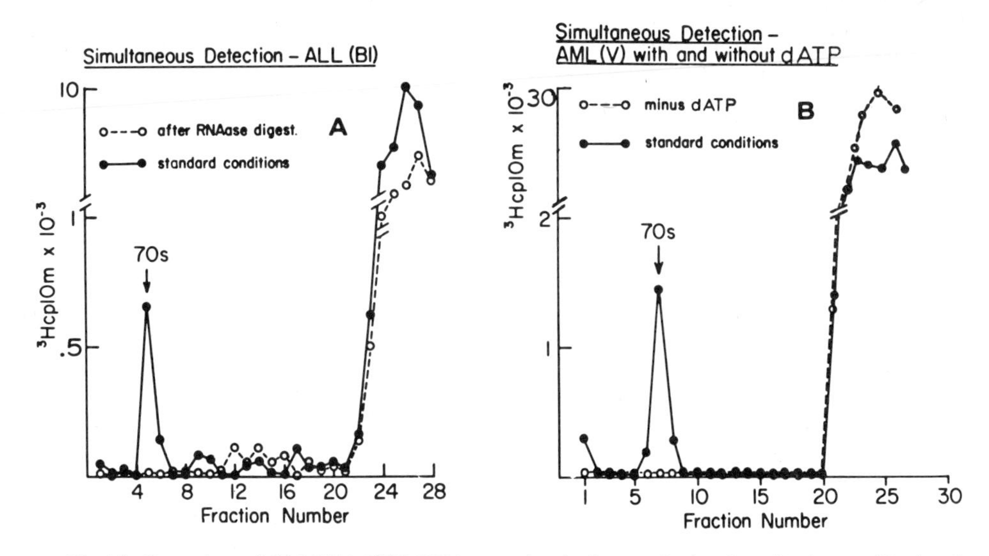

Fig 19. Detection of 70S-RNA-[^{3}H]-DNA complex in human leukemic cells. 1 gm of leukemic WBC was washed in 5 ml of 0.01 M NaCl, 0.01 M Tris-HCl, pH 7.4, resuspended in 4 ml of 5% sucrose, 0.005 M EDTA, 0.01 M Tris-HCl, pH 8.3, and ruptured with three strokes of a Dounce homogenizer. The nuclei were removed by low speed centrifugation (2000 g, 5 min, 2°C). The supernatant was brought to a final concentration of 1 mg/ml trypsin (Worthington) and incubated at 37°C for 30 min. A ten-fold excess of lima bean trypsin inhibitor (Worthington) was added (final concentration 3 mg/ml) and the solution again centrifuged at 2000 g for 5 min at 2°C. The supernatant was then centrifuged at 45,000 rpm for 60 min at 2°C. The resulting cytoplasmic pellet was resuspended in 0.5 ml of 0.01 M Tris-HCl, pH 8.3, brought to 0.1% Nonidet P-40 (Shell Chemical Co.) and incubated at 0°C for 15 min. DNA was synthesized in a typical reverse transcriptase reaction mixture (final volume 1 ml) containing: 50 μmol of Tris-HCl, ph 8.3, 20 μmol NaCl, 6 μmol $MgCl_2$, 100 μmol each of dATP, dGTP, dCTP, and 50 μmol-[^{3}H]-dTTP (Schwarz Biochemical, 8000 cpm per pmol). 50 μg/ml actinomycin D was added to inhibit DNA-instructed DNA synthesis. After incubation at 37°C for 15 min, the reaction was adjusted to 0.2M NaCl and 1% SDS, and deproteinized by phenol-cresol extraction. The aqueous phase was layered on a 10 to 30% gradient of glycerol in TNE buffer (0.01 M Tris-HCl, pH 8.3, 0.1 M NaCl, 0.003 M EDTA) and centrifuged in a SW-41 rotor Spinco at 40,000 rpm for 180 min at 2°C. Fractions were collected from below and assayed for TCA-precipitable radioactivity. In this, as in all sedimentation analysis (Table II), 70S RNA of the avian myeloblastosis virus was used as a marker.

A) One aliquot of product was run on the gradient as a control and the other was pretreated with 20 μg of RNase 1 (Worthington) for 15 min at 37°C prior to sedimentation analysis.

B) Reactions with and without dATP (29).

product and template in a glycerol gradient sedimentation analysis. The split peak seen here is occasionally observed and may be due to conformational changes in the RNA template. The material included within the brackets of Figure 20A was pooled and the complex re-isolated by precipitation from 0.4 M NaCl with alcohol. An aliquot was analyzed on a Cs_2SO_4 gradient with results as shown in Figure 20B. It will be seen that a small amount of free DNA is released during the manipulation, but most of the product bands in the RNA density region, indicating that the DNA molecules remain complexed to large RNA molecules. The remainder of the DNA complex isolated from the glycerol gradient of Figure 20A was subjected to alkali digestion to remove the RNA. The resulting purified [^{3}H]-DNA was then annealed to 70S RNAs prepared

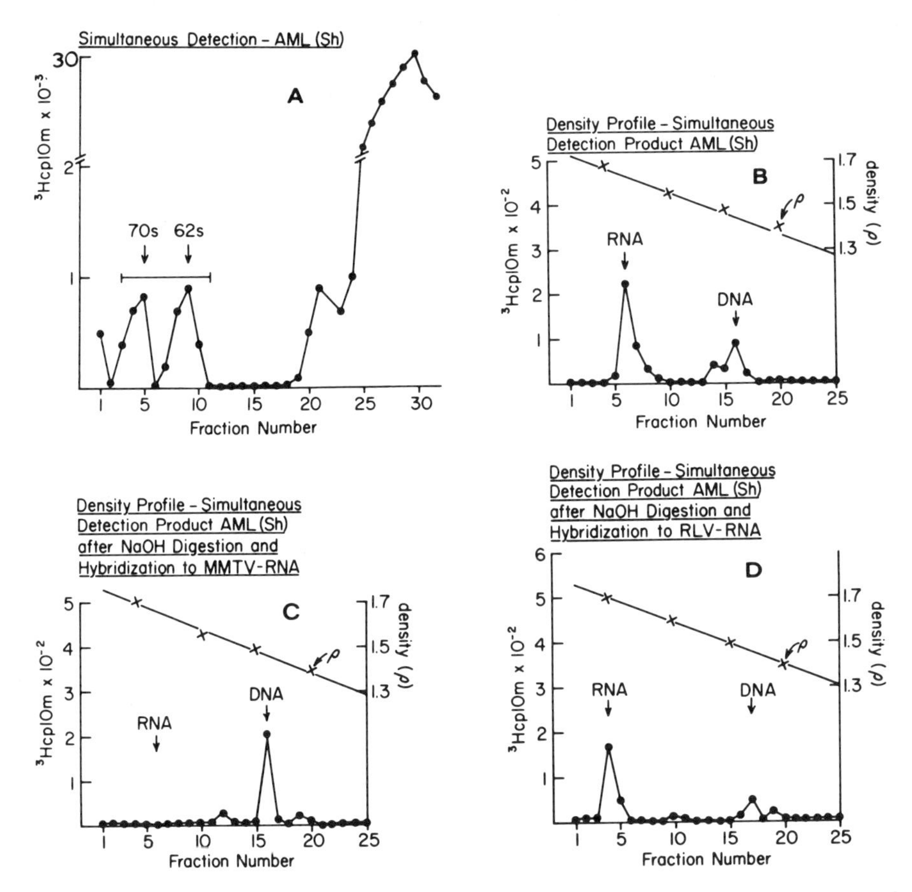

Fig 20. Assay for 70S-RNA-[^{3}H]-DNA complex in human leukemic cells (20A) with subsequent Cs_2SO_4 gradient analysis (20B-D). 10% of the glycerol gradient fractions (20A) was assayed for TCA-precipitable counts. The remaining 90% of the bracketed fractions was pooled and precipitated with two volumes of ethanol in the presence of yeast RNA carrier (15 μg/ml final concentration). After centrifugation at 10,000 g for 10 min at -20°C, the precipitate was resuspended in 0.1 ml TNE buffer and divided into three aliquots. Aliquot 1 (Fig 20B) was added directly to 11 ml of half-saturated cesium sulphate (Gallard-Schlessinger) in 0.003 M EDTA (ρ=1.52 g/ml) and centrifuged in a 50 Ti rotor (Spinco) at 44,000 rpm for 60 h, at 15°C. Aliquots 2 and 3 were digested in 0.4 M NaOH at 43°C for 24 h, neutralized with HCl in the presence of 0.01 M Tris-HCl, pH 7.4, and brought to 50% formamide (Eastman). After heat denaturation for 10 min at 85°C, aliquot 2 was hybridized to 0.2 μg MMTV-RNA (20C), aliquot 3 to 0.2 μg RLV-RNA (20D) in the presence of 0.4 M NaCl at 37°C for 8 hr and then submitted to Cs_2SO_4 density analysis (29).

from mouse mammary tumor virus (MMTV) and Rauscher leukemia virus (RLV). The outcomes of the annealing reactions were analyzed in Cs_2SO_4 gradients. It is clear from Figure 20C that the ^{3}H-DNA product shows no ability to hybridize with MMTV-RNA, a result consistent with our findings (14) that the RNA from human leukemic cells does not hybridize to MMTV-DNA. In sharp contrast are the results shown in Figure 20D with RLV-RNA. Here 75% of the leukemic ^{3}H-DNA is found complexed to the RLV-RNA, again a result logically expected from our earlier demonstration that leukemic RNA hybridizes to synthetic DNA complementary to RLV-RNA. In the experiments of Figure 20 the background was 14 cpm, which was subtracted.

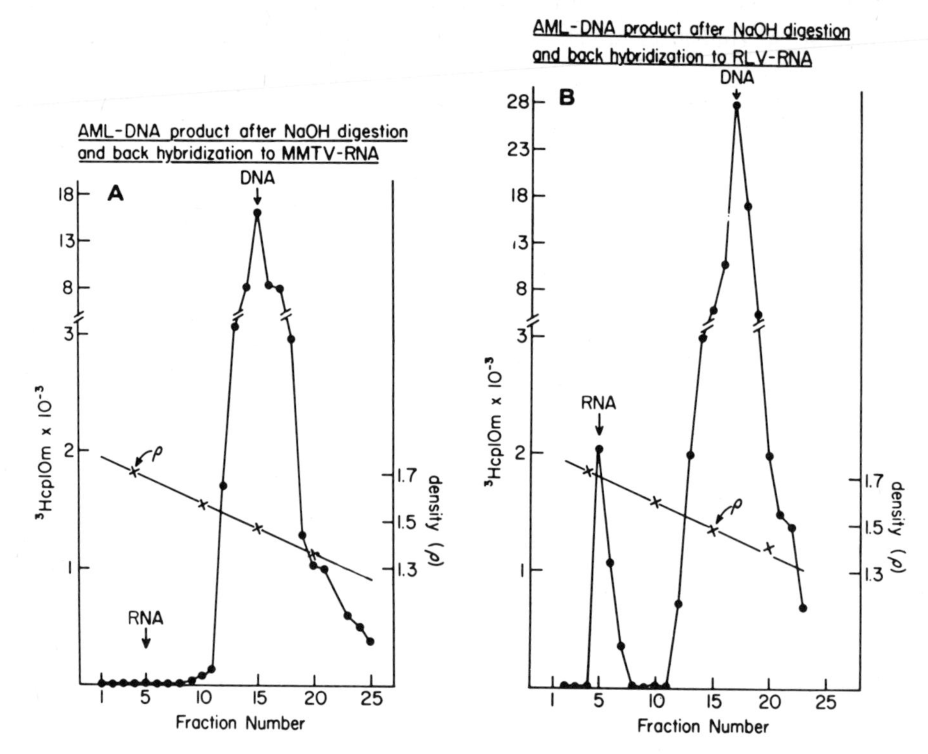

Fig 21. Cs_2SO_4 density profiles of human leukemic endogenous (^{3}H]-DNA product hybridized to MMTV-RNA(21A) and RLV-RNA(21B). 0.1 g of leukemic white blood cells was treated as described in Figure 19, except that the reverse transcriptase reaction was run in the presence of 100 μg of actinomycin D and the glycerol gradient sedimentation step was omitted. After phenol-cresol extraction, the aqueous phase was precipitated with two volumes ethanol. The precipitate was digested in 0.4 M NaOH for 24 hr at 43°C, neutralized, and divided into two aliquots. Aliquot 1 was hybridized to MMTV-RNA (A), aliquot 2, to RLV-RNA (B) (29).

Hybridizability of the DNA product to relevant viral RNAs provides the most revealing information and also requires more product than for a simple simultaneous detection test. We consequently modified our procedure to maximize our yield by avoiding the loss that attends isolation of the DNA synthesized from its 70S complex in a glycerol gradient, a procedure followed in all the experiments described thus far. Instead, the total DNA product of an endogenous reaction was recovered free of protein and the RNA removed by alkali treatment. Annealing reactions of such a product with RLV and MMTV-RNAs are shown in Figure 21. The pattern is the same as with the ^{3}H-DNA isolated from the 70S complex, hybridization being observed only with RLV-RNA and not with the unrelated MMTV-RNA.

The experiments thus far detailed involved 4 patients, one with acute lymphoblastic leukemia (ALL) and the other 3 with acute myelogenous leukemia (AML). It was of obvious interest to see how widespread these findings were. Table II summarizes the results obtained on examination of the peripheral white blood cells of 23 patients all in the active phases of their disease. In addition to the acute leukemias, 4 cases are also included involving chronic lymphocytic leukemia (CLL) and chronic myelogenous leukemia (CML).

TABLE II

Simultaneous Detection of 70S RNA and Reverse Transcriptase in Leukemic Cells

Leukemias	Simul. det. cpm	RNase sensitivity	Hybrid. to RLV-RNA	Hybrid. to AMV- or MMTV-RNA
Acute lymphatic				
Od	400	+	NT	NT
Sm	95	+	NT	NT
Acute lymphatic/ lymphosarcoma				
Hl	805	+	+	-
Hz	200	NT	+	-
Bx	185	+	NT	NT
St	105	+	NT	NT
Acute myelogenous				
Ge	170	+	NT	NT
De	985	+	+	-
Be	305	+	NT	NT
Vi	1295	+	+	-
Sh	1010	NT	+	-
S	115	NT	+	-
Du	415	NT	+	-
He	400	NT	+	-
Bl	605	+	NT	NT
Ga	215	+	+	-
C	285	+	NT	NT
Pe	0	NT	NT	NT
Ne	1400	NT	+	-
Chronic lymphatic				
Ri	200	+	NT	NT
Chronic myelogenous				
Ec	405	NT	+	-
Har	390	NT	+	-
An	600	NT	+	-

NT = Not Tested.

Note:

Column 1 depicts acid-prepipitable radioactivity found in the 70S region of the glycerol gradients, as described in the legend to Figure 19.

Column 2 depicts abolition by RNase 1 of acid-precipitable radioactivity in the 70S region of glycerol gradients, as described in the legend to Figure 19.

Columns 3 and 4 depict hybridization of leukemic [^{3}H]-DNA product to RLV-RNA and MMTV-RNA, as described in the legend to Figure 20. Plus values indicate a significance of greater than 999 out of a thousand.

Of the 23 leukemic patients examined, 22 showed clear evidence in their peripheral white blood cells of a reaction leading to the appearance of endogenously synthesized DNA in the 70S region of a glycerol gradient sedimentation analysis. Nine of these were tested for ribonuclease sensitivity of the 70S-DNA complex, and in all 9 the complexes were destroyed. In 9 others, the DNA was recovered from the complex and annealed to RLV-RNA and to either MMTV- or AMV-RNA. In all 9, hybridization occurred with the RLV-RNA and not to either of the unrelated MMTV- or AMV-RNAs. In 4 patients enough DNA complex was formed to permit a complete characterization of the product. In all 4, the DNA complex was destroyed by ribonuclease and the purified DNA hybridized to RLV-RNA and not to either MMTV- or AMV-RNA. The absence of complexes with MMTV- and AMV-RNA eliminates the possibility that the hybridization with RLV-RNA is due solely to stretches of A. All tumor viral RNAs contain such stretches (17 - 19).

In summary, with one exception, the 23 leukemic patients examined showed clear evidence that their white blood cells contained a reverse transcriptase associated with a 70S RNA that was used as a template to generate DNA complements. It was further shown in the 13 out of 13 instances in which the tests could be carried out, that the DNA synthesized hybridized specifically to the RNA of the mouse leukemogenic agent.

In contrast with these results are those obtained with a control series of 18 white blood samples from nonleukemic patients. These included 9 with elevated white blood cell counts (in the range of 25,000 per mm^3) due to a variety of disorders and 9 normal blood donors. None of these showed evidence of the synthesis of a DNA complex that was ribonuclease-sensitive and contained DNA that could hybridize specifically to RLV-RNA. The necessity of carrying out the supplementary tests on the "70S DNA" is well illustrated by our experience with the controls. Three of the nonleukemic patients and 3 of the normal individuals did yield enough labeled DNA in the 70S region to warrant further investigation. However, in every case these nonleukemic reactions were either RNase-resistant, insensitive to the omission of one of the required deoxyriboside triphosphates, or nonhybridizable to RLV-RNA. Thus, none of the 18 nonleukemic individuals yielded evidence of the specific DNA synthesis found in almost all of the leukemic patients examined.

It is important to emphasize that the negative outcomes with the control group do not mean that a similar reaction is nonexistent in nonleukemic individuals. Our procedure would not have detected viral agents that do not have sequence homology to at least one of the three (i.e., RLV, MMTV, and AMV) oncogenic viruses employed as a source of RNA in the hybridization tests. Further, it should be noted that it would not be terribly surprising if occasional healthy individuals showed evidence of a positive reaction leading to formation of 70S-DNA complex homologous to RLV-RNA. In this connection, it is relevant to recall that we reported (11,27) the presence of 70S RNA and reverse transcriptase in the milk of perfectly normal disease-free women.

CONCLUSIONS

The experiments just described were designed to probe further the etiological significance of the earlier (14) findings in human leukemic cells of RNA homologous to that of the Rauscher leukemia virus. The data obtained show that at least a portion of the viral-related RNA we were detecting is in the form of a 70S-RNA template physically associated with a reverse transcriptase, two of the diagnostic features of the RNA tumor viruses. Further, the DNA synthesized by the leukemic cell reverse transcriptase on its own endogenous template is related in sequence to RLV-RNA. Note that this last result is complementary to and completes the logic of our earlier experiments in which the DNA synthesized on RLV-RNA was used as a probe to find the viral-related information in leukemic cells.

It is noteworthy that positive outcomes were observed in more than 95% of the leukemic patients, whether they were acute or chronic, lymphoblastic or myelogenous. Thus, despite their disparate clinical pictures and differing cellular pathologies, these various types of leukemias have very similar viral-related information.

The presence in human neoplasias of viral-related RNA or reverse transcriptase are individually suggestive of a viral agent. The data just described carry this implication much further. They demonstrate that human leukemic cells contain a 70S viral-related RNA associated with a reverse transcriptase, the two uniquely identifying characteristics of the RNA tumor viruses. Ultimately, it will be necessary to determine whether the complexes identified are infectious and transforming particles.

REFERENCES

1. Spiegelman, S.: DNA and the RNA viruses. Proc. Royal Society of London B., 177:87, 1971.
2. Temin, H. M., and Mizutani, S.: RNA-dependent DNA polymerase in virions of Rous sarcoma virus. Nature, 226:1211, 1970.
3. Baltimore, D.: RNA-dependent DNA polymerase in virions of RNA tumor viruses. Nature, 226:1209, 1970.
4. Spiegelman, S., and Schlom, J.: Reverse transcriptase in oncogenic RNA viruses. In: Virus Cell Interactions and Viral Antimetabolites. Edited by D. Shugar. Academic Press, London, p 115 (FEBS Symposium), 1972.
5. Schlom, J., Harter, D. H., Burny, A., and Spiegelman, S.: DNA polymerase activities in virions of visna virus, a causative agent of a "slow" neurological disease. Proc. Nat. Acad. Sci USA, 68:182, 1971.
6. Spiegelman, S., Burny, A., Das, M. R., Keydar, J., Schlom, J., Travnicek, M., and Watson, K.: Characterization of the products of RNA-directed DNA polymerases in oncogenic RNA viruses. Nature, 227:563, 1970.
7. Hall, B. D., and Spiegelman, S.: Sequence complementarity of T2-DNA- and T2-specific RNA. Proc. Nat. Acad. Sci. USA, 47:137, 1961.
8. Fujinaga, K., and Green, M.: Mechanism of viral carcinogenesis by the deoxyribonucleic acid mammalian viruses; IV Related virus-specific ribonucleic acids in tumor cells induced by "highly" oncogenic adenovirus types 12, 18, and 31. J. Virol., 1:576, 1967.
9. Lindberg, U., and Darnell, J. E.: SV-40-specific RNA in the nucleus and polyribosomes of transformed cells. Proc. Nat. Acad. Sci. USA, 65:1089, 1970.
10. Green, M., Rokutanda, H., and Rokutanda, M.: Virus-specific RNA in cells transformed by RNA tumor viruses. Nature New Biology, 230:229, 1971.
11. Schlom, J., Spiegelman, S., and Moore, D. H.: RNA-dependent DNA polymerase activity in virus-like particles isolated from human milk. Nature, 231:97, 1971.
12. Axel, R., Schlom, J., and Spiegelman, S.: Evidence for translation of viral-specific RNA in cells of a mouse mammary carcinoma. Proc. Nat. Acad. Sci. USA, 69:535, 1972.
13. Axel, R., Schlom, J., and Spiegelman, S.: Presence in human breast cancer of RNA homologous to mouse mammary tumor virus RNA. Nature, 235:32, 1972.

14. Hehlmann, R., Kufe, D., and Spiegelman, S.: RNA in human leukemic cells related to the RNA of a mouse leukemia virus. Proc. Nat. Acad. Sci. USA, 69:435, 1972.
15. Kufe, D., Hehlmann, R., and Spiegelman, S.: Human sarcomas contain RNA related to the RNA of a mouse leukemia virus. Science, 175:182, 1972.
16. Hehlmann, R., Kufe, D., and Spiegelman, S.: Viral-related RNA in Hodgkins' disease and other human lymphomas. Proc. Nat. Acad. Sci., USA, 69:1727, 1972.
17. Gillespie, D., Marshall, S., and Gallo, R. C.: RNA of RNA tumor viruses contains poly A. Nature New Biology, 236:227, 1972.
18. Green, M., and Cartas, M.: The genome of RNA tumor viruses contains polyadenylic acid sequences. Proc. Nat. Acad. Sci. USA, 69:791, 1972.
19. Lai, M. M. C., and Duesberg, P. H.: Adenylic acid-rich sequence in RNAs of Rous sarcoma virus and Rauscher mouse leukemia virus. Nature, 235:383, 1972.
20. Robinson, W. S., Robinson, H. L., and Duesberg, P. H.: Tumor virus RNA's. Proc. Nat. Acad. Sci. USA, 58:825, 1967.
21. Rokutanda, M., Rokutanda, H., Green, M., Fujinaga, K., Ray, R. K., and Gurgo, C.: Formation of viral RNA-DNA hybrid molecules by the DNA polymerase of sarcoma-laeukemia viruses. Nature, 227:1026, 1970.
22. Bishop, D. H. L., Ruprecht, R., Simpson, R. W., and Spiegelman, S.: Deoxyribonucleic acid polymerase of Rous sarcoma virus: reaction conditions and analysis of the reaction product nucleic acids. J. Virology, 8:730, 1971.
23. Schlom, J., and Spiegelman, S.: Simultaneous detection of the reverse transcriptase and high molecular weight RNA unique to oncogenic RNA viruses. Science, 174:840, 1971.
24. Duesberg, P. H., and Cardiff, R. D.: Structural relationships between the RNA of mammary tumor virus and those of other RNA tumor viruses. Virology, 36:696, 1968.
25. Feller, W. F., and Chopra, H. C.: Studies of human milk in relation to the possible viral etiology of breast cancer. Cancer, 24:1250, 1969.
26. Moore, D. H., Charney, J., Kramarsky, B., Lasfargues, E. Y., Sarkar, N. H., Brennan, M. J., Burrows, J. H., Sirsat, S. M., Paymaster, J. C., and Vaidya, A. B.: Search for a human breast cancer virus. Nature, 229:611, 1970.
27. Schlom, J., Spiegelman, S., and Moore, D. H.: Detection of high molecular weight RNA in particles from human milk. Science, 175:542, 1972.
28. Gulati, S. C., Axel, R., and Spiegelman, S.: The detection of reverse transcriptase and high molecular weight RNA in malignant tissue. Proc. Nat. Acad. Sci., USA, 69:2020, 1972.
29. Baxt, W., Hehlmann, R., and Spiegelman, S.: Human leukemic cells contain reverse transcriptase associated with a high molecular weight viral-related RNA. Nature New Biology, 240:72, 1972.

22. DISCUSSION

Dr. Beers:

All four of these papers have called serious attention to the problem of Koch's postulates. Each one of them tried to attack this problem with a different methodology, ranging from hybridization to immunological testing to actual demonstration of a transmissible agent. I wonder if we could start the discussion off by considering some of the technical problems in trying to fulfill Koch's four postulates for an etiological agent in a malignancy.

Dr. Spiegelman:
(New York)

Well, the last step, the demonstration of a transmissible agent, has not been done and, of course, it would be in many ways the most convincing. It is obvious that we are going to have to be lucky and find a susceptible animal. Remember we have had the mouse mammary agent with us for more than forty years and to this day nobody has ever caused breast cancer in any other animal but the mouse with that. So we don't know how long it is going to take us to be lucky, but I think the attempt has to be made.

The obvious thing that a number of laboratories are trying to do now with the agents from breast cancers and leukemia is to get them growing in a tissue culture situation in a laboratory so that we can study them. That, I think, we can be more hopeful will work.

Dr. Todaro:
(Bethesda, Md.)

I think in the mouse model system with the induced viruses it is possible to test and it is testable. In fact, one can get the virus that appears in spleens of normal mice and show that the virus is oncogenic when put into other mice.

Now, as Dr. Spiegelman pointed out, there is also the model in the avian system in which the virus from chick embryo cells seems not to be able to grow in chickens, but is able to grow in Japanese quail. I would fully agree that at some point we do need an infectious transmissible agent to begin to work with. It may be that we will have to experiment with various primates to find a susceptible host to these agents that are found in the human material.

Dr. Aurelian:
(Baltimore, Md.)

I think there is another way in which one could approach the problem. This would be by prospective studies if we know what agent we are looking for.

Dr. Spiegelman:

These are going on, as you know, at least with breast cancer. Prospective studies, though, don't ever prove. I mean, they make it likely and more and more likely, but you still have to have an experimental situation in which you can do a Koch-type experiment.

Dr. Beers: I would like to raise an additional question with respect to demonstrating a transmissible agent. Is a tissue culture or a cell clone a suitable model as opposed to the entire animal?

Dr. Spiegelman: Well, I think at least you will prove that you have an infectious agent and that it does something to cells which is oncogenic-like, but ultimately you really want to have an animal system in which you produce the disease.

Dr. Beers: You certainly can demonstrate it in the tissue culture, but you might not be able to demonstrate it in the intact animal. Then where are you?

Dr. Todaro: The Wooly monkey Sarcoma virus has been demonstrated to transform cells in tissue culture and also to produce tumors in marmosets, another primate species. The point I was trying to make is that if we could find a primate model system first looking in tissue culture for transmissible virus, that also may be a host that will allow direct testing of the oncogenic potential.

Dr. Pollard: (Notre Dame, Ind.) Dr. Todaro, with respect to reverse transcriptase, can mice be immunized against their own transcriptase and possibly provide a means of control of leukemia?

Secondly, can you administer antiserum to reverse transcriptase and prevent leukemia in these animals?

Dr. Todaro: Well, with regard to the first question, no. In fact, if you look at spontaneous tumors in mice, they do not have antibody to their own reverse transcriptase. It does appear you have to put it in a foreign species.

Now, you have to remember that a reverse transcriptase is a minor component of the virus and represents half to one percent of the total viral protein. It is not surprising that it doesn't induce antibody formation in its own host. The mouse tumors that are transmitted in rats do make antibody to the mouse polymerase. They don't in the mouse, at least in the limited number of tests that we have looked at.

Now, the second question: It might be possible to protect animals passively if, as I suggested, there really was a shower of virus that follows specific sequences, hormonal changes, or radiation or chemical carcinogens. I feel it would be much more likely that this protection would make more sense to be directed against a viral surface component and not against the reverse transcriptase.

I find it quite unlikely that antibody to the reverse transcriptase would have any practical therapeutic value. We are using it as a diagnostic procedure.

Dr. Ryser:
(Baltimore, Md.)

My question is addressed to that part of Dr. Todaro's talk that suggests that chemical carcinogens activate a C virus oncogene. However appealing that unitarian view is, it is in a way rather difficult to reconcile with some of the biological properties of the chemically induced tumors. There is the long period of latency that follows initiation by chemical carcinogens. It has been suggested that this initiation knocks out oncogene repressor. It would not be readily understandable how knocking out that repressor would be followed by a long latency. The second property is the extraordinary variability that is seen among chemical cancers. If the C-oncogenes were to play an overriding role, one would expect some uniformity in the biological properties of the chemical tumors. As you know, if one looks at the antigenicity of chemical tumors, one finds that two tumors produced by the same carcinogen in the same animal differ widely in their antigenic properties and that there are differences in antigenic properties even among different cells from the same tumor.

I suspect that there might be alternative ways to implicate an oncogene or C-particle in chemical carcinogenesis and I would be very happy to suggest one, but before I go much farther I would like to have your comments as to these reservations.

Dr. Todaro:

We would suggest that the first step, which involves the activation of viral information from a single cell in the body, is only one of the steps that will eventually lead to a tumor. There are going to be control mechanisms at the level of the single cell and at the level of groups of cells. Certainly the immune system is going to play an important role. I don't think the long latent period argues either for or against. There are certainly enough systems in which when one adds a virus there is a very long latent period before the expression of some tumors. Which are the factors that offer the best place to interrupt the cycle? Now, it may be at the level of the single cell or it may be by enhanced immune defense. We really don't know yet.

With respect to the second part, that about different tumors having slightly different antigens, I think a point should be clear. If the tumors have antibody to the viral-specific antigens, surface antigens, themselves, these will be recognized by the animal as foreign and will be rejected. The whole concept of immune surveillance is based on this.

The tumors that come through are going to be the tumors with the minimal viral expression. You can do experiments with a nonproducer tumor cell which is transformed by sarcoma virus and is making no virus. That cell is very capable of producing tumors, even in immunologically competent adult mice. But if you add virus to it or if you induce the virus out, that animal now rejects the tumor, because it recognizes the cell surface antigens and the budding viruses as foreign.

Therefore, there should be a selection against extreme viral expression. I think this is part of the reason that I would expect that except in highly inbred strains you are not going to get lots of virus being produced in a tumor or in very immumodeficient animals.

Dr. Ryser: Do you believe that this answer accounts for the already known difference in the homogeneity of the antigenicity of the virally induced tumors and the chemically induced tumors? This is a fairly well-recognized fact.

Dr. Todaro: I think the point is that if there are strong antigens, they would be rejected by the animal. We never would see them. What you are seeing, I think, in the chemically induced tumors are subtle, small changes which reflect genetic drift more than anything else. The fact that the one tumor can differ from another in an antigen certainly doesn't mean that they have a different mechanism of induction.

Dr. Ryser: One of the trends in chemical carcinogenic research is examination of the bonding of the carcinogen to one of the molecules like DNA, RNA, and protein. Some carcinogens are known to bind particularly to RNA.

I wonder whether one could not invoke the reverse transcriptase in this case as a possible means of starting with these modified RNAs and transcribing them into modified DNAs, which amounts to a kind of somatic mutation. Would you comment on this possibility, at least?

Dr. Todaro: It is possible.

Dr. Bader: (Bethesda, Md.) I would like to make a statement which is relevant to much of the work that has been discussed this morning and this afternoon. I don't mean it to be a criticism of any specific aspect of these discussions, but I would like to make a plea for a documentation of purity of reagents that have been used in many of the studies described.

To be more specific, the experimental approach in many of the studies discussed today begins with a virus preparation, the endogenous reverse transcriptase, the synthesis of a DNA product, the isolation of the DNA, and the use of this DNA to examine the RNA somewhere else. Unless the whole system is described very carefully, it is very difficult to assess just what it is you are studying here.

For instance, the typical virus preparation not only contains viral 70-S RNA but also many cellular nucleic acids, including a lot of cellular DNA and many RNA species. Those same preparations will contain a large variety of enzymes. Some dozen to fifteen have been described. You might consider that among them would be normal DNA polymerase and some of the other cellular DNA polymerases which have been discussed here.

So, clearly, if you had a little detergent and deoxyribonucleotide triphosphates, the DNA that comes out of this sytem would very often not only be virus-specific but it will be cellular or, that is, DNA which has been transcribed from cellular material.

Dr. Spiegelman: Well, we go to great pains to avoid all these complications which you have mentioned. In the first place, wherever possible you first purify out the RNA, get rid of the DNA, and the virion, and anything else. Then you give that clean RNA to a clean enzyme and give it the subtrate to synthesize. But you don't stop there, because when you try to clean anything you do the best you can and you may not have really eliminated all of the contaminants. So even with that you have to check the specificity of the products. If we make a product which hybridizes equally well to cellular RNA as to viral RNA, we throw it out the window, or down the drain, really.

All of our products are monitored, and they are also checked against unrelated viral RNA to make sure that we do have a probe which is going to give us sensible information.

Dr. Bader: Let me ask, specifically in your experience, Dr. Spiegelman, have you isolated the 70-S viral RNA and then made a product from that and used that in your hybridization?

Dr. Spiegelman: Yes; we do that routinely.

Dr. Bader: I wasn't clear on that point.

Dr. Spiegelman: I couldn't give you all the details. You mean the 70-S RNA from what?

Dr. Bader: From virions?

Dr. Spiegelman: From MMTV, yes. When we mate with Rauscher, yes.

Dr. Bader: The problem is not just to tell a virus preparation, but it is to take the 70-S RNA and specifically make DNA product.

Dr. Spiegelman: We do that.

Dr. Bader: Not everyone does.

Dr. Spiegelman: No, but if you are forced on occasion, you may not have enough virus to go through all of that and end up with enough RNA to do anything. If you are forced to use a crude preparation, then you have to be even more careful and monitor the product to make sure you have not made some nonsense, but there are ways of checking it. You don't have to be licked by it.

Dr. Kallen:
(Phila., Pa.) Dr. Spiegelman, the E.B. virus has been implicated in infectious mononucleosis, and my understanding is that those cells are available in culture. Have you looked at those?

Dr. Spiegelman: No, we would very much like to. We don't have them. We have been very anxious actually to look at patients with infectious mononucleosis directly, because I am a little bit afraid of tissue culture lines.

Dr. Kallen: Dr. Spiegelman, with respect to getting only 70% positives in your Hodgkins' disease series, I wondered whether you have looked at the histologic breakdown of subtypes, lymphocytic depletion phenomena, and so forth, because only some of those have been implicated with respect to E.B. virus.

Dr. Spiegelman: I am afraid that we did not look at Hodgkins' with the E.B. at all. The 70% positives is just a matter of sensitivity of the method. If you increase the sensitivity, I am sure that number will go up to virtually 100%, as it has in the leukemics.

Dr. Kallen: My point is, I guess, that there is some evidence that Hodgkins' disease doesn't have a single etiology. It is heterogenous with respect to age distribution, sex, and histologic type.

Dr. Spiegelman: I am afraid I would have to have much more compelling evidence than exists to accept the hypothesis of a multiple etiological cause of Hodgkins' disease or any other kind of cancer.

Dr. Lin:
(Staten Island, N.Y.) Dr. Todaro, we have been working on a visina virus, and we feel very strongly about the transformation of mouse cells by visna virus. This work has been confirmed by other laboratories. Dr. Strong has been trying to do just that for about one year, and he has been unable to get any of these cells transformed, including the mouse cells provided by Dr. Takamoto. I recall you mentioned the mouse, the transformed cell, and induced tumor. Would you like to elaborate on this point or do you have any additional data?

Dr. Todaro: It is in the literature. We haven't done any ourselves. I think the point I was trying to make was that there were reverse transcriptase-containing viruses in which there was less than complete evidence that those viruses could cause tumors in their natural host.

For example, visina has not been shown to produce tumors in sheep, but there was evidence which I thought was quite substantial that visina could transform mouse cells and the mouse cells in turn would produce tumors.

It is not quite the same thing. But that suggests that, as with Mason-Pfizer virus, these are potential oncogenic agents, and I am not sure whether to classify them as oncorna viruses, which I don't like because it does suggest a common property of all of these viruses, their oncogenic capacity. The point I was suggesting for calling them protoviruses is that it might make more sense to define them in terms of their mode of replication than a particular pathologic property. Now, visna does share a lot of properties with C-type viruses.

Dr. Spiegelman: I would like to make a comment about the negative responses in another sense. I think you people must realize that we were extraordinarily lucky to get any reaction at all using a mouse agent to look at a human tumor. The fact is we now are beginning to see that the amount of homology that does exist between the mouse agents and the putative human agent is rather small. When we used the product made by the material taken from a human tumor, we get much more hybridization. We can hybridize essentially virtually all of our input. Instead of dealing with 2% hybrid we can deal with 80% hybrid.

Now, using that material as a probe to examine the tumors, we are finding many more positives than we did before. So I am sure that when we learn how to make this human material in adequate amounts, we will have a much more sensitive probe than we have been forced to use in our earlier investigation.

Dr. Silber: (New York) Dr. Spiegelman, since the level of ribonuclease in normal leukocytes is over ten-fold greater than that of leukemic leukocytes, I wonder if the higher ribonuclease activity than normal would interfere with the detection of the 70-S RNA? Have you tested cells from leukemics and from normals, mixing them and then extracting them to see whether you could still detect the 70-S RNA from the leukemic?

Dr. Spiegelman: We have looked at normals and we can't find any of this sort of reaction. We have done mixture experiments to make sure that it wasn't a problem of destruction. The frequency of positives among the various types of leukemias doesn't depend upon whether they are lymphocytic or myelogenous.

The ribonuclease contents are rather different but that doesn't seem to bother us.

Dr. Silber: Actually, you have very similar nuclease contents in acute lymphoblasts and acute myeloblasts.

Dr. Spiegelman: How about the CMLs?

Dr. Silber: They are slightly higher, but still considerably lower than the normals.

Dr. Spiegelman: I see.

Dr. Gallo: (Bethesda, Md.) I think there are a couple of points that need emphasizing. The ribonuclease-sensitive DNA syntheses that is in the particulate fractions that Dr. Spiegelman described and that we described in the earlier part of the talk are crude systems. We know that ribonuclease-sensitive DNA synthesizing systems can be found in normal cells, as Temin showed with the chick embryo and as we showed with the normal blood lymphocytes stimulated with phytohemoglutinin. Therefore, that kind of a test is not sufficient for any aspect of a role of a virus or a viral-type enzyme.

What makes Dr. Spiegelman's approach more definite is hybridization. I think he would agree with me on that. What we tried to show is, to answer Dr. Bader's question, perhaps, that we were purifying the enzyme from that pellet and simply showing that it had characteristics like the viral enzyme, that it could copy pure 70-S DNA with a purified enzyme. But just getting a ribonuclease-sensitive DNA synthesis with a particulate fraction, I don't think shows anything at this time except that you have a ribonuclease-sensitive DNA synthesis. It may be of some biological interest for other reasons than the viral enzyme. It may mean the viral enzyme, but that is not documentation, really, of anything at that stage.

Dr. Spiegelman: I'm glad you emphasized that, Bob, because that is a key issue. You have to show that the nucleic acid that you synthesized is relative to the disease that you think you are studying, and there is only one way to do it and that is to show homology with a relevant oncogenic piece of information.

Dr. Gallo: Could I have one more question, George? On the five or six items that you proposed for experiment that could lend support to the oncogene theory, could you pick out one or two that might help to distinguish it from the protovirus theory?

Dr. Todaro: Yes. I think actually a good many of them do. The fact of complete DNA copy in all normal cells is not a part of the protovirus theory. The protovirus theory says this evolves and can eventually, in certain circumstances, lead to a complete virus which can get out. Not every single cell has a complete copy. And also, as you point out, the protovirus theory doesn't really say anything about the oncogenic capacity. The question of what is the mechanism of cancer is really what the oncogene theory is about.

The second critical point is concerned with the question of whether there are specific RNA sequences found in transformed and tumor cells which are not found in normal cells.

Dr. Gallo: Would you worry about the data like Zerhausen's with E.B. virus in which, perhaps, information in Burkett's cells is more than in the control cells? If the DNA virus were an activator of an oncogene, what does this mean?

Dr. Todaro: I think one possibility which is certainly far from proven is that it does activate the endogenous C-type information. I think Dr. Spiegelman's data, although he might not agree with me, would lend support to that idea. In fact, I would like to ask him if this information is found in all the leukemics and lymphoma patients, how would he guess it got there? Did it get there by a horizontal transmission?

Dr. Spiegelman: I don't think it is too profitable to guess. I think the nature of the experiments that have to be performed to make a decision are fairly obvious. Once they are done, we will know.

But is it, in fact, true that the oncogene hypothesis demands one full copy of all the information required to synthesize the virus particle, or does the oncogene hypothesis say that all you need is information to make a cancer cell? These two are different.

Dr. Todaro: No, there are two aspects to it. The virogene hypothesis says that all the information needed to make a complete virus in all–

Dr. Spiegelman: Now there are two hypotheses? The virogene hypothesis and the oncogene hypothesis?

Dr. Todaro: No. They have always been the same. It is only that portion which codes for the transforming protein that renders a cell oncogenic.

Dr. Spiegelman: But no, does the oncogene hypothesis require that all of the virogene be there? You see, that is the thing that makes it difficult to test.

Dr. Todaro: Not that all of the virogene be expressed.

Dr. Spiegelman: But it is silent in the genome?

Dr. Todaro: Yes.

Dr. Spiegelman: Now we can check the hypothesis.

Dr. Olitt: I wonder if normal—that is, mature—leukocytes are the appropriate control for the leukemia leukocytes that you study. Have you studied your normal lymphocytes turned on by PHA or other mitogenic agents, or by specific antigens for the Rauscher complementary RNA or DNA?

Dr. Spiegelman: We have not done enough with PHA stimulated ones. In the number that we did examine, we could not find the synthesis of any DNA which could back-hybridize to the Rauscher RNA so far.

Dr. Olitt: Dr. Aurelian, I was a little bit disturbed by your definition of latency, and I think a distinction has to be made between latency within a single cell and latency within the organism. In the Herpesvirus infection, one may have a latent infection in which the virus is not in the skin, between periods of lesion, but may be found in the nerve root, in the ganglion, in the nerve itself. I think this is different from the concept of latency or integration into the cell itself without productive virus being present.

Dr. Aurelian I agree that what you say is different. I am not sure that you have all the evidence. In the natural infected organism the virus is latent in the ganglion. The ganglion experiments have been done by infecting an animal and following the virus for approximately four weeks or as much as four months in the infected animal.

However, ganglia from humans that carry at a specific sight the latent infection do not reveal the virus. There is a controversy over the mechanism with regard to latency of viruses persistence in cells in humans. One theory is that the virus survives as a genome, be it in the nerve cell or be it in the epithelial cell.

Dr. DeFabo: (Rockville, Md.) Dr. Todaro, what would be the selective advantage for a cell carrying this oncogene?

Dr. Todaro: The selective advantage for the cell would be that it could grow at the expense of other cells. I think the question you are really asking is why has it been preserved in evolution and why is it that there is C-type virus in practically all vertebrates that have been looked at. If it has some growth-stimulating effect, it could have a growth advantage or, as Temin has suggested, the reverse transcriptase could play some role early in differentiation. I don't think there is any strong selective advantage against it, because in natural population cancer is probably not a significant cause of death. Most animals have a lot of other things to worry about besides cancer. It is really a factor only in a civilized population or where modern medicine has reached a point that cancer and heart disease become important problems. In the wild, the tumors would tend to appear after the reproductive age, where there certainly wouldn't be any selection against their appearance.

There is probably very minimal selection against it which is overbalanced by a short selective advantage at some point in development. I don't think it constitutes a major negative factor in natural evolution.

Dr. Gillespie: (Baltimore, Md.) If one finds a cytoplasmic DNA copy of viral information in a spontaneously induced tumor, could you reconcile that with the oncogene theory or would that be a way to distinguish the two?

Dr. Todaro: In a spontaneously induced tumor without the additional exogenous virus, virus could be released if its DNA is synthesized either by the RNA polymerase or by the reverse transcriptase. This really is quite an important point. I think there is rather good indirect evidence that the reverse transcriptase is involved in initiating an infection. Whether it is involved in the release process we really have no idea, but I would say it need not be involved.

Dr. Gillespie: You emphasized that it must come out as normal cellular expression.

Dr. Todaro: Well, eventually what comes out is a 70-S RNA which is properly packaged and synthesized, presumably from DNA. That could involve the making and packaging of the 70-S RNA or it could involve the reverse transcriptase, even in the release step.

Dr. Spiegelman: The experiment that Montagnier described to us today, which is really quite a fascinating one, it seems to me is not readily consistent with the oncogene hypothesis. If the oncogene hypothesis is true, you should in fact be able to use DNA from normal cells for infection and obtain virus particles. The information is there.

Now, you could say, in the transformed cell you may have expanded the information. Maybe that is the difference. Then the effort should really be made with normal DNA. This test should be sensitive enough, because the expansion can't be more than a factor of, let's say, 50 from our experience.

Dr. Todaro: It could be in a more accessible form when it is infected from the outside. Yes, I would certainly agree, certainly with the GS-positive chick which is making GS antigen, but presumably not making virus.

Dr. Winker: (Wash., D.C.) Perhaps of importance might be the site in which viruses are incorporated into the genome. There may be effects of parts of viruses or the whole virus on the functionality of the genome and the oncogene, depending on the location within the chromosome of the oncogenic virus. I have heard nothing about the possible positions of these viruses.

How redundant is the virus information in the genome? This may be important in terms of a possible spectrum of tumors produced by viruses. Perhaps, the more the virus incorporated the wider the variety of redundancy, the wider the spectrum of tumors that can be produced.

In chemical carcinogenesis, we have the problem of a spectrum of tumors. We think that has nothing to do with viruses. When we ask the question the other way around, do the virologists have a spectrum of tumors produced by a single virus?

One could postulate a relationship between chemical and viral carcinogenesis and reconcile the two processes.

The chemical carcinogene may help to spring the virus loose at one point. Then after a long latency it may aid in the reinsertion of the viral genome in a critical point in the chromosome.

Dr. Todaro: Yes, I think that is really quite an important point. Although I think it is very doubtful that there is horizontal transmission in the sense of conventional "you-cough-on-me-and-I-catch-your-virus" kind of thing, it might well be that in a single animal, virus is induced from a cell. For example, we are inducing virus from fibroblasts. Now, it may be that it has to get into another cell under a situation where it is not controlled for it to have its effect. This is why I was suggesting there may well be a role for antiviral therapy within a given individual.

Dr. Winker: Unless the virus remains in the same cell.

Dr. Todaro: Yes, if it reinserts within the same cell.

Dr. Sear: (Wash., D.C.) Dr. Todaro, you briefly mentioned antiviral agents. There have been experiments with rifampicin drugs against tissue culture systems. Is there any future for this type of drug, rifampicin, in patient therapy?

Dr. Todaro: Dr. Gallo is right there, and he will be glad to answer that.

Dr. Gallo: They were used in more than tissue culture. They were used in fresh leukemic and normal cells. We find with some rifampicin derivatives that one could get selective cytotoxicity with some rifampicin derivatives on cells.

I don't think inhibiting a reverse transcriptase specifically with it would be the cure of cancer or have a potent antitumor effect unless there are active viral particles or reverse transcriptase systems with nucleic acids without a whole virus that can infect neighboring cells after the disease occurs.

There is some suggestive clinical evidence that an infectious type of agent, whatever it be, viral or not, may still be present after a patient has the disease.

There are now at least three reported cases of people with leukemia transplanted with donor cells, after radiation to ablate their bone marrow.

A month of two later, transformation of donor cells occurred in the recipient. This would support the argument that some agent may still be there. In that case, the specific inhibitor of a reverse transcriptase, if RNA virus is the inciting agent, would be theoretically useful in combination with cytotoxic agents.

Dr. Beers: I want to thank our speakers for the excellent series of papers they have given in these two days. I also want to thank our sponsors of this symposium, Miles Laboratories, for their generosity and the great help, as I said before, of Dr. O'Donovan, the coordinator, and I also express thanks to the Johns Hopkins Medical Institutions for providing the facilities for the symposium.

INDEX

Library of Congress Cataloging in Publication Data

International Symposium on Molecular Biology, 6th, Baltimore, 1972.
Cellular modification and genetic transformation by exogenous nucleic acids.

(The Johns Hopkins medical journal. Supplement no. 2)
"Sponsored by Miles Laboratories, inc."
Includes bibliographical references.
1. Genetic transformation--Congresses. 2. Viral genetics--Congresses. 3. Oncogenic viruses--Congresses. 4. Cytogenetics--Congresses. 5. Nucleic acid synthesis--Congresses. I. Beers, Roland F., ed. II. Tilghman, R. Carmichael, 1904– ed. III. Title. IV. Series. [DNLM: 1. Cell transformation,

Neoplastic--Congresses. 2. Nucleic acids--Congresses. 3. Transformation, Genetic--Congresses. W1 J0158H no. 2 1973. XNLM: [QW 51 I601c 1972]]
QH448.4.I57 1972 574.8′732 735909
ISBN 0-8018-1511-8